Nonlinear Circuit Simulation and Modeling

Discover the nonlinear methods and tools needed to design real-world microwave circuits with this tutorial guide. Balancing theoretical background with practical tools and applications, it covers everything from the basic properties of nonlinear systems such as gain compression, intermodulation and harmonic distortion, to nonlinear circuit analysis and simulation algorithms, and state-of-the-art equivalent circuit and behavioral modeling techniques. Model formulations discussed in detail include time-domain transistor compact models and frequency-domain linear and nonlinear scattering models. Learn how to apply these tools to the design of real circuits with the help of a power amplifier design example, which covers all stages from active device model extraction and the selection of bias and terminations, through to performance verification. Realistic examples, illustrative insights, and clearly conveyed mathematical formalism make this an essential learning aid for both professionals working in microwave and RF engineering, and graduate students looking for a hands-on guide to microwave circuit design.

José Carlos Pedro is Professor of Electrical Engineering at the Universidade de Aveiro and a senior researcher at Instituto de Telecomunicações, Portugal, and a fellow of the IEEE.

David E. Root is Keysight Research Fellow at Keysight Technologies, Inc., USA, and a fellow of the IEEE. He is a co-author of *X-Parameters* (Cambridge University Press, 2013) and a co-editor of *Nonlinear Transistor Model Parameter Extraction Techniques* (Cambridge University Press, 2011).

Jianjun Xu is a senior device modeling R&D engineer at Keysight Technologies, Inc., USA.

Luís Cótimos Nunes is a radio-frequency research assistant at Instituto de Telecomunicações, Universidade de Aveiro, Portugal.

The Cambridge RF and Microwave Engineering Series

Series Editor
Steve C. Cripps, Distinguished Research Professor, Cardiff University

Editorial Advisory Board
James F. Buckwalter, UCSB
Jenshan Lin, University of Florida
John Wood, Obsidian Microwave, LLC.

"This is a remarkable book in every way that is important: clear, professional, and with good coverage of its subject. With the importance of nonlinearity in modern wireless circuit and system design, this book should be on every engineer's bookshelf."

Dr Steve Maas, *Nonlinear Technologies, Inc.*

"A much needed foundational and pedagogical treatment of how modern nonlinear approaches include and transcend their more familiar but overly restrictive linear precursors. Presented from multiple mathematical and conceptual perspectives, this book opens the door for both a richer theoretical understanding and a more confident application to real-world design challenges confronting today's engineers."

Jay Alexander, Chief Technology Officer, *Keysight Technologies*

"With the advent of more powerful computers, the importance of models in almost every sphere of engineering and science has continued to grow rapidly. The promise of being able to more easily and more accurately represent the non-linear effects of the real world and therefore design higher performing products more quickly is within our grasp. But it is rare to find a book which brings together the collection of knowledge to both understand and practice the new skills required to embrace and succeed with non-linear modeling - but this one does. The use of practical examples and the many visualizations of non-linear behaviour further help to cement fundamental conclusions. While the theory is presented, the book clearly demonstrates how this can be transferred into practical application with modern CAD tools. In an age where artificial intelligence and machine learning are all in vogue, linking models to physical processes helps engineers to pinpoint and address area for improvement in the real world."

Dr Mark Pierpoint, Senior Vice President, *Keysight Technologies*

Nonlinear Circuit Simulation and Modeling

Fundamentals for Microwave Design

JOSÉ CARLOS PEDRO
Universidade de Aveiro

DAVID E. ROOT
Keysight Technologies, Inc.

JIANJUN XU
Keysight Technologies, Inc.

LUÍS CÓTIMOS NUNES
Universidade de Aveiro

CAMBRIDGE
UNIVERSITY PRESS

CAMBRIDGE
UNIVERSITY PRESS

University Printing House, Cambridge CB2 8BS, United Kingdom

One Liberty Plaza, 20th Floor, New York, NY 10006, USA

477 Williamstown Road, Port Melbourne, VIC 3207, Australia

314-321, 3rd Floor, Plot 3, Splendor Forum, Jasola District Centre, New Delhi - 110025, India

79 Anson Road, #06-04/06, Singapore 079906

Cambridge University Press is part of the University of Cambridge.

It furthers the University's mission by disseminating knowledge in the pursuit of education, learning and research at the highest international levels of excellence.

www.cambridge.org
Information on this title: www.cambridge.org/9781107140592
DOI: 10.1017/9781316492963

© Cambridge University Press 2018

First published 2018

A catalogue record for this publication is available from the British Library

ISBN 978-1-107-14059-2 Hardback

"X-parameters" is a trademark of Keysight Technologies, Inc.

José Carlos Pedro: To Maria João
David E. Root: To Marilyn, Daniel, and Alex
Jianjun Xu: To Cynthia (Yang), Lynna (Mengmeng), and Lexie (Yoyo)
Luís Cótimos Nunes: To Catarina

Contents

Preface

The idea for this book had its genesis in the positive reception accorded to the short courses "Fundamentals of Device Modeling for Nonlinear Circuit Simulation and Microwave Design" and "The Basics of Computer Aided Nonlinear Microwave Circuit Design," which José Pedro and David Root presented for three years at two major international conferences. At that time, what moved us toward this endeavor was the shared view that there was a significant set of computer-aided design methods and techniques for nonlinear RF/microwave circuits that had not yet become standard practice for the vast majority of RF circuit designers. Despite the fact that most of the basic scientific results of nonlinear RF computer-aided engineering-based circuit design had already appeared in the late 1980s and had its boom of article publications in the 1990s and the beginning of the 2000s, the practicing engineer could not yet take full advantage of the enormous potential provided by this knowledge.

There are undoubtedly several reasons that can be advanced to explain the existing gap between the academic or industrial researcher and the practicing circuit designer. Among the most important reasons, we feel, is the engineering curriculum itself. In fact, even today, most bachelor- and master-level electrical engineering programs are still focused on linear analysis and design methods. So after graduation, when microwave engineers must face nonlinear circuits, they tend to base their designs on limited experience, heuristic extrapolations of familiar linear concepts beyond their domains of validity, and many trial-and-error iterations in the lab.

This book is intended to change this situation, filling the gap between the classical but restricted domain of linear microwave circuit design and the modern requirements for knowledgeable fundamentals of nonlinear circuit simulation and nonlinear device modeling. Being directed to both graduate students and RF design engineers, it adopts a hybrid approach in which basic mathematical foundations are presented along with many real examples. This way, this text goes all the way from the basics to the state-of-the-art in RF/microwave circuit simulation and device modeling. Since the mathematical formalism of nonlinear systems can be quite formidable, and even intimidating, it is always presented as a comparison to and an extension beyond the already known analysis of linear systems, and accompanied by simple illustrative application examples and exercises.

With this objective in mind, an introductory chapter starts the book, Chapter 1 – Linear and Nonlinear Circuits, in which the basics of nonlinear systems' responses to various types of stimuli are introduced as the natural result of lifting the linearity restriction. Many important concepts, unknown in linear circuit design, such as gain compression, intermodulation, and harmonic distortion, are defined and then systematized using some simple and intuitive nonlinear mathematical models.

Then, Chapter 2 – Basic Nonlinear Microwave Circuit Analysis Techniques continues to the time-, frequency- and mixed-domain nonlinear circuit analysis and simulation algorithms, explaining the fundamental principles of the classic SPICE-like transient analysis and the workhorses of modern microwave circuit simulators: harmonic-balance and shooting (also known as the periodic-steady-state) methods. Finally, this chapter concludes with simulation techniques devoted to modern wireless signals such as envelope-following simulation engines and their enabled circuit-level/system-level co-simulation capabilities.

Because simulation requires numerical algorithms and device models, the next part of the book is devoted to the various equivalent-circuit and behavioral model formulations and extraction procedures. So, Chapter 3 – Linear Behavioral Models in the Frequency Domain: S-parameters starts with the linear behavioral model of microwave components, the S-parameter matrix, as an introduction to its extension to the nonlinear realm, to be treated in Chapter 4 – Nonlinear Frequency Domain Behavioral Models. Using a rigorous mathematical formalism, but also many practical measurements, the modern general framework of X-parameters is defined and carefully explained, with attention to pedagogy, from multiple perspectives, along with some simple, but insightful, examples.

Following this structure, Chapter 5 – Linear Device Modeling and Chapter 6 – Nonlinear Device Modeling address equivalent circuit models of microwave active devices with a particular emphasis on field-effect transistors. Discussing first the equivalent circuit concept and its limitations, the text then moves on to treat many different modeling approaches, such as arbitrary function approximation using ad hoc basis functions or systematic artificial neural-networks. However, such a theme would not be complete without many other practically important and scientifically challenging topics such as electro-thermal and trapping modeling, equivalent circuit model scaling, symmetry and charge and energy conservation principles.

Finally, to facilitate the comprehension of all these concepts and illustrate their application in a real engineering environment, the book concludes with Chapter 7 – Nonlinear Microwave CAD Tools in a Power Amplifier Design Example. This consists of the discussion of a practical nonlinear microwave circuit design, in which all design steps, from the active device model extraction, to the selection of bias and terminations, up to load- and source-pull prediction and the performance verification are addressed.

After collecting all material and then turning it into a coherent technical text, the authors hope that our readers will find it as exciting to read as it was for us to write. Actually, if our book causes more practicing engineers to use these computer-aided nonlinear microwave circuit design methods and techniques, then we will have fulfilled our initial goal.

Acknowledgments

The first author thanks his wife, Maria João, for the many hours that were due to her and were instead directed to this work.

The second author thanks his dear wife, Marilyn, for her patience over a time-period well beyond that estimated!

The fourth author would like to thank his girlfriend and family for their patience and understanding during the time taken to prepare this book.

The second and third authors thank their many past and present colleagues at HP, Agilent, and Keysight Technologies for collaboration, inspiration, and encouragement. We thank Keysight management for support.

The first and fourth authors thank their colleagues of the Wireless Circuits and Systems' Group for their encouragement and the many technical discussions that, directly or indirectly, contributed to the material of this book.

The first and fourth authors also thank Instituto de Telecomunicações – Universidade de Aveiro for all its available facilities and support.

We are grateful to many outstanding technical professionals in academia and industry for their collegial collaboration, and in several cases (specifically referenced), for allowing us to publish joint results obtained by accessing their technologies.

We thank Dr. Julie Lancashire and her staff at Cambridge University Press for their efforts.

1 Linear and Nonlinear Circuits

This chapter has a two-fold objective. First, it introduces the nomenclature that will be used throughout the book. Second, it presents the basic mathematical theory necessary to describe nonlinear systems, which will help the reader to understand their rich set of behaviors. This will clarify several important distinctions between linear and nonlinear circuits and their mathematical representations.

We shall start with a brief review of linearity and linear systems, their main properties and underlying assumptions. A reader familiarized with the linear system realm can understand the limitations of the theoretical abstraction framed in the linearity mathematical concept, realizing its validity borders and so be prepared to cross them, i.e., to enter the natural world of nonlinearity. We will then introduce nonlinear systems and the responses that we should expect from them. After this, we will study one static, or memoryless, nonlinearity and a dynamic one, i.e., one that exhibits memory. This will then establish the foundations of nonlinear static and dynamic models and their basic extraction procedures.

The chapter is presented as follows: Section 1.1 is devoted to nomenclature and Section 1.2 reviews linear system theory. Sections 1.3 and 1.4 illustrate the types of behaviors found in general nonlinear systems and, in particular, in nonlinear RF and microwave circuits. Then, Sections 1.5 and 1.6 present the theory of nonlinear static and dynamic systems that will be useful to understand the nonlinear circuit simulation algorithms treated in Chapter 2 and the device modeling techniques of Chapters 3–6. Mathematics of nonlinear systems, and in particular dynamic ones, is not easy or trivial. So, we urge you to not feel discouraged if you do not understand it after your first read. What you will find in the next chapters will certainly help provide a physical meaning and practical usefulness to most of these sometimes abstract mathematical formulations. Finally, Section 1.7 closes this chapter with a brief conclusion.

1.1 Basic Definitions

We will frequently use the notion of model and system, so it is convenient to first identify these concepts.

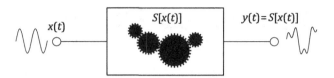

Figure 1.1 Illustration of the system concept.

1.1.1   Model

A model is a ***mathematical description***, or representation, of a set of particular features of a physical entity that combines the observable (i.e., measurable) magnitudes and our previous knowledge about that entity. Models enable the simulation of a physical entity and so allow a better understanding of its observed behavior and provide predictions of behaviors not yet observed. As models are simplifications of the physically observable, they are, by definition, an approximation and restricted to represent a subset of all possible behaviors of the physical device.

1.1.2   System

As depicted in Figure 1.1, a system is a model of a machine or mechanism that transforms an input (excitation, or stimulus, usually assumed as a function of time), $x(t)$, into an output (or response, also varying in time), $y(t)$. Mathematically, it is defined as the following operator: $y(t) = S[x(t)]$, in which $x(t)$ and $y(t)$ are, themselves, mathematical representations of the input and output measurable signals, respectively. Please note that, contrary to ordinary mathematical functions, which operate on numbers (i.e., that for a given input number, x, they respond with an output number, $y = f(x)$), mathematical operators map functions, such as $x(t)$, onto other functions, $y(t)$. So, they are also known as *mathematical function maps*. And, similar to what is required for functions, a particular input must be mapped onto a particular, unique, output.

When the operator is such that its response at a particular instant of time, $y(t_0)$, is only dependent on that particular input instant, $x(t_0)$, i.e., the system transforms each input value onto the corresponding output value, the operator is reduced to a function and the system is said to be ***static or memoryless***. When, on the other hand, the system output cannot be uniquely determined from the instantaneous input only but depends on $x(t_0)$ and its $x(t)$ past and future values, $x(t \pm \tau)$, i.e., the system is now an operator of the whole $x(t)$ onto $y(t)$, we say that the system is ***dynamic*** or that it exhibits ***memory***. (In practice, real systems cannot depend on future values because they must be causal.) For example, resistive networks are static systems, whereas networks that include energy storage elements (memory), such as capacitors, inductors or transmission lines, are dynamic.

Defined this way, this notion of a system can be used as a representation, or model, of any physical device, which can either be an individual component, a circuit or a set of circuit blocks. An interesting feature of this definition is that a system is nestable, i.e., it is such that a block (circuit) made of interconnected individual systems (circuit elements or components) can still be treated as a system. So, we will use this concept of system whenever we want to refer to the properties that we normally observe in components or circuits.

1.1.3   Time Invariance

Although the system response, $y(t)$, varies in time, that does not necessarily mean that the system varies in time. The change in time of the response can be only a direct consequence of the input variation with time. This time-invariance of the operator is expressed by stating that the system reacts exactly in the same way regardless at which time it is subjected to the same input. That is, if the response to $x(t)$ is $y(t) = S[x(t)]$, and another test is made after a certain amount of time, τ, then the response will be exactly the same as before, except that now it will be naturally delayed by that same amount of time $y(t - \tau) = S[x(t - \tau)]$. This defines a ***time-invariant*** system. If, on the other hand, $y(t - \tau) \neq S[x(t - \tau)]$, then the system is said to be ***time-variant***.

The vast majority of physical systems, and thus of electronic circuits, are time-invariant. Therefore, we will assume that all systems referred to in this and succeeding chapters are time-invariant unless otherwise explicitly stated.

After finalizing the study of this chapter, the reader may try Exercise 1.5 which constitutes a good example of how we can make use of this time-variance property for enabling us to treat, as a much simpler linear time-variant system, a modulator that is inherently nonlinear and time-invariant.

1.2   Linearity and the Separation of Effects

Now we will define a linear system as one that obeys superposition and recall how we use this property to determine the response of a linear system to a general excitation.

1.2.1   Superposition

A system is said to be linear if it obeys the principle of superposition, i.e., if it shares the properties of additivity and homogeneity.

The additivity property means that if $y_1(t)$ is the system response to $x_1(t)$, $y_1(t) = S[x_1(t)]$, $y_2(t)$ is the system's response to $x_2(t)$, $y_2(t) = S[x_2(t)]$, and $y_T(t)$ is the response to $x_1(t) + x_2(t)$, then

$$y_T(t) = S[x_1(t) + x_2(t)] = S[x_1(t)] + S[x_2(t)] = y_1(t) + y_2(t) \qquad (1.1)$$

The additivity property is the mathematical statement that affirms that a linear system reacts to an additive composition of stimuli as an additive composition of responses, as if the system could distinguish each of the stimuli and treat them separately. In practical terms, this would mean that, if, in the lab, the result of an experiment with a cause $x_1(t)$ would produce an effect $y_1(t)$, and another, independent, experiment, on another cause $x_2(t)$, would produce $y_2(t)$, then, a third experiment, now made on a third stimulus $x_1(t) + x_2(t)$, would produce a response that is the numerical summation of the two previously obtained effects $y_1(t) + y_2(t)$.

On the other hand, the homogeneity property means that if α is a constant, then the response to $\alpha x(t)$ will be $\alpha y(t)$, i.e.,

$$S[\alpha x(t)] = \alpha S[x(t)] = \alpha y(t) \tag{1.2}$$

The homogeneity property is the mathematical description of proportionality that says that an α times larger cause produces an α times larger effect. However, it does not necessarily state that the effects are proportional to their corresponding causes. For example, although the current and the voltage in a constant (linear) capacitance obey the homogeneity principle, they are not proportional to each other. In fact, since the current in a capacitor is given by (1.3), the current to a twice as large $v_c(t)$ will be twice as large as $i_c(t)$. However, that does not mean that $i_c(t)$ is proportional to $v_c(t)$, as can be readily noticed when $v_c(t)$ is a ramp in time and $i_c(t)$ is a constant.

$$i_c(t) = C\frac{dv_c(t)}{dt} \tag{1.3}$$

In summary, *linear systems* obey the principle of *superposition*,

$$\begin{aligned}S[\alpha_1 x_1(t) + \alpha_2 x_2(t)] &= S[\alpha_1 x_1(t)] + S[\alpha_2 x_2(t)] = \alpha_1 S[x_1(t)] + \alpha_2 S[x_2(t)]\\ &= \alpha_1 y_1(t) + \alpha_2 y_2(t)\end{aligned} \tag{1.4}$$

1.2.2   Response of a Linear System to a General Excitation

Superposition has very useful consequences that we now briefly review. They all revolve around that idea of the separation of effects, whereby we can expand any previously untested stimulus into a summation of previously tested excitations, making general predictions about the system responses.

1.2.2.1   Linear Response in the Time Domain

In the time domain, this means that, if we represent any input, $x(t)$, as composed of the succession of its time samples, taken at regular intervals, T_s, of a constant sampling frequency $f_s = 1/T_s$, so that they asymptotically produce the same effect of $x(t)$, $x(nT_s)T_s$,

$$x(t) \approx \sum_{n=-N}^{N} x(nT_s)T_s\delta(t - nT_s) \tag{1.5}$$

in which $\delta(t - nT_s)$ is the Dirac delta, or impulse, function centered at nT_s, where n is the number of samples, (see Figure 1.2(a)), and we know the response of the system to one of these impulse functions of unity amplitude, $h(t) = S[\delta(t)]$ (see Figure 1.2(b)), then we can readily predict the response to any arbitrary input $x(t)$ as

$$\begin{aligned}y(t) \equiv S[x(t)] &\approx S\left[\sum_{n=-N}^{N} x(nT_s)T_s\delta(t - nT_s)\right] = \sum_{n=-N}^{N} S[x(nT_s)T_s\delta(t - nT_s)]\\ &= \sum_{n=-N}^{N} x(nT_s)T_s S[\delta(t - nT_s)] = \sum_{n=-N}^{N} x(nT_s)h(t - nT_s)T_s\end{aligned} \tag{1.6}$$

(a)

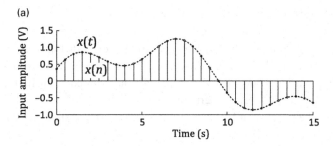

(b)

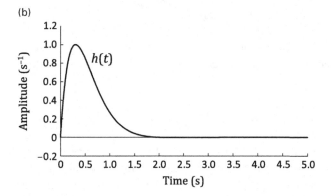

(c)

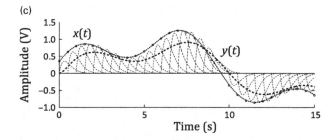

Figure 1.2. Response, $y(t)$, of a linear dynamic and time-invariant system to an arbitrary input, $x(t)$, when this stimulus is expanded in a summation of Dirac delta functions. (a) Input expansion with the base of delayed Dirac delta functions $x(n) = x(nT_s)\delta(t - nT_s)$. (b) Impulse response of the system, $h(t) = S[\delta(t)]$. (c) Response of the system to $x(t)$, $y(t) = S[x(t)]$.

by simply making use of the additivity and homogeneity properties (as shown in Figure 1.2(c)). Expression (1.6) is exact in the limit when the sampling interval, T_s, tends to zero and N tends to infinity, becoming the well-known convolution integral:

$$y(t) \equiv S[x(t)] = \int_{-\infty}^{\infty} x(\tau)h(t - \tau)d\tau = \int_{-\infty}^{\infty} h(\tau)x(t - \tau)d\tau \qquad (1.7)$$

1.2.2.2   Linear Response in the Frequency Domain

So, in time domain, we only needed to know the system response to one input basis function – the impulse response, $h(t) = S[\delta(t)]$, to be able to predict the response to any other arbitrary input. Similarly, in the frequency domain we only need to know the response to one input basis function, the cosine, although tested at all frequencies, to predict the response to any arbitrary periodic input.

Actually, since the cosine can be given as the additive combination of two complex exponentials

$$A \cos(\omega t) = A \frac{e^{j\omega t} + e^{-j\omega t}}{2} \qquad (1.8)$$

from a mathematical viewpoint, we only need to know the response to that basic complex exponential. This response can be obtained from (1.7) as

$$\int_{-\infty}^{\infty} h(\tau)e^{j\omega(t-\tau)}d\tau = e^{j\omega t}\int_{-\infty}^{\infty} h(\tau)e^{-j\omega\tau}\,d\tau = H(\omega)e^{j\omega t} \qquad (1.9)$$

in which $H(\omega)$ is the Fourier transform of $h(\tau)$. This is an interesting result that tells us that the response to an arbitrary $x(t)$ can be easily computed by summing up the Fourier components of that input scaled by the system's response to each particular frequency. Indeed, if $R(\omega)$ is the frequency-domain Fourier representation of a time-domain signal $r(t)$, so that

$$R(\omega) = \int_{-\infty}^{\infty} r(t)e^{-j\omega t}dt \qquad (1.10a)$$

and

$$r(t) = \frac{1}{2}pi\int_{-\infty}^{\infty} R(\omega)e^{j\omega t}d\omega \qquad (1.10b)$$

then, the substitution of (1.10) into (1.7) would lead to

$$Y(\omega) = H(\omega)X(\omega) \qquad (1.11)$$

where $Y(\omega)$ can be related to $y(t)$ – as $X(\omega)$ is related to $x(t)$ – by the Fourier transform of (1.10). This expression tells us the following two important things.

First, the time-domain convolution of (1.7) between the input, $x(t)$, and the impulse response, $h(\tau)$, becomes the product of the frequency-domain representation of these two entities, $X(\omega)$ and $H(\omega)$, respectively.

Second, the response of a linear time-invariant system to a continuous-wave (CW) signal (an unmodulated carrier of frequency ω, specifically $\cos(\omega t)$) is another CW signal of the same frequency with, possibly, different amplitude and phase. Consequently, the response to a signal of complex spectrum will only have frequency-domain components at the frequencies already present at the input. A time-invariant linear system is incapable of generating new frequency components or of performing any qualitative transformation of the input spectrum.

Finally, equation (1.11) tells us that, in the same way we only needed to know the system's impulse response to be able to predict the response to any arbitrary stimulus in the time domain, we just need to know $H(\omega)$ to predict the response to any arbitrary periodic input described in the frequency domain. As an illustration, Figure 1.3 depicts the measured transfer function $S_{21}(\omega)$, in amplitude and phase, of a microwave filter.

(a)

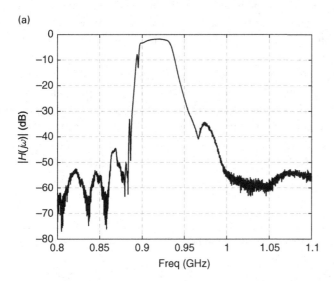

(b)

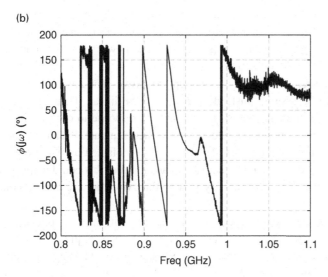

Figure 1.3 Example of the frequency-domain transfer function of a linear RF circuit, $H(\omega)$: measured forward gain, $S_{21}(\omega)$, in amplitude – (a) and phase – (b), of a microwave filter.

1.3   Nonlinearity: The Lack of Superposition

As all of us have been extensively taught and trained in working with linear systems, and with the additivity and homogeneity properties being so intuitive, we may easily fall into the trap of believing that these should be properties naturally inherent to all physical systems. But this is not the case. In fact, most of macroscopic physical systems behave very differently from linear systems, i.e., they are not linear. Actually, we use the term nonlinear systems to identify them.

Since we have been making the effort to define all important concepts used so far, we should start by defining a nonlinear system. But that is not a straightforward task as there is no general definition for these systems. There is only the unsatisfying definition of defining something by what it is not: a nonlinear system is one that is not linear, i.e., a nonlinear system is one that does not obey the principle of superposition. This is an intriguing, but also revealing, situation, which tells us that if linear systems are the ones that obey a precise mathematical principle, nonlinear systems are all the other ones. Hence, from an engineering standpoint the relevant question to be answered is: Are nonlinear systems often seen, or used, in practice? To demonstrate their importance, let us try a couple of very common, RF electronic examples. But, before these, the reader may want to try the two simpler examples discussed in Exercises 1.1–1.4.

Example 1.1 Active Devices and Amplifiers   In this example we will show that any active device must be nonlinear.

As a first step, we will show that all active devices depend on two different excitations. One is the input signal and the other is the dc power supply. This means, as illustrated in Figure 1.4, that amplifiers are transducers that convert the power supplied by a dc power source into output signal power, i.e. they convert dc into RF power.

Now, as the second step in our attempt to prove that any active device must be nonlinear, let us assume, instead, that it could be linear. Then, it would have to obey the additivity property, which means that the response to each of the inputs, the signal and the power supply, should be determined separately. That is, the response to the auxiliary supply and to the signal should be obtained as if the other stimulus would not exist. And we would come back to an amplifier that could amplify the signal power without requiring any auxiliary power, thus violating the energy conservation principle.

Although this argument seems quite convincing, it raises a puzzling question, because, if it is impossible to produce amplifiers without requiring nonlinearity, we should be magicians as we all have already seen and designed linear amplifiers. So, how can we overcome this paradox?

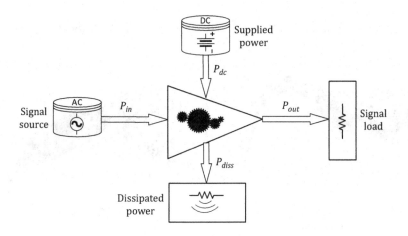

Figure 1.4 Illustration of the power flow in a transducer or amplifier.

According to the power flow shown in Figure 1.4, where P_{in}, P_{out}, P_{dc} and P_{diss} are, respectively, the signal input and output powers, the supplied dc power and the dissipated power (herein assumed as all forms of energy that are not correlated with the information signal, such as heat, harmonic generation, intermodulation distortion, etc.), the amplifier gain, G, can be defined by

$$G \equiv \frac{P_{out}}{P_{in}} \tag{1.12}$$

And this G must be constant and independent of P_{in} for preserving linearity.

Imposing the energy conservation principle to this transducer results in

$$P_{out} + P_{diss} = P_{in} + P_{dc} \tag{1.13}$$

from which the following constraint can be found for the gain:

$$G(P_{in}) = 1 + \frac{P_{dc} - P_{diss}}{P_{in}} \tag{1.14}$$

Since P_{diss} cannot decrease below zero (100% dc-to-RF conversion efficiency) and P_{dc} must be limited (as is proper from real power sources), $G(P_{in})$ cannot be kept constant but must decrease beyond a certain maximum P_{in}.

In RF amplifiers, this gain decrease with input signal power is called gain compression. In practice, amplifiers not only exhibit a gain variation when their input amplitude changes, but also an input-dependent phase shift. This is particularly important in RF amplifiers intended to process amplitude modulated signals as this input modulation is capable of inducing nonlinear output amplitude and phase modulations. These are the well-known AM/AM and AM/PM nonlinear distortions, often plotted as shown in Figure 1.5(a) and (b), respectively.

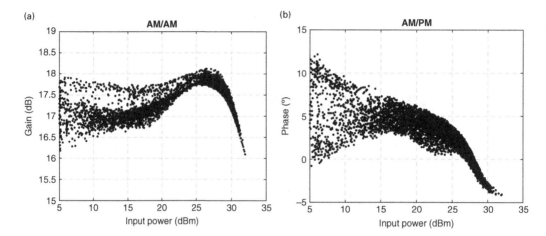

Figure 1.5 Illustration of measured (a) amplitude – AM/AM – and (b) phase-shift – AM/PM – gain variations as a function of input signal amplitude. Please note how these plots are not any idealized lines, but a cloud of dots that reveal hysteretic trajectories.

This analysis shows that linearity can only be obeyed at sufficiently small signal levels, and that it is only a matter of excitation amplitude to make an apparently linear amplifier expose its hidden nonlinearity.

Actually, this study provided us a much deeper insight of linearity and linear systems. Linearity is what we obtain when looking only at the system's input to output signal mapping (leaving aside the dc-to-RF energy conversion process) and when the signal is a very small perturbation of the dc quiescent point. So, linear systems are the conceptual mathematical model for the behaviors obtained from analytic operators (i.e., that are continuous and infinitely differentiable mappings), when these are excited with signals whose amplitudes are infinitesimally small as compared with the magnitude of the quiescent points. And it is under this small-signal operation regime that the linear approximation is valid. We will come back to this important concept later.

Example 1.2 A Sinusoidal Oscillator   A sinusoidal oscillator is another system that depends on nonlinearity to operate. Although in basic linear system analysis we learned how to predict the stable and unstable regimes of amplifiers, and so to predict oscillations, we were not told the complete story. To understand why, we can just use the above results on the analysis of the amplifier and recognize that, by definition, an oscillator is a system that provides an output even without an input. That is, contrary to an amplifier that is a nonautonomous, or forced, system, an oscillator is an autonomous one. So, if it would not rely on any external source of power, it would violate the energy conservation principle. Like an amplifier, it is, instead, a transducer that converts energy from a dc power supply into signal power at some frequency ω. Hence, like the amplifier, it must rely on some form of nonlinearity. But, unlike the amplifier, in which we have shown that, seen from the input signal to the output signal, it could behave in an approximately linear way, we will now show that not even this is possible in an oscillator.

To see why, consider the following linear differential equation of constant (i.e., time-invariant) coefficients – one of the most common models of linear systems:

$$LC\frac{d^2 i(t)}{dt^2} + (R_S + R_L + R_A)C\frac{di(t)}{dt} + i(t) = C\frac{dv_s(t)}{dt} \tag{1.15}$$

which describes the loop current, $i(t)$, sinusoidal oscillations of a series RLC circuit when the excitation vanishes, $v_s(t) = 0$. The proof that this equation is indeed the model of a linear time-invariant system is left as an exercise for the reader (see Exercise 1.6).

This RLC circuit is assumed to be driven by an active device whose model for the power delivered to the network is the negative resistance R_A, and to be loaded by the load resistance R_L and the inherent LC tank losses R_S. It can be shown that the solution of this equation, when $v_s(t) = 0$, is of the form

$$i(t) = Ae^{-\lambda t}\cos(\omega t) \tag{1.16}$$

where $\lambda = \frac{R_S + R_L + R_A}{2L}$ and $\omega = \sqrt{\frac{1}{LC} - \lambda^2}$.

The first curious result of this linear oscillator model is that it does not provide any prediction for the oscillation amplitude A as if A could be any arbitrary value. The second is that, to keep a steady-state oscillation, i.e., one whose amplitude does not decay or increase exponentially with time, λ must be exactly (i.e., with infinite precision) zero, or $R_A = -(R_S + R_L)$ something our engineering common sense finds hard to believe. Both of these unreasonable conditions are a consequence of the absence of any energy constraint in (1.15), which, itself, is a consequence of the performed linearization. In practice, what happens is that the active device is nonlinear, its negative resistance is not constant but an increasing function of amplitude, $R_A(A) = -f(A)$, so that a negative feedback process keeps the oscillation amplitude constant at $A = f^{-1}(R_S + R_L)$, in which $f^{-1}(.)$ represents the inverse function of $f(.)$.

Although nonlinearity is often seen as a source of perturbation, referred to with many terms, such as *harmonic distortion, nonlinear cross-talk, desensitization,* or *intermodulation distortion,* it plays a key role in wireless communications. As a matter of fact, these two examples, along with the amplitude modulator of Exercises 1.3–1.5, show that nonlinearity is essential for amplifiers, oscillators, modulators, and demodulators. And since wireless telecommunication systems depend on these devices to generate RF carriers, translate information base-band signals back and forth to radio-frequency frequencies (reducing the size of the antennas), and provide amplification to compensate for the free-space path loss, we easily conclude that without nonlinearity wireless communications would be impossible. As an illustration, Figure 1.6 shows the block diagram of a wireless transmitter where the blocks from which nonlinearity should be expected are put in evidence.

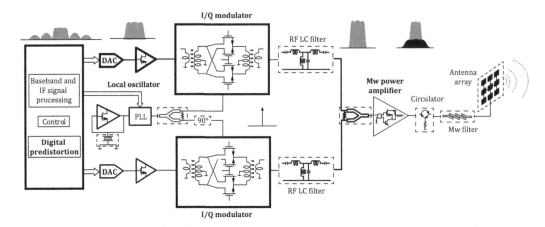

Figure 1.6 Block diagram of a wireless transmitter where the blocks from which nonlinearity is expected are highlighted: linear blocks are represented within dashed line boxes whereas nonlinear ones are drawn within solid line boxes.

1.4   Properties of Nonlinear Systems

This section illustrates the multiplicity of behaviors that can be found in nonlinear dynamic systems. Although a full mathematical analysis of those responses does not constitute a key objective of this chapter, we will nevertheless base our tests in a simple circuit so that each of the observed behaviors can be approximately explained by relating it to the circuit topology and components.

The analyses will be divided in forced and autonomous regimes, like the ones found in amplifiers or frequency multipliers and oscillators, respectively. However, to obtain a more applied view of these responses we will further group forced regimes in responses to CW and modulated excitations.

1.4.1   An Example of a Nonlinear Dynamic Circuit

To start exploring some of the basic properties of nonlinear systems, we will use the simple (conceptual) amplifier shown in Figure 1.7.

In order to preserve the desired simplicity, enabling us to qualitatively relate the obtained responses to the circuit's model, we will assume that the input and output block capacitors, C_B, and the RF bias chokes, L_{Ch}, are short-circuits and open circuits to the RF signals, respectively. Therefore, the dc blocking capacitors can be simply replaced by two ideal dc voltage sources, the gate RF choke can be neglected, and the drain choke must be replaced by a dc current source that equals the average $i_{DS}(t)$ current, I_{DS}. This is illustrated in Figure 1.7(b). Furthermore, we will also assume that the FET's feedback capacitor, C_{gd}, can be replaced by its input and output reflected Miller capacitances, which value $C_{gd\_in} = C_{gd}(1 - A_v)$ and $C_{gd\_out} = C_{gd}(A_v - 1)/A_v$, respectively. Assuming that the voltage gain, A_v, is negative and much higher than

(a)

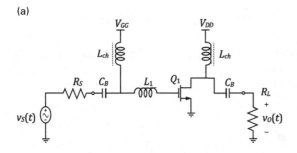

(b)

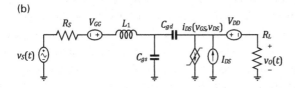

(c)

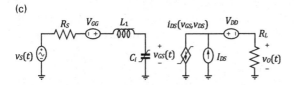

Figure 1.7 (a) Conceptual amplifier circuit used to illustrate some properties of forced nonlinear systems. (b) Equivalent circuit when the dc block capacitors and the dc feed inductances are substituted by their corresponding dc sources. (c) Simplified unilateral circuit after C_{gd} was reflected to the input via its Miller equivalent.

one (rigorously speaking, much smaller than minus one) the total FET's input capacitance, C_i, will be approximately given by $C_i = C_{gs} + C_{gd}|A_v|$ while the output capacitance, C_o, will equal C_{gd}, being thus negligible. Under these conditions, the schematic of Figure 1.7(b) becomes the one shown in Figure 1.7(c), whose analysis, using Kirchhoff's laws, leads to

$$v_S(t) = R_S i_G(t) - V_{GG} + L_1 \frac{d i_G(t)}{dt} + v_{GS}(t) \tag{1.17}$$

and

$$v_{DS}(t) = V_{DD} - R_L [i_{DS}(t) - I_{DS}] \tag{1.18}$$

in which $i_G(t) = C_i \frac{d v_{GS}(t)}{dt}$ is the gate current flowing through C_i and $i_{DS}(t)$ is the FET's drain-to-source current. This drain current is assumed to be some suitable static nonlinear function of the gate-to-source voltage, $v_{GS}(t)$, and drain-to-source voltage, $v_{DS}(t)$.

In case $v_{DS}(t)$ is kept sufficiently high so that $v_{DS}(t) \gg V_K$, the FET's knee voltage, $i_{DS}(t)$ can be considered only dependent on the input voltage, $i_{DS}(v_{GS})$. Using these results in (1.17), the differentiation chain rule leads to a second order differential equation

$$v_S(t) = R_S C_i \frac{dv_{GS}(t)}{dt} - V_{GG} + L_1 C_i \frac{d^2 v_{GS}(t)}{dt^2} + v_{GS}(t) \tag{1.19}$$

whose solution, $v_{GS}(t)$, allows the determination of the amplifier output voltage as

$$v_o(t) = v_{DS}(t) - V_{DD} = -R_L\{i_{DS}[v_{GS}(t)] - I_{DS}\} \tag{1.20}$$

1.4.2   Response to CW Excitations

Because of its central role played in RF circuits, we will first start by identifying the responses to sinusoidal, or CW, (plus the dc bias) excitations. Hence, our amplifier is described by a circuit that includes two nonlinearities: $i_{DS}(v_{GS}, v_{DS})$ that is static and C_i, a nonlinear capacitor, which, depending on the voltage gain, evidences its nonlinearity when the amplifier suffers from $i_{DS}(v_{GS}, v_{DS})$ induced gain compression.

Figure 1.8 shows the drain-source voltage evolution in time, $v_{DS}(t)$, for three different CW excitation amplitudes, while Figure 1.9 depicts the respective spectra. Under small-signal regime, i.e, small excitation amplitudes, in which the FET is kept in the saturation region and v_{gs} (the signal component of the composite v_{GS} voltage defined by $v_{gs} \equiv v_{GS} - V_{GS}$ and, in this case, $V_{GS} = V_{GG}$) is so small that $i_{DS}(v_{GS}, v_{DS}) \approx I_{DS} + g_m v_{gs} + g_{m2} v_{gs}^2 + g_{m3} v_{gs}^3 \approx I_{DS} + g_m v_{gs}$, the amplifier presents an almost linear response without any other harmonics than the dc and the fundamental component already present at the input. As we increase the excitation amplitude, the $v_{GS}(t)$ and $v_{DS}(t)$ voltage swings become sufficiently large to excite the FET's $i_{DS}(v_{GS}, v_{DS})$ cutoff ($v_{GS}(t) < V_T$, the FET's threshold voltage) and knee voltage nonlinearities ($v_{DS}(t) \approx V_K$), and the amplifier starts to evidence its nonlinear behavior, producing other frequency-domain harmonic components or time-domain distorted waveforms.

In a cubic nonlinearity as the one above used for the dependence of i_{DS} on v_{GS}, this means that a sinusoidal excitation, $v_s(t) = A\cos(\omega t)$, produces a gate-source voltage of $v_{gs}(t) = |V_{gs}(\omega)| \cos(\omega t + \phi_i)$, in which

$$V_{gs}(\omega) = \frac{A}{1 - \omega^2 L_1 C_i + j\omega R_S C_i} = H_i(\omega)A = |H_i(\omega)|e^{j\phi_i}A \tag{1.21}$$

produces the following $i_{ds}(t)$ response:

$$
\begin{aligned}
i_{ds}(t) \approx\ & g_m |H_i(\omega)| A\cos(\omega t + \phi_i) + \frac{1}{2} g_{m2} |H_i(\omega)|^2 A^2 \\
& + \frac{1}{2} g_{m2} |H_i(\omega)|^2 A^2 \cos(2\omega t + 2\phi_i) \\
& + \frac{3}{4} g_{m3} |H_i(\omega)|^3 A^3 \cos(\omega t + \phi_i) \\
& + \frac{1}{4} g_{m3} |H_i(\omega)|^3 A^3 \cos(3\omega t + 3\phi_i)
\end{aligned}
\tag{1.22}
$$

which evidences the generation of a linear component, proportional to the input stimulus, a quadratic dc component and second and third harmonics, beyond a cubic term at the fundamental. Actually, it is this fundamental cubic term that is responsible

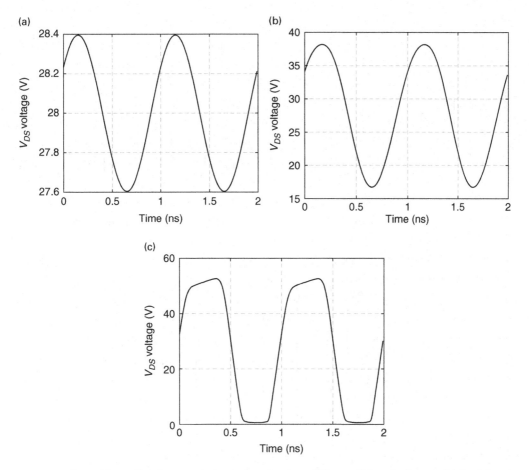

Figure 1.8 $v_{DS}(t)$ voltage evolution in time for three different CW excitation amplitudes. Note the distorted waveforms arising when the excitation input is increased.

for modeling the amplifier's gain compression since the equivalent amplifier transconductance gain is

$$G_m(A) \equiv \frac{I_{ds}(\omega)}{A} \approx g_m H_i(\omega) + \frac{3}{4} g_{m3} H_i(\omega)|H_i(\omega)|^2 A^2 \tag{1.23}$$

i.e., is dependent on the input amplitude.

In communication systems, in which an RF sinusoidal carrier is modulated with some amplitude and phase information (the so-called complex envelope [1]), such a gain variation with input amplitude has always deserved particular attention, since it describes how the input amplitude and phase information are changed by the amplifier. This defines the so-called AM/AM and AM/PM distortions. That is what is represented in Figures 1.10 and 1.11, in which the input–output fundamental carrier amplitude and phase (with respect to the input phase reference) is shown

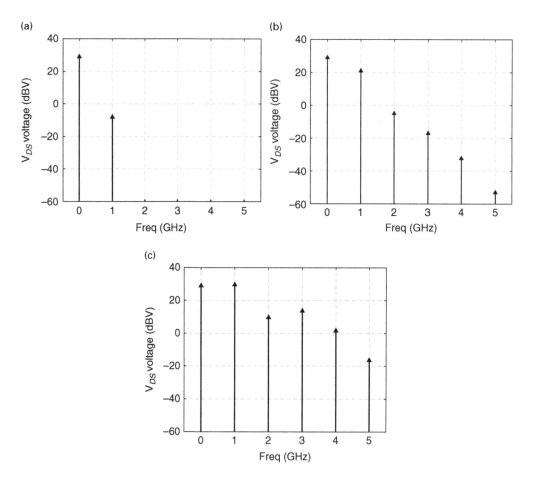

Figure 1.9 $V_{ds}(\omega)$ spectrum for three different CW excitation amplitudes. Note the increase in the harmonic content with the excitation amplitude.

versus the input amplitude. When the output voltage waveform becomes progressively limited by the FET's nonlinearities, its corresponding gain at the fundamental component gets compressed and its phase-lag reduced. Indeed, when the voltage gain is reduced, so is the C_{gd} Miller reflected input capacitance, and thus C_i. Hence, the phase of $H_i(\omega)$ increases, and consequently, the $v_{GS}(t)$ fundamental component shows an apparent phase-lead (actually a reduced phase-lag), revealed as the AM/PM of Figure 1.11.

What happens in real amplifiers (see, for example, [2]) is that, because of the circuit nonlinear reactive components, or, as was here the case, because of the interactions between linear reactive components (C_{gd}) and static nonlinearities ($i_{DS}(v_{GS}, v_{DS})$, which manifests itself as a nonlinear voltage gain, A_v), the equivalent gain suffers a change in both magnitude and phase manifested under CW excitation as AM/AM and AM/PM.

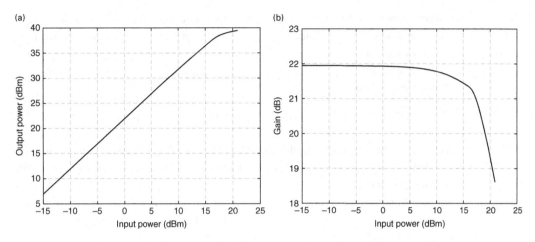

Figure 1.10 Response to CW excitations: (a) AM/AM characteristic and (b) Gain characteristic.

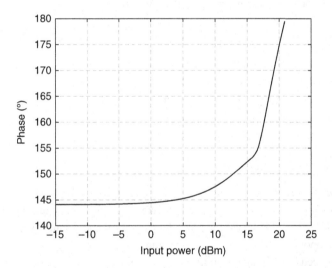

Figure 1.11 Response to CW excitations: AM/PM characteristic.

1.4.3   Response to Multitone or Modulated Signals

The extension of the CW regime to a modulated one is trivial, as long as we can assume that the circuit responds to a modulated signal like

$$v_S(\tau, t) = a(\tau) \cos \left[\omega_c t + \phi(\tau) \right] \tag{1.24}$$

without presenting any memory to the envelope. This means that the circuit treats our modulated carrier as a succession of independent CW signals of amplitude $a(\tau)$ and phase $\phi(\tau)$. That is, we are implicitly assuming that the envelope varies in a much slower and uncorrelated way – as compared to the carrier – (the narrow bandwidth approximation), as if the RF carrier would evolve in a fast time t while the base-band envelope would evolve in a much slower time τ. In that case, for example, a two-tone signal of frequencies $\omega_1 = \omega_c - \omega_m$ and $\omega_2 = \omega_c + \omega_m$, i.e., whose frequency separation is $2\omega_m$ and is centered at ω_c, and in which $\omega_c \gg \omega_m$, can be seen as a double sideband amplitude modulated signal of the form

$$v_S(\tau, t) = a(\tau) \cos(\omega_c t) = A\cos(\omega_m \tau) \cos(\omega_c t) \tag{1.25}$$

Applied to our nonlinear circuit, this modulated signal will generate several out-of-band and in-band distortion components centered at the carrier harmonics and the fundamental carrier, respectively.

This is illustrated in Figure 1.12, where both the time-domain envelope waveform and frequency-domain spectral components of our amplifier response to a two-tone signal of $f_c = 1$ GHz and $f_m = 10$ kHz, are represented.

In fact, still using the cubic nonlinearity of $i_{DS}(v_{GS})$ and the input transfer function of (1.21), we would obtain an output response of the form

$$
\begin{aligned}
i_{ds}(\tau, t) \approx {} & g_m |H_i(\omega_c)| a(\tau) \cos(\omega_c t + \phi_i) + \frac{1}{2} g_{m2} |H_i(\omega_c)|^2 a^2(\tau) \\
& + \frac{1}{2} g_{m2} |H_i(\omega_c)|^2 a^2(\tau) \cos(2\omega_c t + 2\phi_i) \\
& + \frac{3}{4} g_{m3} |H_i(\omega_c)|^3 a^3(\tau) \cos(\omega_c t + \phi_i) \\
& + \frac{1}{4} g_{m3} |H_i(\omega_c)|^3 a^3(\tau) \cos(3\omega_c t + 3\phi_i)
\end{aligned}
\tag{1.26}
$$

which is composed of the following terms:

(i) $g_m |H_i(\omega_c)| a(\tau) \cos(\omega_c t + \phi_i) = g_m |H_i(\omega_c)| A\cos(\omega_m \tau) \cos(\omega_c t + \phi_i)$ is the two-tone fundamental linear response located at $\omega_1 = \omega_c - \omega_m$ and $\omega_2 = \omega_c + \omega_m$;

(ii) $\frac{1}{2} g_{m2} |H_i(\omega_c)|^2 a^2(\tau)$ is located near the 0-order harmonic frequency of the carrier, and is composed of a tone at dc, of amplitude $\frac{1}{4} g_{m2} |H_i(\omega_c)|^2 A^2$ – representing the input-induced bias change from the quiescent point to the actual dc bias operating point – and another tone at $\omega_2 - \omega_1$ or $2\omega_m$, of amplitude $\frac{1}{4} g_{m2} |H_i(\omega_c)|^2 A^2$, representing the dc bias fluctuation imposed by the time-varying carrier amplitude.

(iii) $\frac{1}{2} g_{m2} |H_i(\omega_c)|^2 a^2(\tau) \cos(2\omega_c t + 2\phi_i)$ is the second harmonic distortion of the amplifier and is composed of three tones at $2\omega_1 = 2\omega_c - 2\omega_m$ and $2\omega_2 = 2\omega_c + 2\omega_m$, of amplitude $\frac{1}{8} g_{m2} |H_i(\omega_c)|^2 A^2$, and a center tone at $\omega_1 + \omega_2 = 2\omega_c$, of amplitude $\frac{1}{4} g_{m2} |H_i(\omega_c)|^2 A^2$;

(iv) $\frac{3}{4} g_{m3} H_i(\omega_c) |H_i(\omega_c)|^2 a^3(\tau) \cos(\omega_c t)$ are again fundamental components located at $\omega_1 = \omega_c - \omega_m$ and $\omega_2 = \omega_c + \omega_m$, of amplitude $\frac{9}{32} g_{m3} |H_i(\omega_c)|^3 A^3$, plus two in-band distortion sidebands located at $2\omega_1 - \omega_2 = \omega_c - 3\omega_m$ and $2\omega_2 - \omega_1 = \omega_c + 3\omega_m$, of amplitude $\frac{3}{32} g_{m3} |H_i(\omega_c)|^3 A^3$;
      and

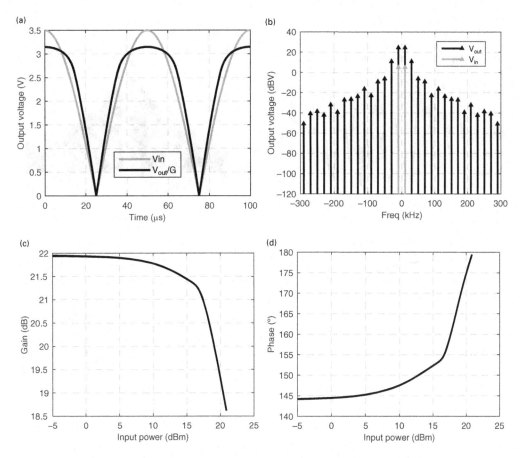

Figure 1.12 (a) Time-domain envelope waveform and (b) frequency-domain envelope spectra of our nonlinear circuit's response to a two-tone signal; (c) and (d) represent, respectively, the amplitude and phase profiles of the amplifier gain defined as the ratio of the instantaneous output to the instantaneous input complex envelopes.

(v) $\frac{1}{4}g_{m3}H_i^3(\omega_c)a^3(\tau)\cos(3\omega_c t)$ the third harmonic distortion of the amplifier, composed of four tones at $2\omega_1 + \omega_2 = 3\omega_c - \omega_m$ and $2\omega_2 + \omega_1 = 3\omega_c + \omega_m$, of amplitude $\frac{3}{32}g_{m3}|H_i(\omega_c)|^3A^3$, and two sidebands at $3\omega_1 = 3\omega_c - 3\omega_m$ and $3\omega_2 = 3\omega_c + 3\omega_m$, of amplitude $\frac{1}{32}g_{m2}|H_i(\omega_c)|^3A^3$.

As the modulation complexity is increased from a sinusoid to a periodic signal, the amplifier response would still be composed by these frequency clusters located at each of the carrier harmonics. A more detailed treatment of this multitone regime can be found in the technical literature, e.g., [3–5]. Because the common frequency separation of the tones of these clusters is inversely proportional to the modulation period, $2\pi/\omega_m$, a real modulated signal – which is necessarily aperiodic or with an infinite period – produces a spectrum that is composed of continuous bands centered at each of the carrier harmonics. Figure 1.13 illustrates a response to such a real wireless communications signal.

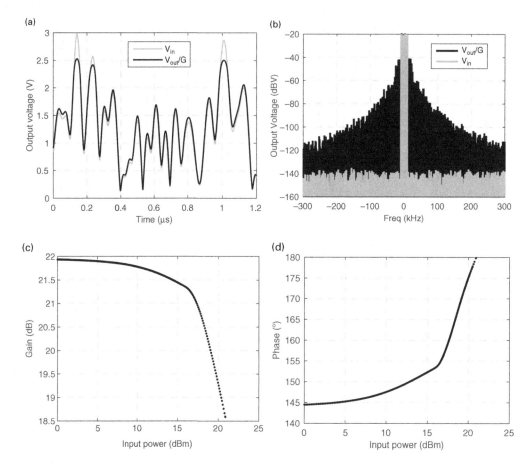

Figure 1.13 (a) Time-domain envelope waveform and (b) frequency-domain envelope spectrum of our nonlinear circuit's response to a 20 kHz bandwidth OFDM wireless signal; (c) and (d) represent, respectively, the amplitude and phase profiles of the amplifier gain defined as the ratio of the instantaneous output to the instantaneous input complex envelopes.

If we now consider a progressively faster envelope, in which the amplifier has not enough time to reach the steady-state before its excitation amplitude or phase changes significantly, this narrowband approximation ceases to be valid and the amplifier exhibits memory to the envelope. This may happen either because the input transfer function $H_i(\omega)$ varies significantly within each of the harmonic clusters, or because the L_{Ch} and C_B bias network elements can no longer be considered short and open circuits, respectively, for the base-band components (of type (ii) above) of this wideband envelope. This means that the response to a particular CW excitation is no longer defined by its instantaneous (to the envelope) amplitude or phase, but also depends on their past envelope states. For example, as the amplitude and phase changes can no longer be entirely defined by the input excitation amplitude, but are different if the envelope stimulus is rising or falling, the AM/AM and AM/PM exhibit hysteresis loops

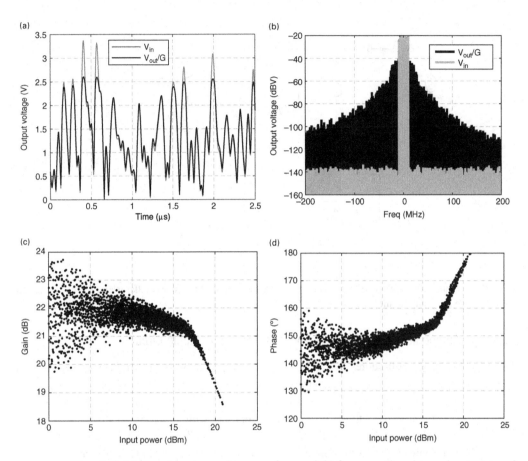

Figure 1.14 (a) Time-domain envelope waveform and (b) frequency-domain envelope spectra of our nonlinear circuit's response to a OFDM wireless signal but now with a much wider bandwidth of 20 MHz; (c) and (d) represent, respectively, the amplitude and phase profiles of the amplifier gain defined as the ratio of the instantaneous output to the instantaneous input complex envelopes.

as the ones previously shown in the measured data of Figure 1.5 and now illustrated again in Figure 1.14. Envelope memory effects is a nontrivial topic, in particular concerning nonlinear power amplifier distortion and linearization studies as can be seen in the immense literature published on this subject [4], [6–9].

In our simple circuit, the main source of these envelope memory effects is the $\frac{1}{2} g_{m2} |H_i(\omega_c)|^2 a^2(\tau) \, i_{ds0}(\tau)$ bias current fluctuation in the drain RF choke inductance, which generates a $V_{ds0}(\tau)$ drain bias voltage variation, and thus gain change as is shown in Figure 1.14 [10]. However, as this $V_{ds0}(\tau)$ variation is given by

$$V_{ds0}(\tau) = V_{DD} - L_{Ch} \frac{di_{ds0}(\tau)}{d\tau} \tag{1.27}$$

there is a gain reduction (increase) when the amplitude envelope is rising (falling) in time, with respect to what would be expected from a memoryless, or static, response.

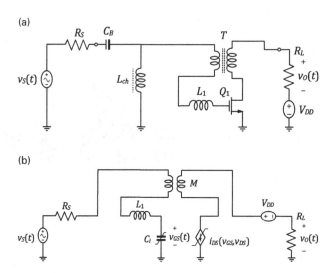

Figure 1.15 (a) Forced transformer-based Armstrong or Meissner oscillator and (b) its simplified equivalent circuit.

1.4.4   Autonomous and Forced Responses from an Unstable Circuit

Now, we will address some of the properties of another important class of nonlinear regimes: the autonomous ones, i.e., regimes capable of generating new frequencies without any excitation. For that, we will use the forced Armstrong or Meissner oscillator [11] because it combines a surprising simplicity with an enormous diversity of effects. Actually, this simple oscillator circuit was invented in the first years of radio, when electronic circuits were implemented with triode tubes. When triodes were replaced by solid-state devices, such as the field-effect transistor, this oscillator acquired the topology shown in Figure 1.15.

This oscillator can be easily distinguished from the above studied amplifier since feedback is no longer any undesirable or parasitic effect but constitutes an essential ingredient of its operation. In this case, the feedback path is provided by the mutual inductor M that couples the input and output circuits of what can be seen as our previous ordinary amplifier.

There are two different types of autonomous regimes, periodic and aperiodic ones. Autonomous periodic regimes are what we normally expect from RF oscillators, while chaotic regimes are strange oscillatory phenomena that are aperiodic, but deterministic, i.e., they cannot be attributed to random noise fluctuations [12–14].

Then, a steady-state periodic oscillatory regime can still be perturbed by an external forcing signal to create a multitude of new phenomena, from phase-locked oscillations, analog frequency division, or self-oscillating mixing [15].

1.4.4.1   Autonomous Periodic Regime

Setting $v_S(t) = 0$, i.e., eliminating any possible external excitation, we obtain the periodic (nearly sinusoidal) autonomous regime depicted in Figure 1.16. In Figure 1.16 (a),

(a)

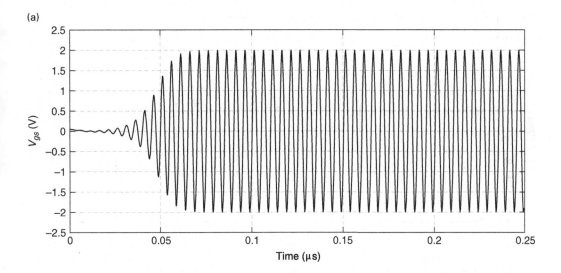

(b)

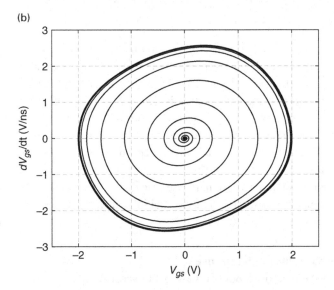

Figure 1.16 (a) $v_{gs}(t)$ voltage in time domain and (b) its corresponding phase-portrait, when the circuit operates in an oscillatory autonomous regime.

we can see the $v_{gs}(t)$ voltage evolution with time, which shows that, because the circuit is unstable in its quiescent point, it evolves from its dc quiescent point toward its sinusoidal steady-state regime.

Figure 1.16 (b) constitutes another commonly adopted alternative representation known as the phase-portrait. It is a graphic of the trajectory, or orbit, of the circuit or system state in the phase-space, i.e., in the state-variables that describe the system. In our second order system, these are the two state-variables of the differential equation that governs this response, $v_{gs}(t)$ and its first derivative in time, $dv_{gs}(t)/dt$, of Eq. (1.18).

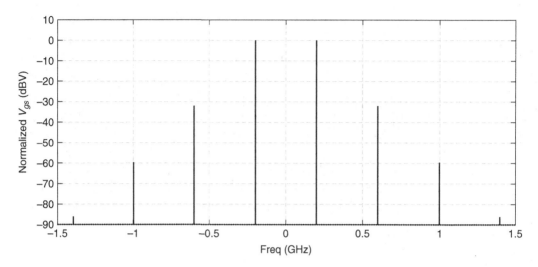

Figure 1.17 Spectrum of the $v_{gs}(t)$ voltage when the circuit operates in an oscillatory autonomous regime.

In the phase-portrait, you can see that the system evolution in time is such that it starts with a null $v_{gs}(t)$, with null derivative (a dc point) and evolves to a cyclic trajectory, also known as a limit cycle, of 2 V amplitude and about $2.5 \times 10^9 \text{ Vs}^{-1}$ $dv_{gs}(t)/dt$ amplitude. This phase-portrait also indicates that the steady state is periodic – as it tends to a constant trajectory, also known as an attractor – and nearly sinusoidal, but not purely sinusoidal. Indeed, if the steady state was a pure sinusoid, i.e., without any harmonics, the steady-state orbit would have to be a circle or an ellipse, depending on the x, $v_{GS}(t)$, and y, $dv_{GS}(t)/dt$, relative scales.

Figure 1.17 presents the corresponding spectrum. Please note that, in the steady state, this periodic waveform is composed of discrete lines at multiples of the fundamental frequency, 200 MHz.

1.4.4.2   Forced Regime over an Oscillatory Response

To conclude this section on the type of responses we should expect from a nonlinear circuit, we now reduce the amount of feedback, i.e., M. As the feedback is now insufficient to develop an autonomous regime, the circuit is stable at its quiescent point. However, when forced with a sufficiently large excitation, it may develop parasitic oscillations as shown in Figures 1.18 and 1.19.

As the circuit is on the edge of instability, it can only develop and sustain the parasitic oscillation for a very particular excitation amplitude. However, there are also cases in which both the forced and autonomous regimes can be permanent and the amplifier/oscillator becomes an injection locked oscillator or a self-oscillating mixer depending on the relation of the frequencies of the excitation and of the oscillatory autonomous regime.

However, intended to illustrate the variety of different response types possible in nonlinear dynamic circuits, Figures 1.20 and 1.21 correspond to none of the

(a)

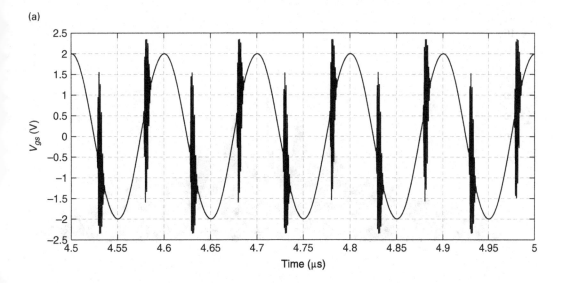

(b)

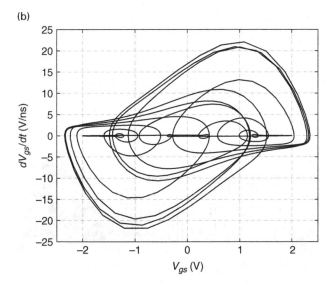

Figure 1.18 (a) $v_{gs}(t)$ voltage in the time domain and (b) its corresponding phase-portrait when the circuit is under a sinusoidal excitation but is evidencing parasitic oscillations.

referred cases. Now, the interaction between the amplitude and frequency of the excitation and of the oscillator itself makes it develop a regime that is neither periodic (indicative of injection locking) nor double periodic (indicative of mixing), but completely aperiodic: a chaotic regime. As shown in Figures 1.20 and 1.21, the regime is nearly periodic but not exactly periodic, as the attractor is not a closed curve that repeats itself indefinitely. Consequently, the spectrum is now filled in with an infinite number of spectral lines spaced infinitesimally closely, i.e., it is a continuous spectrum.

(a)

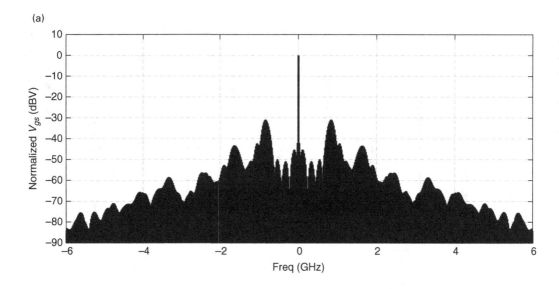

(b)

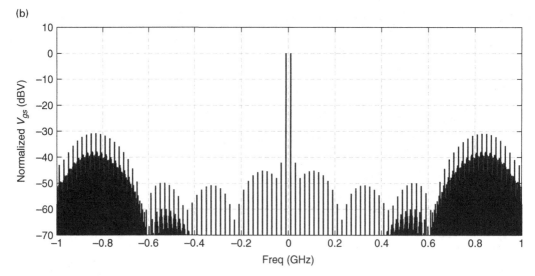

Figure 1.19 (a) Wide span and (b) narrow span spectrum of the $v_{gs}(t)$ voltage when the circuit is operating as an amplifier but exhibiting parasitic oscillations.

1.5   Example of a Static Transfer Nonlinearity

In the previous section, we illustrated the large variety of responses that can be expected from a nonlinear system. Now, we will try to systematize that, treating nonlinear systems according to their response types. That is what we will do in this and the next sections, starting from a static system to then evolve to a dynamic one. Because of the importance played by frequency-domain responses in RF and microwave electronics,

(a)

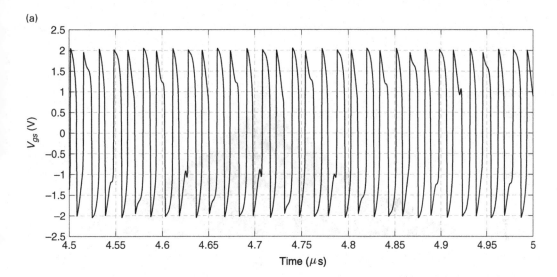

(b)

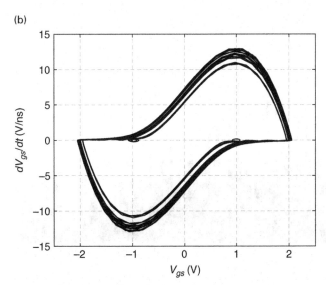

Figure 1.20 (a) $v_{gs}(t)$ voltage in time domain and (b) phase-portrait of the unstable amplifier in chaotic regime.

we will start to study the system's responses in the time domain and then observe their responses in the frequency domain.

1.5.1   The Polynomial Model

One of the simplest static nonlinear models we can think of is the polynomial because it can be interpreted as the natural extension of a static linear operator. Actually, the general Pth degree polynomial form of

(a)

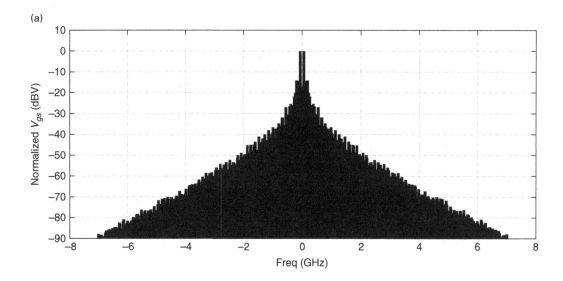

(b)

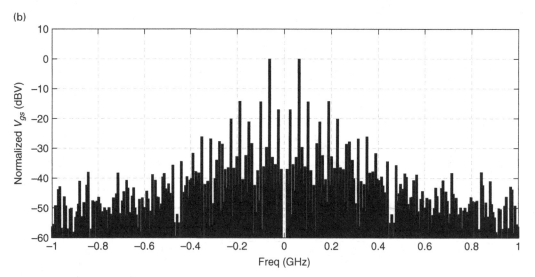

Figure 1.21 (a) Wide span and (b) narrow span spectrum of the $v_{gs}(t)$ voltage of the unstable amplifier in chaotic regime.

$$y(x) \approx \hat{y}(x) = k_0 + \sum_{p=1}^{P} k_p x^p \qquad (1.28)$$

is a natural extension of the linear system model defined by $\hat{y}(x) = k_1 x$.

Moreover, polynomials have universal approximating properties, as they are known to approximate any analytic function, $y(x)$, (one that is smooth, i.e., that is continuous and its derivatives with respect to the input are also continuous), as well as we want, in a certain input amplitude domain.

There is, however, something subtle hidden behind this apparent simplicity that we should point out. As happens to any other model, (1.28) does not define any particular model, but a model format, i.e., a mathematical formulation, or a family of models of our nonlinear static $y(x)$ function, each one specified by a particular set of coefficients $\{k_0, k_1, k_2, \ldots, k_p, \ldots, k_P\}$. To define a particular polynomial approximation, $\hat{y}(x)$, for a particular function, $y(x)$, we need to find a procedure to identify, or extract, the polynomial model's coefficients.

1.5.1.1   The Taylor Series Approximation

For example, we could start thinking of what would be the polynomial approximation that would minimize the error in the vicinity of a particular measured point $[x_0, y(x_0)]$ and then let this error increase smoothly outside that region. Actually, this means that this polynomial approximation should be exact (provide null error) at this fixed point, x_0, and thus it would have the following form:

$$\hat{y}(x) = y(x_0) + k_1(x - x_0) + k_2(x - x_0)^2 + k_3(x - x_0)^3 + \ldots + k_P(x - x_0)^P \quad (1.29)$$

To identify these polynomial coefficients, we could start by realizing that differentiating (1.29) once with respect to x we obtain

$$k_1 = \left.\frac{d\hat{y}(x)}{dx}\right|_{x=x_0} \quad (1.30)$$

Then, performing successive differentiations we would be easily led to conclude that

$$k_p = \frac{1}{p!}\left.\frac{d^p\hat{y}(x)}{dx^p}\right|_{x=x_0} \quad (1.31)$$

and that the polynomial approximation is the Taylor series given by

$$\hat{y}(x) = y(x_0) + \sum_{p=1}^{P}\frac{1}{p!}\left.\frac{d^p y(x)}{dx^p}\right|_{x=x_0}(x - x_0)^p \quad (1.32)$$

Note that the information that must be measured for extracting the $P+1$ coefficients of this polynomial is $y(x_0)$ plus the first P derivatives of $y(x)$ in x_0.

This Taylor series approximation is thus the best polynomial approximation in the vicinity of x_0, or the best small-signal polynomial approximation in that vicinity. So, in our $i_D(v_{GS})$ function example, x_0 can be regarded as the quiescent point V_{GS}, k_1 is the device's transconductance in that quiescent point, and $(x-x_0) = (v_{GS}-V_{GS})$ becomes the small-signal deviation [4], [16–17].

1.5.1.2   The Least Squares Approximation

Unfortunately, as this polynomial approximation directs all the attention to minimize the error in the vicinity of the fixed point, it cannot control the error outside that region, i.e., it may provide a good small-signal, or quasi-linear, approximation, but it will surely be very poor in making large-signal predictions. For these large-signal predictions, we have to find another coefficient extraction method that can take care of the $y(x)$ function behavior in a wider region of its input and output domain than simply the vicinity of a particular fixed point.

For that, we will rely on one of the most attractive features of polynomial models, their linearity in the parameters. Although polynomial responses are obviously non-linear to the input, they are linear-in-the-parameters, which means that, for example, the pth order polynomial response, $k_p x^p$, varies nonlinearly with x, but linearly with its coefficient, or parameter, k_p. And this is what allows the result in discrete time, $y(nT_s)$, of a polynomial approximation as (1.28) to be written as the following linear system of equations

$$
\begin{bmatrix} y(0) \\ \vdots \\ y(n) \\ \vdots \\ y(N) \end{bmatrix} \approx \begin{bmatrix} 1 & x(0) & \cdots & x^p(0) & \cdots & x^P(0) \\ \vdots & \vdots & & \vdots & & \vdots \\ 1 & x(n) & \cdots & x^p(n) & \cdots & x^P(n) \\ \vdots & \vdots & & \vdots & & \vdots \\ 1 & x(N) & \cdots & x^p(N) & \cdots & x^P(N) \end{bmatrix} \begin{bmatrix} k_0 \\ k_1 \\ \vdots \\ k_p \\ \vdots \\ k_P \end{bmatrix} \tag{1.33a}
$$

or in matrix-vector form

$$
\mathbf{y} \approx \mathbf{X}\mathbf{k} \tag{1.33b}
$$

and so to obtain the coefficients' set as the solution of the following linear system of equations:

$$
\begin{bmatrix} k_0 \\ k_1 \\ \vdots \\ k_p \\ \vdots \\ k_P \end{bmatrix} = \begin{bmatrix} 1 & x(0) & \cdots & x^p(0) & \cdots & x^P(0) \\ \vdots & \vdots & & \vdots & & \vdots \\ 1 & x(n) & \cdots & x^p(n) & \cdots & x^P(n) \\ \vdots & \vdots & & \vdots & & \vdots \\ 1 & x(N) & \cdots & x^p(N) & \cdots & x^P(N) \end{bmatrix}^{-1} \begin{bmatrix} y(0) \\ \vdots \\ y(n) \\ \vdots \\ y(N) \end{bmatrix} \tag{1.34a}
$$

or

$$
\mathbf{k} = \mathbf{X}^{-1}\mathbf{y} \tag{1.34b}
$$

in which $\mathbf{X}^{-1}$ stands for the inverse of the matrix $\mathbf{X}$, and $y(0),\ldots,y(N)$ are the $N+1 = P+1$ measured input and output data points $[x(n), y(n)]$.

Although, at this stage, we might think we have just said everything we could have said on polynomial model extraction, that is far from being the case. What we described was a procedure for finding just two specific polynomials. The first one, the Taylor series, focuses all its attention on one fixed point, providing a good small-signal prediction in that quiescent point, but discarding any specific details that may exist outside the vicinity of that point. The second passes through $N+1 = P+1$ measured points, distributed in a wider input domain, but still pays no attention to how the approximation behaves in between these measured data points. Actually, specifying that the polynomial approximation should pass exactly through each of the measured points may not be a good policy, simply because of a fundamental practical reason: you may not be approximating the function, but a corrupted version of it due to the inevitable measurement errors. Fortunately, none of these can be said to be *the* polynomial approximation of $y(x)$ in the defined input domain, but *one of many* possible polynomial approximations of that function in that domain. Different sets of coefficients result in different polynomial approximations, and to say that one is better than the other we should first establish an error criterion.

The most widely accepted criterion is the **normalized mean square error** (NMSE), a metric of the normalized (to the output average power) distance between the actual response $y(n)$ and the approximated one, $\hat{y}(n)$,

$$NMSE \equiv \frac{\dfrac{1}{N+1}\sum_{n=0}^{N}|y(n) - \hat{y}(n)|^2}{\dfrac{1}{N+1}\sum_{n=0}^{N}|y(n)|^2} \tag{1.35}$$

in which n, or nT_s, is again the index of the discrete time samples. Please note that, expressed this way, the model is not approximating the function but its response to a particular input, $x(nT_s)$, which indicates that you may obtain a different set of coefficients, for a different realization of the input. Actually, this is true for any model or model extraction procedure: we never approximate the system operator, but its observable behavior under a certain excitation or group of excitations, which reveals the importance of the excitation selection. Therefore, to find a model that takes care of the error outside the previously mentioned $P+1$ observation points, we have to give it information on the function behavior between these points. In other words, we have to increase the number of excitation points.

At this point it should be obvious that to extract $P + 1$ coefficients we need, at least, $P+1$ realizations, or $N+1 = P+1$ time samples of the input, $x(n)$, and the output, $y(n)$. However, nothing prohibits the use of more data points, and it should be intuitively clear that if we could provide more information about the system behavior (i.e., measured data) than the required minimum, it should not do any harm. On the contrary, it should even be better as we could, for example, compensate for the above referred measurement errors as finite resolution errors (or finite precision number representation) or random errors (due to instrumentation noise or other random fluctuations).

This is what is depicted in Figure 1.22, in which, beyond the ideal $P+1$ [$x(n)$, $y(n)$] point set, we have represented another $P+1$ point set obtained from measurements, and thus corrupted by measurement errors. Please note that, although there is an ideal polynomial that passes exactly through the ideal $P+1$ points (in this case $P=12$), the polynomial that passes through the measured ($P+1 = 13$) points is a different one.

To overcome this modeling problem and move toward our objective of enlarging the approximation domain and providing an extraction method that is more resilient to data inaccuracies, we could think of obtaining a much richer data set to reduce the measurement error through some form of averaging. The question to be answered is: How can we use a larger number of measured data points, say 200 points, to extract the desired

(a)

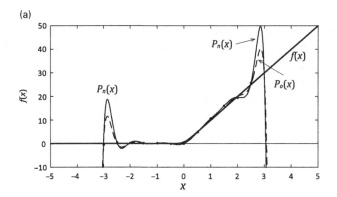

(b)

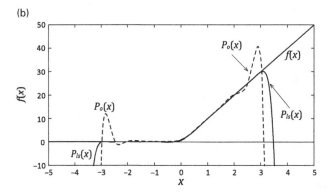

Figure 1.22 The polynomial approximation. (a) $P_o(x)$ is a 12'th degree polynomial that passes through 13 ideal, or error free, data points of $f(x)$ in the $[-3, +3]$ input domain, while $P_n(x)$ is a similar polynomial but now extracted from 13 noisy measured data points. (b) $P_o(x)$ is the same 12'th degree polynomial of (a) extracted from error free data, while $P_{ls}(x)$ is another 12'th degree polynomial that approximates 200 error corrupted measured data points, in the least squares sense, in the same $[-3, +3]$ input domain.

12'th degree polynomial, taking advantage of the extra data points to correct the measurement errors? The answer to this question relies on augmenting the number of lines of (1.33). By doing that, we will relax the need of the 12'th degree polynomial to pass through all 200 points; we will only require that the model approximation passes close to the measured data in the minimum mean squared error sense of (1.35). That is, we will find the polynomial coefficient set that obeys

$$\min\left\{\sum_{n=0}^{N}\left[y(n) - \sum_{p=0}^{P} k_p x^p(n)\right]^2\right\} \tag{1.36}$$

Since it can be shown that there is only one minimum and no maximum, such a solution can be obtained by equating to zero the derivative of the error function with respect to the vector, **k**, of the polynomial coefficients, making

$$\sum_{n=0}^{N} \frac{d}{d\mathbf{k}} \left[y(n) - \sum_{p=0}^{P} k_p x^p(n) \right]^2 = 0 \tag{1.37}$$

This equation can be expressed as

$$\sum_{n=0}^{N} \frac{d}{dk_0} \left[y(n) - \sum_{p=0}^{P} k_p x^p(n) \right]^2 = \sum_{n=0}^{N} 2 \left[y(n) - \sum_{p=0}^{P} k_p x^p(n) \right] = 0$$

$$\sum_{n=0}^{N} \frac{d}{dk_1} \left[y(n) - \sum_{p=0}^{P} k_p x^p(n) \right]^2 = \sum_{n=0}^{N} 2 \left[y(n) - \sum_{p=0}^{P} k_p x^p(n) \right] x(n) = 0$$

$$\vdots$$

$$\sum_{n=0}^{N} \frac{d}{dk_p} \left[y(n) - \sum_{p=0}^{P} k_p x^p(n) \right]^2 = \sum_{n=0}^{N} 2 \left[y(n) - \sum_{p=0}^{P} k_p x^p(n) \right] x^p(n) = 0 \tag{1.38}$$

$$\vdots$$

$$\sum_{n=0}^{N} \frac{d}{dk_P} \left[y(n) - \sum_{p=0}^{P} k_p x^p(n) \right]^2 = \sum_{n=0}^{N} 2 \left[y(n) - \sum_{p=0}^{P} k_p x^p(n) \right] x^P(n) = 0$$

which can also be written as

$$\sum_{n=0}^{N} y(n) = \sum_{n=0}^{N} \left[\sum_{p=0}^{P} k_p x^p(n) \right]$$

$$\sum_{n=0}^{N} y(n)x(n) = \sum_{n=0}^{N} \left[\sum_{p=0}^{P} k_p x^p(n) \right] x(n) = 0$$

$$\vdots$$

$$\sum_{n=0}^{N} y(n)x^p(n) = \sum_{n=0}^{N} \left[\sum_{p=0}^{P} k_p x^p(n) \right] x^p(n) = 0 \tag{1.39}$$

$$\vdots$$

$$\sum_{n=0}^{N} y(n)x^P(n) = \sum_{n=0}^{N} \left[\sum_{p=0}^{P} k_p x^p(n) \right] x^P(n) = 0$$

or, in matrix form, as

$$\mathbf{X}^T \mathbf{y} = \mathbf{X}^T \mathbf{X} \, \mathbf{k} \tag{1.40}$$

which can be solved for $\mathbf{k}$ through

$$\mathbf{k} = \left(\mathbf{X}^T \mathbf{X} \right)^{-1} \mathbf{X}^T \mathbf{y} \tag{1.41}$$

in which $\mathbf{X}^T$ stands for the transpose matrix of $\mathbf{X}$. This minimum squared error solution is also depicted in our polynomial approximation example of 1.22.

There is an interpretation of this derivation that is worth discussing. It arises when we realize that the left-hand side of (1.39) and (1.40), $\mathbf{X}^T\mathbf{y}$, is the cross-correlation at zero of the measured output, $y(n)$, with each of the polynomial model kernels, $x^p(n)$, while their right-hand side is the product of the cross-correlation of all model kernel responses with themselves, the so-called regression matrix $(\mathbf{X}^T\mathbf{X})$ times the model coefficients, $\mathbf{k}$ [18]. Indeed, the discrete cross-correlation between two time functions $x(n)$ and $y(n)$ is calculated by

$$R_{xy}(m) = \frac{1}{N+1} \sum_{n=0}^{N} x(n)y(n-m) \tag{1.42}$$

If the excitation were selected so that each model kernel response was independent, or uncorrelated, of all the others, i.e., if

$$\sum_{n=0}^{N} x^{p_1}(n)x^{p_2}(n) = 0 \tag{1.43}$$

except when $p_1 = p_2$, we would say that the model kernels' responses to that particular excitation were orthogonal, and the regression matrix $(\mathbf{X}^T\mathbf{X})$ would be a diagonal matrix composed by the responses' auto-correlations at zero. In that case, the coefficients were simply the cross-correlation of the output to its corresponding model kernel, normalized by the power (auto-correlation at zero) of the corresponding model kernel response. Hence, each of the model coefficients, k_p, would be a measure of "how much" of its corresponding model kernel, $x^p(n)$, is in the observed output, $y(n)$, in a similar way as the coefficients of the Fourier series of a signal constitute a metric of "how much" of the corresponding harmonic components are in that signal[1].

In the opposite extreme case, i.e., when the excitation is poorly selected so that two or more kernels cannot be independently exposed, the regression matrix has a nearly zero determinant, the linear system of equations of (1.40) is then said to be ill-conditioned, and the coefficients cannot be correctly extracted.

1.5.2   Time-Domain Response of a Static Transfer Nonlinearity

The calculation of the response of our (static) polynomial to a general excitation in time domain is a trivial task. Since a static system responds to an excitation in an instantaneous way (i.e., it only reacts to the excitation at the present time), the system response at any arbitrary instant t_0, $y(t_0)$, can be given by the algebraic computation of the model of our system at that instant t_0, $y(t_0) = S[x(t_0)]$. Thus, the response to a general excitation $x(t)$ can be given by the successive application of this rule for each time instant: $y(t) = S[x(t)] = k_0 + k_1 x(t) + k_2 x^2(t) + k_3 x^3(t) + \dots$.

---

[1] The interested reader can, for example, consult [19] on orthogonal polynomials for various excitations, such as the sinusoid (the Chebyshev polynomial). the Gaussian noise (the Hermite polynomial), or the triangular wave (the Legendre polynomial).

1.5.3   Frequency-Domain Response of a Static Transfer Nonlinearity

To express the response of our polynomial static nonlinearity in the frequency domain, we will start again by stating that $x(n)$ should be expressed as the discrete Fourier series

$$x(n) = \sum_{q=-Q}^{Q} X_q e^{jq\omega nT_s} \tag{1.44}$$

where X_q is the abbreviated form of $X(q\omega)$, which we will then substitute in (1.28) to obtain

$$
\begin{aligned}
y(n) = k_0 &+ k_1 \sum_{q=-Q}^{Q} X_q e^{jq\omega nT_s} + k_2 \sum_{q_1=-Q}^{Q} \sum_{q_2=-Q}^{Q} X_{q_2} X_{q_1} e^{j(q_1+q_2)\omega nT_s} \\
&+ k_3 \sum_{q_1=-Q}^{Q} \sum_{q_2=-Q}^{Q} \sum_{q_3=-Q}^{Q} X_{q_1} X_{q_2} X_{q_3} e^{j(q_1+q_2+q_3)\omega nT_s} \\
&+ \ldots + k_p \sum_{q_1=-Q}^{Q} \ldots \sum_{q_P=-Q}^{Q} X_{q_1} \ldots X_{q_P} e^{j(q_1+\ldots+q_P)\omega nT_s}
\end{aligned} \tag{1.45}
$$

This time-domain response can then be rewritten to give

$$
\begin{aligned}
Y_q = k_1 X_q &+ k_2 \sum_{q_1=-Q}^{Q} X_{q_1} X_{q-q_1} + k_3 \sum_{q_1=-Q}^{Q} \sum_{q_2=-Q}^{Q} X_{q_1} X_{q_2} X_{q-(q_1+q_2)} \\
&+ \ldots + k_P \sum_{q_1=-Q}^{Q} \ldots \sum_{q_{P-1}=-Q}^{Q} X_{q_1} \ldots X_{q_{P-1}} X_{q-(q_1+\ldots+q_{P-1})}
\end{aligned} \tag{1.46}
$$

in which Y_q, or $Y(q\omega)$, stands again for the $q\omega$ frequency component of $y(t)$ and constitute the desired frequency-domain response model. A close look into the second order response of (1.46) reveals that the Fourier component at, for example, the second harmonic, $+2\omega$, results from the products of the component pairs at $[-(Q-2),Q]$, $[-(Q-3),(Q-1)]$, $\ldots$, $[-1,3]$, $[0,2]$, $[1,1]$, $[2,0]$, $[3,-1]$, $\ldots$, $[Q,-(Q-2)]$. In general, as expressed in (1.46), the result at $q\omega$ results from all possible summations of terms arising from products of the components at any $q_1\omega$ and $(q-q1)\omega$, i.e., (1.46) is a discrete convolution. As a matter of fact, originating, in the time domain, from the product of $x(t)$ by itself, in the frequency domain it should correspond to a convolution of $X(\omega)$ by itself. Consequently, the p'th order term of (1.46), which arises from the time-domain product of $x(t)$ by itself p times, results in the convolution of $X(\omega)$ by itself p times.

What we have just found leads us to an important property of static nonlinear systems, which we will find many other times during the course of our study. Because the response at the time sample n, $y(n)$, only depends on the excitation of that time sample, $x(n)$, i.e., it is a local operation, it can be calculated point by point, regardless of the excitation at any other time instant. However, that same static nonlinear function becomes a nonlocal operation, when computed in the frequency domain, since the response at each $q\omega$ depends, in general, of the excitation's input frequency content at all frequencies.

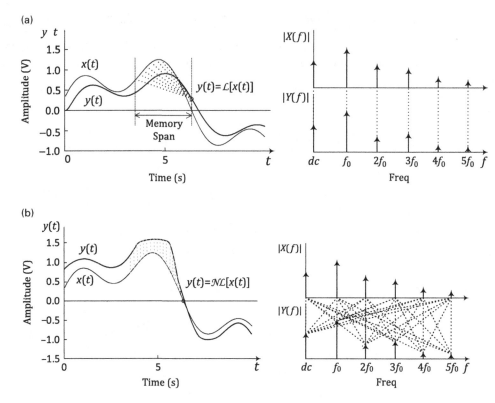

Figure 1.23 Local and nonlocal responses in time and frequency domains: (a) depicts the time and frequency mappings found in linear dynamic systems, which are nonlocal in time but local in frequency, while (b) illustrates the local in time and nonlocal in frequency mappings of nonlinear static systems.

The property we have just described is a consequence of the fact that the Fourier transform of a local function of time (frequency), such as a product, has its support spread throughout all frequencies (times), since it is converted in a convolution. Actually, we had already seen a nonlocal operation in time, which was the convolution that arose as the response of dynamic linear systems. And, in that case, it was the frequency-domain response that was local, as it was given by the product of the Fourier-transform of the impulse response function by the Fourier transform of the input stimulus.

These important conclusions are summarized in Figure 1.23.

1.6   Example of a Dynamic Transfer Nonlinearity

Now that we have already addressed the example of a static transfer nonlinearity, we can pass to a dynamic one. For that, we will review the linear case and explain how we can generalize it to the nonlinear domain. Similarly to what we have done with the static

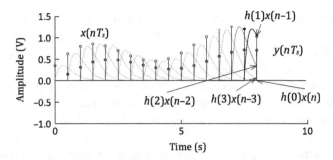

Figure 1.24 The convolutional response of a linear system to an input described as the summation of several impulse responses that extend from the observation time nT_s up to the system's memory span $(n-M)T_s$.

case, we will first obtain the time-domain response of our dynamic system and then move to derive its frequency-domain counterpart.

1.6.1 Time-Domain Response of a Dynamic Transfer Nonlinearity

When, in Section 1.2, we were reviewing the time-domain response of a linear time-invariant dynamic system to an input $x(t)$, we concluded that it should have the form of a convolution of this stimulus $x(t)$ with the impulse response function $h(t)$. If the system were assumed stable, causal, and of finite memory, the impulse response function would start at $t=0$ and last up to some memory span $t=T$. In the discrete time nT_s, we would say that, if $h(mT_s)$ is the response to a unity impulse centered at nT_s, and the input can be expressed as a combination of successive past impulses appearing at nT_s, $(n-1)T_s$, ..., $(n-M)T_s$, or simply, n, $(n-1)$, ..., $(n-M)$, then the system response at nT_s, $y(n)$, can be expressed as the summation, or linear combination, of $h(0)x(n)$, $h(1)x(n-1)$, ..., $h(M)x(n-M)$, which constitutes the discrete convolution, as illustrated in Figure 1.24:

$$y(n) = \sum_{m=0}^{M} h(m)x(n-m) \tag{1.47}$$

Therefore, in linear systems, we can assume that the response to a summation of one unity impulse at nT_s, $x(n)$, and another at $(n-1)T_s$, $x(n-1)$, is the summation of their responses, as if they were applied separately, as shown in Figure 1.24. In nonlinear systems that is not the case; superposition no longer applies, and (1.47) becomes invalid. In fact, we can even question the usefulness of having $x(t)$ expanded as a summation of some predefined basis function as the Dirac delta function.

Nevertheless, we need to start from something to be able to proceed, and so we will keep this expansion of the input stimulus. Hence, we will assume that, beyond the linear response, the system responds to a pair of impulses, say $x(n)$ and $x(n-1)$, with some unknown response that depends on the magnitude of the two impulses and on their relative location. Unfortunately, this still involves such a high level of generality that impedes any progress. However, we can try to move another small step forward if we assume that this two-impulse response is similar to the one that we could expect from a polynomial, i.e., something like $k_2x^2(n)$, $2k_2x(n)x(n-1)$ and $k_2x^2(n-1)$. Unfortunately, our intuition tells us that this cannot yet be right as we know that, if the system has memory, its response to the two impulses must depend on the relative position of the two impulses and present a tail in time. So, we should consider different coefficients and different contributions for any of the two different impulse times.

For example, if the two impulses were both present at the same time, we should have something like

$$\sum_{m=0}^{M} k_2(m)x^2(n-m) \tag{1.48}$$

But, if the two impulses appeared at different instants in time, we should now have the following more general bidimensional form

$$\sum_{m_1=0}^{M}\sum_{m_2=0}^{M} h_2(m_1,m_2)x(n-m_1)x(n-m_2) \tag{1.49}$$

in which $h_2(m_1,m_2)$ plays a similar role as the linear impulse response $h(m)$, but now for the second order response. If $h(m)$, or $h_1(m)$, is the linear, or first order, impulse response function, $h_2(m_1,m_2)$ is the system's nonlinear second order impulse response function.

So, the generalization of (1.47) from linear to nonlinear systems and the generalization of (1.28) from a static polynomial to one with memory must be given by [18]

$$
\begin{aligned}
y(n) = h_0 &+ \sum_{m=0}^{M} h_1(m)x(n-m) + \sum_{m_1=0}^{M}\sum_{m_2=0}^{M} h_2(m_1,m_2)x(n-m_1)x(n-m_2) \\
&+ \sum_{m_1=0}^{M}\sum_{m_2=0}^{M}\sum_{m_3=0}^{M} h_3(m_1,m_2,m_3)x(n-m_1)x(n-m_2)x(n-m_3) \\
&+ \ldots + \sum_{m_1=0}^{M}\ldots\sum_{m_P=0}^{M} h_P(m_1,\ldots,m_P)x(n-m_1)\ldots x(n-m_P)
\end{aligned}
\tag{1.50}
$$

Expressed in continuous time, each of these discrete multidimensional convolutions becomes a continuous convolution in a similar way as the first order one of (1.6) turned into (1.7), and the summations of (1.50) become multidimensional integrals (see, e.g., [19]).

This model plays an important role in nonlinear dynamic systems since it is the simplest one can conceive to represent, simultaneously, nonlinear and dynamic behavior. Furthermore, its derivation can give important hints on the inherent complexity necessary to model these nonlinear dynamic devices. Nevertheless, it should be noted that, although (1.28) is a universal approximator of static analytic functions, (1.50) is restricted to dynamic operators that, beyond being analytic, must also be causal and stable and possess finite memory. In fact, the summations of (1.50) have support on only the present and the past time (the summation indexes m can only take positive integer values) (causality), the model cannot mimic any unstable behavior as it does not incorporate feedback (stability), and the present output does not depend on the inputs that appeared in an infinitely remote past, but are limited to $x(n-MT_s)$ (finite memory).

Similarly to what we said about the static polynomial of (1.28), this dynamic polynomial is only a general form and its complete specification depends on the particular way the coefficients are determined.

If the coefficients of (1.50) are selected as multidimensional partial derivatives in a predefined fixed point, $[x_0, h_0 = y(x_0)]$, (1.50) becomes a Taylor series with memory, also known as the Volterra series [19]. The coefficients that define the first order impulse response function of (1.47) and (1.50) constitute the linear dynamic response in the vicinity of the quiescent point, and all the other higher order coefficients acquire a similar role. If, on the other hand, an approximation in the minimum mean square error sense is desired, then the coefficients must be determined through a generalization of what was already discussed for the static case. Indeed, since (1.50) can be expressed as

$$
\begin{bmatrix} y(0) \\ \vdots \\ y(n) \\ \vdots \\ y(N) \end{bmatrix} = \begin{bmatrix} 1 & \dots & x(0-m) & \dots & x(0-m_1)\dots x(0-m_p) & \dots & x(0-M)\dots x(0-M) \\ \vdots & & \vdots & & \vdots & & \vdots \\ 1 & \dots & x(n-m) & \vdots & x(n-m_1)\dots x(n-m_p) & \dots & x(n-M)\dots x(n-M) \\ \vdots & & \vdots & & \vdots & & \vdots \\ 1 & \dots & x(N-m) & \dots & x(N-m_1)\dots x(N-m_p) & \dots & x(N-M)\dots x(N-M) \end{bmatrix}
$$

$$
\cdot \begin{bmatrix} h_0 \\ h_1(0) \\ \vdots \\ h_1(m) \\ \vdots \\ h_1(M) \\ h_2(0,0) \\ \vdots \\ h_2(m_1,m_2) \\ \vdots \\ h_2(M,M) \\ \vdots \\ h_p(m_1,\dots,m_p) \\ \vdots \\ h_P(M,\dots,M) \end{bmatrix} \tag{1.51a}
$$

or

$$\mathbf{y} = \mathbf{X}\mathbf{h} \tag{1.51b}$$

then the coefficients that minimize the mean squared error must be given by

$$\mathbf{h} = \left(\mathbf{X}^T\mathbf{X}\right)^{-1}\mathbf{X}^T\mathbf{y} \tag{1.52}$$

Before closing this subject, it is convenient to discuss two issues. The first one is related to an alternative interpretation of (1.50), while the second refers to the inclusion of feedback.

When studying linear system theory, we learned that a system like the one described by (1.47) was named a finite impulse response filter, FIR, and that (1.47) could be derived with a different, but obviously equivalent, reasoning. In short, the idea was to say that if our system has a finite memory, its response, $y(n)$, must be given as a function of the present input, $x(n)$, and all the past samples up to the memory span, $x(n-M)$:

$$y(n) = f[x(n), x(n-1), \ldots, x(n-m), \ldots, x(n-M)] \tag{1.53}$$

Then, if we added that the system is linear, we were saying that this function $f[.]$ was linear, which means that its output should be a linear combination of its inputs, i.e., a summation of all of its $M+1$ inputs, each one multiplied by a coefficient, and the result was the one of (1.47).

If we now take this idea and want to generalize it to a nonlinear system, we would say that, if $f[.]$ was nonlinear, (1.53) would now represent some kind of nonlinear finite impulse response filter, or nonlinear FIR filter. And if, beyond that, we would say that we would approximate the $(M+1)$-to-1 multidimensional function $f[.]$ by a polynomial, (1.50) would become the linear combination of (1.47) for the first order response, the second order combination of (1.49) for the second order response, and so on, ending up with the complete response of (1.50). So, the $h_p(m_1, \ldots, m_p)$ parameters can either be interpreted as groups of coefficients that constitute the system's discrete p'th order nonlinear impulse response functions or seen separately as the coefficients of a P'th order $(M+1)$-to-1 multidimensional polynomial approximation of our nonlinear FIR filter.

Actually, this alternative interpretation of (1.50) can even allow us to generalize the introduced model of our nonlinear dynamic system in two different directions.

One way to generalize this is to say that the polynomial form of (1.50) is only a particular approximation of the desired multidimensional function $f[.]$ and other possibilities could be tried. For example, if $f[.]$ was approximated by a multidimensional table in which the delayed inputs $x(n-m)$ are the indexing data and the table content is some interpolation of the measured $y(n)$ response, we would have a time-delayed look-up table, TD-LUT, model. As another alternative, we could approximate $f[.]$ by an artificial neural network [20] of the form

$$y(n) = b_o + \sum_{p=1}^{P} w_{o,p} f_a \left[b_{i,p} + \sum_{m=0}^{M} w_{i,m,p} x(n-m) \right] \tag{1.54}$$

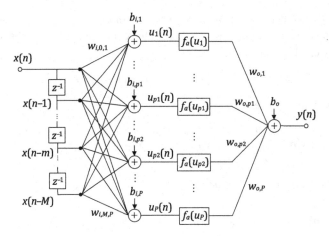

Figure 1.25 The time delay artificial neural network model which can be seen as a generalization of the polynomial FIR filter.

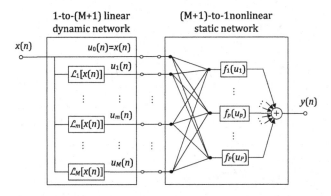

Figure 1.26 Canonic Wiener Model as a generalization of the nonlinear polynomial FIR filter. It is composed of a 1-to-$(M+1)$ linear dynamic filter bank followed by a $(M+1)$-to-1 static, or memoryless, nonlinear function.

in which b_o, the output bias, $w_{o,p}$, the output weights, $b_{i,p}$, the input biases, and $w_{i,m,p}$, the input weights, are fitting coefficients and $f_a[.]$, the activation function, is some appropriate nonlinear function, typically a sigmoidal function like the hyperbolic tangent, $f_a(x) = \tanh(x)$, or a radial-basis function like the Gaussian, $f_a(x) = \exp(-x^2)$. This structure, named the time-delay artificial neural network, TD-ANN, is depicted in Figure 1.25.

Another more elaborate way to generalize (1.50) is to say that the delayed versions of the input are simply one of many possible alternatives of including memory. In general, we could say that a nonlinear FIR model could be represented as a 1-to-$(M+1)$ linear dynamic filter bank followed by a $(M+1)$-to-1 nonlinear static function as shown in Figure 1.26, the so-called Wiener Canonic Model [19].

Another, even more general way to express the response of a nonlinear dynamic system, would be to lift the restriction that the system should be stable and of finite memory. For that, we again recall what we learned from linear systems in that beyond the linear feedforward FIR model, we could also have a recursive, or feedback, infinite impulse response, IIR, filter of the form

$$y(n) = f[x(n), \ldots, x(n-M), y(n-1), \ldots, y(n-L)] \tag{1.55}$$

so that (1.47) would become

$$y(n) = \sum_{m=0}^{M} h_{10}(m)x(n-m) + \sum_{l=1}^{L} h_{01}(l)y(n-l) \tag{1.56}$$

and (1.50) could be generalized to be

$$y(n) = h_0 + \sum_{m=0}^{M} h_{10}(m)x(n-m) + \sum_{l=1}^{L} h_{01}(l)y(n-l) + \ldots$$

$$+ \sum_{m_1=0}^{M} \sum_{m_2=0}^{M} h_{20}(m_1, m_2)x(n-m_1)x(n-m_2)$$

$$+ \sum_{m=0}^{M} \sum_{l=1}^{L} h_{11}(m, l)x(n-m)y(n-l)$$

$$+ \sum_{l_1=1}^{L} \sum_{l_2=1}^{L} h_{02}(l_1, l_2)y(n-l_1)x(n-l_2) \tag{1.57}$$

$$+ \sum_{m_1=0}^{M} \ldots \sum_{m_P=0}^{M} h_{P0}(m_1, \ldots, m_P)x(n-m_1) \ldots x(n-m_P)$$

$$+ \sum_{m_1=0}^{M} \ldots \sum_{m_{P-r}=0}^{M} \sum_{l_1=1}^{L} \ldots \sum_{l_r=1}^{L} h_{(P-r)r}(m_1, \ldots, m_{P-r}, l_1, \ldots, l_r)$$

$$x(n-m_1) \ldots x(n-m_{P-r})y(n-l_1) \ldots y(n-l_r)$$

Please note that this model now involves feedback and so it can be used to model unstable or even autonomous systems. In fact, feeding the input from the past outputs and starting from a nonzero initial output state, the output can either decrease in time toward zero or tend to any steady-state constant, increase or decrease without bound, or tend to a steady-state periodic or chaotic oscillatory behavior, mimicking the wide range of different behaviors we noted in Section 1.4. Unfortunately, this new capability may become a disadvantage when, due to imperfections of the model, we realize that we were using an unstable model to represent a stable system.

Another issue worth noting is that (1.52) can still be used to extract such a model as long as the **y**, **X**, and **h** matrices are augmented accordingly. In particular, beyond the input information, **X** now also includes past output data. However, it is also true that if the feedforward FIR model already involves a huge number of coefficients, this number is further increased in the IIR model of (1.57).

Finally, we would like to make a short note to say that it is not difficult to conceive a structure similar to the Canonic Wiener Model, but now with feedback, and so to imagine corresponding generalizations of the above mentioned TD-LUT and TD-ANN models.

1.6.2   Frequency-Domain Response of a Dynamic Transfer Nonlinearity

Similarly to what we did to obtain the frequency-domain response of first the linear dynamic and then the nonlinear static systems, we will start by stating again the decomposition of the input in the Fourier series of (1.44). Then substituting it in the nonlinear FIR filter of (1.50), we get

$$
\begin{aligned}
y(n) = h_0 &+ \sum_{m=0}^{M} \sum_{q=-Q}^{Q} h_1(m) X_q e^{jq\omega(n-m)T_s} \\
&+ \sum_{m_1=0}^{M} \sum_{m_2=0}^{M} \sum_{q_1=-Q}^{Q} \sum_{q_2=-Q}^{Q} h_2(m_1,m_2) X_{q_1} e^{jq_1\omega(n-m_1)T_s} X_{q_2} e^{jq_2\omega(n-m_2)T_s} \\
&+ \ldots + \sum_{m_1=0}^{M} \cdots \sum_{m_p=0}^{M} \sum_{q_1=-Q}^{Q} \cdots \sum_{q_p=-Q}^{Q} h_P(m_1,\ldots,m_p) X_{q_1} e^{jq_1\omega(n-m_1)T_s} \cdots \\
&\qquad\qquad \ldots X_{q_p} e^{jq_p\omega(n-m_p)T_s}
\end{aligned}
\tag{1.58}
$$

Grouping the summations on the indexes m_p and n, separately, and defining the multidimensional frequency-domain transfer function, $H_P(q_1\omega,\ldots,q_p\omega)$ or $H_P(q_1,\ldots,q_p)$, as the multidimensional Fourier transform of the p'th order nonlinear impulse response function

$$
H_P(q_1,\ldots,q_p) = \sum_{m_1=0}^{M} \cdots \sum_{m_p=0}^{M} h_p(m_1,\ldots,m_p) e^{-j\omega(q_1 m_1 + q_p m_p)T_s}
\tag{1.59}
$$

we get

$$
\begin{aligned}
y(n) = h_0 &+ \sum_{q=-Q}^{Q} H_1(q) X_q e^{jq\omega nT_s} \\
&+ \sum_{q_1=-Q}^{Q} \sum_{q_2=-Q}^{Q} H_2(q_1,q_2) X_{q_1} X_{q_2} e^{j(q_1+q_2)\omega nT_s} \\
&+ \ldots + \sum_{q_1=-Q}^{Q} \cdots \sum_{q_p=-Q}^{Q} H_P(q_1,\ldots,q_P) X_{q_1} \cdots X_{q_p} e^{j(q_1+\ldots+q_p)\omega nT_s}
\end{aligned}
\tag{1.60}
$$

From this expression, we can now derive the desired Fourier coefficients of the $y(n)$ time-domain response, $Y(q\omega)$ or Y_q, as

$$
\begin{aligned}
Y_q = H_1(q) X_q &+ \sum_{q_1=-Q}^{Q} H_2(q_1, q - q_1) X_{q_1} X_{q-q_1} \\
&+ \ldots + \sum_{q_1=-Q}^{Q} \cdots \sum_{q_{P-1}=-Q}^{Q} H_P[q_1,\ldots,q - (q_1 + \ldots + q_{P-1})] \\
&\qquad\qquad X_{q_1} \cdots X_{q-(q_1+\ldots+q_{P-1})}
\end{aligned}
\tag{1.61}
$$

which is again a series of frequency-domain convolutions, but now weighted by the nonlinear transfer functions $H_P(q_1,\ldots,q_p)$. Actually, this expression is a generalization of the one obtained for the static case, Eq. (1.46). Since a static system has no memory, its response is independent of frequency, and therefore $H_p(q_1, \ldots, q_p)$ is a constant, k_p, which can be written outside the summation. This independence of the $H_p(q_1, \ldots, q_p)$ transfer functions on frequency is the mathematical statement of the fact that the system is static or memoryless as, in that case, its response cannot depend on the excitation past but only on the present excitation, which implies that $h_p(m_1, \ldots, m_p)$ must be zero except when all $m_1, \ldots, m_p$ are zero. This means that $h_p(m_1, \ldots, m_p)$ is a multidimensional impulse and so its Fourier transform $H_p(q_1, \ldots, q_p)$ must be constant.

1.7   Summary

The objective of this first chapter was to give the reader a detailed introduction to nonlinear systems.

We started by showing that although nonlinearity is essential to wireless communications and present in the majority of RF and microwave circuits, it is usually not given the necessary attention because we lack the elegant and simple analytical tools we were accustomed to use in linear circuits.

In an attempt to show the relationship between what is known from the linear systems and what is studied for the first time about the nonlinear systems, we tried to show how nonlinear systems, i.e., their characteristics and their mathematical modeling tools, can be understood as a generalization of linear systems. However, we hope we also conveyed the message that because linearity is a very restrictive concept that can be condensed into a simple mathematical relation – the superposition principle – this generalization to nonlinear systems is much more complex in terms of both the type of responses and required mathematical analyses than one might expect. On the other hand, it is exactly this complexity of behaviors, and their importance, that make nonlinear circuits an exciting topic and an attractive area for research and practice.

1.8   Exercises

Exercise 1.1   A voltage amplifier with offset is a circuit that presents an open-circuit output voltage, $v_o(t)$, that can be related to its input voltage $v_i(t)$ as $v_o(t) = A_v v_i(t) + V_{offset}$. Using (1.1) and (1.2), prove that such an amplifier is nonlinear, even if its input–output graphic representation is a straight line.

Exercise 1.2   What could be done to convert our voltage amplifier of Exercise 1.1 into a linear system?

Hint: Interpret the offset, V_{offset}, as the response to a second independent variable, or stimulus.

Exercise 1.3   Prove, using (1.1) and (1.2), that the most basic circuit of wireless communications, the ideal amplitude modulator, a multiplier of two inputs – the modulation envelope and the carrier – that produces one single output – the resulting modulated signal – must be a nonlinear system.

Exercise 1.4 Prove again that the ideal amplitude modulator of Exercise 1.3 must be nonlinear, but now using the result of Section 1.2, which stated that the response of any linear system (dynamic or static) to a sinusoid is another sinusoid of the same frequency.

Exercise 1.5 Show that the ideal amplitude modulator, which, as seen in Exercise 1.3, must inherently be a nonlinear system of two inputs (the modulation envelope and the carrier) and one output (the resulting modulated signal) can be seen as a linear time-variant system of one single input (the modulation envelope) and one output (the same resulting modulated signal).

Exercise 1.6 Prove that the sinusoidal oscillator model of Eq. (1.15) indeed represents a linear system.

References

[1] S. Benedetto, E. Biglieri and R. Daffara, "Modeling and performance evaluation of nonlinear satellite links – a Volterra series approach," *IEEE Trans. Aerospace Electronic Syst.*, vol. AES-15, Apr. 1979, 494–507.

[2] L. C. Nunes, P. M. Cabral and J. C. Pedro, "AM/AM and AM/PM distortion generation mechanisms in Si LDMOS and GaN HEMT based RF power amplifiers," *IEEE Trans. Microw. Theory Techn.*, vol. MTT-62, Apr. 2014, 799–809.

[3] J. C. Pedro and N. B. Carvalho, "On the use of multitone techniques for assessing RF components' intermodulation distortion," *IEEE Trans. on Microw. Theory Techn.*, vol. 47, Dec. 1999, 2393–2402.

[4] J. C. Pedro and N. B. Carvalho, *Intermodulation Distortion in Microwave and Wireless Circuits*, Norwood, MA: Artech House, 2003.

[5] N. Boulejfen, A. Harguem, and F. A. Ghannouchi, "New closed-form expressions for the prediction of multitone intermodulation distortion in fifth-order nonlinear RF circuits/systems," *IEEE Trans. Microw. Theory Techn.*, vol. MTT-52, Jan. 2004, 121–132.

[6] W. Bosch and G. Gatti, "Measurement and simulation of memory effects in predistortion linearizers," *IEEE Trans. Microw. Theory Techn.*, vol. MTT-37, Dec. 1989, 1885–1890.

[7] N. B. Carvalho and J. C. Pedro, "A comprehensive explanation of distortion sideband asymmetries," *IEEE Trans. Microw. Theory Techn.*, vol. MTT-50, Sep. 2002, 2090–2101.

[8] D. R. Morgan, Z. Ma, J. Kim, M. G. Zierdt and J. Pastalan, "A generalized memory polynomial model for digital predistortion of RF power amplifiers," *IEEE Trans. Sig. Process.*, vol. SP-54, Oct. 2006, 3852–3860.

[9] J. Verspecht, J. Horn, L. Betts, D. Gunyan, R. Pollard, C. Gillease and D. E. Root, "Extension of X-parameters to include long-term dynamic memory effects," *IEEE MTT-S Int. Microw. Symp. Dig.*, 2009, 741–744.

[10] T. R. Cunha, E. G. Lima and J. C. Pedro, "Validation and physical interpretation of the power-amplifier polar volterra model," *IEEE Trans. Microw. Theory Techn.*, vol. MTT-58, Dec. 2010, 4012–4021.

[11] A. Meissner, "Production of waves by cathode ray tubes," U.S. Patent 1 924 796 A, Aug. 29, 1933 (filed Mar. 16, 1914).

[12] T. Matsumoto, "Chaos in electronic circuits," *Proc. IEEE*, vol. 75, Aug. 1987, 1033–1057.

[13] L. O. Chua, C. W. Wu, A. Huang and G.-Q. Zhong, "A universal circuit for studying and generating chaos. Part I. Routes to chaos," *IEEE Trans. Circ. Syst.*, vol. CAS-40, Oct. 1993, 732–744.

[14] G. Chen and T. Ueta (Editors), *Chaos in Circuits and Systems*, Singapore: World Scientific Publishing, 2002.

[15] A. Suárez, *Analysis and Design of Autonomous Microwave Circuits*, New York: John Wiley & Sons, 2009.

[16] A. M. Crosmun and S. A. Maas, "Minimization of intermodulation distortion in GaAs MESFET small-signal amplifiers," *IEEE Trans. Microw. Theory Techn.*, vol. MTT-37, Sept. 1989, 1411–1417.

[17] J. C. Pedro and J. Perez, "Accurate simulation of GaAs MESFET's intermodulation distortion using a new drain-source current model," *IEEE Trans. Microw. Theory Techn.*, vol. MTT-42, Jan. 1994, 25–33.

[18] V. J. Mathews and G. L. Sicuranza, *Polynomial Signal Processing*, New York, NY: John Wiley & Sons, 2000.

[19] M. Schetzen, *The Volterra and Wiener Theories of Nonlinear Systems*, New York, NY: John Wiley & Sons, 1980.

[20] M. Gupta, L. Jin and N. Homma, *Static and Dynamic Neural Networks*, New York, NY: John Wiley & Sons, 2003.

2 Basic Nonlinear Microwave Circuit Analysis Techniques

The main goal of this chapter is to review the various nonlinear simulation techniques available for RF and microwave circuit analysis, and to present the basic operation and properties of those techniques. Although this is not intended to be a course on nonlinear analysis or simulation, which is far beyond the scope of this book, we believe it is still important for the reader to have a basic idea of how these simulation tools function. This way, he or she can select the most appropriate technique for their particular task and be aware of its limitations.

Although mathematical details will be kept to a minimum, they can still intimidate many microwave engineers. If this is the case, do not worry, stick to the qualitative explanations and the physical interpretations provided to describe the equations. Ultimately, this is more important than the mathematical derivations themselves.

Before we move on to discuss the details of each of the simulation techniques, it is convenient to first state a set of rules common to all of them. So, we start by defining what is meant by circuit or system simulation.

As explained in Chapter 1, the concept of system can be used in its most general form to represent either circuit elements, circuits, or complete RF systems. Therefore, we will use the notion of system whenever we want to achieve this level of generality, and we will adopt the designation of circuit element, component, circuit, or RF system otherwise.

Nowadays, circuit or RF system simulation is a process in which the behavior of a real circuit or RF system is mimicked, solving, in a digital computer, the equations that – we believe – describe the signal excitation and govern the physical operation of that circuit. Hence, the success of simulation should be evaluated comparing the real circuit or RF system behavior with the one predicted by the simulator. The simulation accuracy depends on both the adopted equations and the numerical techniques used to solve them. The set of equations that represent the behavior of each element constitutes the model of the element. The set of equations that describe how these are connected (the circuit or RF system topology) defines the model of the system. The numerical techniques, or numerical methods, used to solve these equations are known as the numerical solvers.

Contrary to linear systems, most nonlinear systems found in practice are described by complex nonlinear equations that do not have any closed form solution. Hence, there is no way to get a general understanding of the system's behavior and we have to rely on partial views of its response to a few particular excitations. This is the underlying reason why simulation plays a much more important role in nonlinear circuit or RF system design than in its linear counterpart.

Currently, the discipline of mathematics on numerical methods is so advanced that the accuracy limitation of modern circuit simulators resides almost entirely on the models' accuracy. But, since the circuit's equations are often derived from nodal analysis (a statement of the universal law of charge conservation), we would not be exaggerating if we said that circuit simulation accuracy limitations must be solely attributed to the circuit components models, or to a possible misuse of the simulator (for example, expecting it to do something for which it was not intended).

Because it is the users who select the simulator, and most of the time, it is their responsibility to provide the simulator with the appropriate models, it is essential that they have, at least, a basic understanding of what the simulator does, so that they can a priori build a rough estimate of the simulation result. Otherwise, the entire simulation effort can be reduced to a frustrating "garbage in, garbage out."

Actually, the main objective of this book is to provide the knowledge to the RF/microwave engineer that prevents falling into any of these traps. The present chapter aims at a correct use of the simulator, while the next four address the components' models.

This chapter on simulation is organized in four major sections. Section 2.1 deals with the mathematical representation, or model, of signals and systems. Section 2.2 treats time-domain simulation tools for aperiodic and periodic regimes. Hence, its main focus is the transient analysis of SPICE-like simulators. Although Section 2.2 already addresses the steady-state periodic regime in the time-domain, via the so-called periodic steady-state, or PSS, algorithm, this is the core of the harmonic-balance, HB, frequency-domain technique of Section 2.3. Finally, Section 2.4 describes a group of algorithms especially devoted to analyze circuits and telecommunication systems excited by modulated RF carriers. From these, special attention is paid to the envelope-transient harmonic-balance, ETHB, a mixed-domain – time and frequency – simulation tool.

2.1   Mathematical Representation of Signals and Systems

As said above, simulation tools comprise the mathematical representation – equations or models – of the system whose behavior is to be predicted, and the numerical solvers for these equations. Therefore, this subsection is intended to introduce general forms for these representations. For that, we will consider the wireless transmitter depicted in Figure 2.1.

2.1.1   Representation of Circuits and Systems

In the wireless transmitter example in Figure 2.1 we can recognize a very heterogeneous set of circuits such as digital signal-processing and control, mixed digital-analog, analog IF or baseband, and RF. Considering only this latter subset, we can note circuits that operate in an *autonomous* way, like the local oscillator, and ones that are *driven*, and

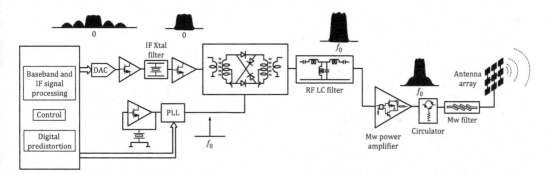

Figure 2.1 The block diagram of a wireless transmitter used to illustrate the disparate nature of circuits and signals dealt with by RF circuit designers.

thus present a *forced* behavior. In addition, we may also divide them into circuits solely composed of *lumped* elements (like ideal resistors, capacitors, inductors, and controlled sources) or into ones that include *distributed* elements (such as transmission lines, microstrip discontinuities, waveguides, etc.).

2.1.1.1 Circuits of Lumped Elements

As seen in Chapter 1, the response of dynamic circuits or systems cannot be given as a function of solely the instantaneous input, but depends also on the input past, stored as the circuit's state. For example, in lumped circuits, i.e., circuits composed of only lumped elements, their state comprises the electric charge stored in the capacitors and the magnetic flux stored in the inductors, the so-called system's *state-variables* or the *state-vector*. Assuming that these capacitor charges and inductor fluxes are memoryless functions of the corresponding voltages and currents (the quasi-static assumption in which accumulated charges, fluxes, or currents are assumed to be instantaneous functions of the applied fields), the *state-equation* is thus the mathematical rule that describes how the state-vector evolves in time. It is an ordinary differential equation, ODE, in time that expresses the time-rate of change of the state-vector, i.e., its derivative, with respect to time, as a function of the present state and excitation:

$$\frac{d\mathbf{s}(t)}{dt} = \mathbf{f}[\mathbf{s}(t), \mathbf{x}(t)] \tag{2.1}$$

where $\mathbf{s}(t)$ is state-vector, $\mathbf{x}(t)$ is the vector of inputs, or *excitation-vector*, (independent current and voltage sources) and $\mathbf{f}[\mathbf{s}(t), \mathbf{x}(t)]$ is a set of functions, each one derived from applying Kirchhoff's currents law, KCL, to the nodes, and voltages law, KVL, to the branches.

With this state-equation, one can calculate the trajectory of the state-vector over time, $\mathbf{s}(t)$. Then, it is easy to compute the *output-vector*, $\mathbf{y}(t)$, i.e., the set of observable currents, voltages, or any algebraic operation on these (such as power, for example) from the following *output-equation*:

$$\mathbf{y}(t) = \mathbf{g}[\mathbf{s}(t), \mathbf{x}(t)] \tag{2.2}$$

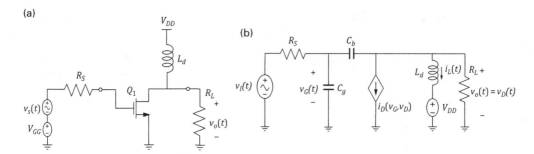

Figure 2.2 Simplified FET amplifier circuit (a) and its equivalent circuit (b) intended to illustrate the analysis's algorithms. $v_I(t) = V_{GG} + v_S(t)$ where $v_S(t)$ is the excitation, $V_{GG} = 3\,$V, $V_{DD} = 10\,$V, $R_s = 50\,\Omega$, $C_g = 50\,$pF, $C_b = 4.4\,$pF, $L_d = 120\,$nH, and $R_L = 6.6\,\Omega$. The MOSFET has a transconductance of $g_m = 10\,$A/V, a threshold voltage of $V_T = 3\,$V, and a knee voltage of $V_K = 1\,$V.

Example 2.1 The State- and Output-Equations   For an illustrative example of this state- and corresponding output-equations, let us use the circuit of Figure 2.2. The application of Kirchhoff currents law to the input and output nodes, leads to

$$\begin{cases} \dfrac{v_G(t) - v_I(t)}{R_s} + C_g \dfrac{dv_G(t)}{dt} + C_b \dfrac{d}{dt}[v_G(t) - v_D(t)] = 0 \\[3mm] C_b \dfrac{d}{dt}[v_D(t) - v_G(t)] + i_D(v_G, v_D) + i_L(t) + \dfrac{v_D(t)}{R_L} = 0 \\[3mm] v_D(t) - V_{DD} = L_d \dfrac{di_L(t)}{dt} \end{cases} \quad (2.3)$$

which can be rewritten as

$$\begin{cases} \dfrac{dv_G(t)}{dt} = \dfrac{1}{C_g}\left[\dfrac{v_I(t) - v_G(t)}{R_s} - i_D(v_G, v_D) - i_L(t) - \dfrac{v_D(t)}{R_L}\right] \\[3mm] \dfrac{dv_D(t)}{dt} = \dfrac{1}{C_g}\dfrac{v_I(t) - v_G(t)}{R_s} - \dfrac{C_b + C_g}{C_b C_g}\left[i_D(v_G, v_D) + i_L(t) + \dfrac{v_D(t)}{R_L}\right] \\[3mm] \dfrac{di_L(t)}{dt} = \dfrac{v_D(t) - V_{DD}}{L_d} \end{cases} \quad (2.4)$$

Defining now $s_1(t) \equiv v_G(t)$, $s_2(t) \equiv v_D(t)$, $s_3(t) \equiv i_L(t)$, $x_1(t) \equiv v_I(t)$ and $x_2(t) \equiv V_{DD}$ allows us to express these nodal equations in the canonical state-equation form of (2.1) as

$$\dfrac{ds_n(t)}{dt} = f_n[s_1(t), \ldots, s_n(t), \ldots, s_N(t), x_1(t), \ldots, x_m(t), \ldots, x_M(t)] \quad (2.5)$$

where $N = 3, M = 2$ and

$$\begin{cases} f_1[s(t), \mathbf{x}(t)] = \dfrac{1}{C_g}\left[\dfrac{x_1(t) - s_1(t)}{R_s} - i_D[s_1(t), s_2(t)] - s_3(t) - \dfrac{s_2(t)}{R_L}\right] \\[4mm] f_2[s(t), \mathbf{x}(t)] = \dfrac{1}{C_g}\dfrac{x_1(t) - s_1(t)}{R_s} - \dfrac{C_b + C_g}{C_b C_g}\left[i_D[s_1(t), s_2(t)] + s_3(t) + \dfrac{s_2(t)}{R_L}\right] \\[4mm] f_3[s(t), \mathbf{x}(t)] = \dfrac{s_2(t) - x_2(t)}{L_d} \end{cases} \quad (2.6)$$

Assuming that the desired observable is the output voltage, $v_O(t)$, the canonical output equation can be obtained following a similar procedure, defining $y(t) \equiv v_O(t) = v_D(t) = s_2(t)$

$$y(t) = g[s_1(t), s_2(t), s_3(t), x_1(t), x_2(t)] \quad (2.7)$$

which, in this case, is simply $y(t) = s_2(t)$.

Because this system is a third order one, its behavior is completely described by a three-dimensional state-vector, i.e., involving three state-variables. So, as explained in Section 1.4.4, its phase-portrait should be represented in a volume whose axis are s_1, s_2, and s_3, or v_G, v_D, and i_L, as is shown in Figure 2.3. Please note that the state-variables correspond to the expected circuit's three independent energy storing elements: $s_1(t) \equiv v_G(t)$ is a measure of the accumulated charge (electric field) in the input capacitor C_g, $s_2(t) \equiv v_D(t)$ is needed to build $v_{GD}(t) = v_G(t) - v_D(t) = s_1(t) - s_2(t)$, a measure of the accumulated charge in the feedback capacitor C_b, and $s_3(t) \equiv i_L(t)$ stands for the accumulated magnetic flux in the output drain inductor L_d.

Although the phase-portrait is a parametric curve in which the evolving parameter is time, it does not allow a direct observation of the trajectory of each of the state-variables over time. In fact, a graph like Figure 2.3 (a) cannot tell us when the system is more active or, on the contrary, when it is almost latent. Such a study would require separate graphs of $s_1(t)$, $s_2(t)$ or $s_3(t)$ versus time as is exemplified in Figure 2.4 for $s_1(t)$ and $s_2(t)$.

2.1.1.2 Circuits with Distributed Elements

Contrary to circuits that include only lumped elements, circuits or systems that involve distributed components, such as transmission lines or other distributed electromagnetic multi-ports, cannot be described as ODEs in time, requiring a description in both time and space. That is the case of the power divider/combiner, the circulator or the output microwave filter of Figure 2.1, and of the transmission line of Figure 2.5. As an example, the transmission line is described by the following partial differential equations, PDEs, in time, t, and space, x.

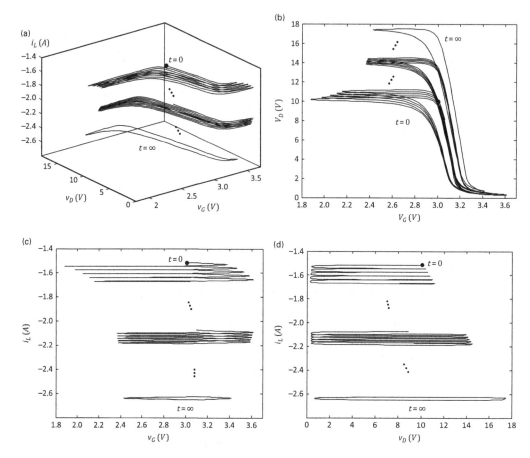

Figure 2.3 (a) Illustrative trajectory of the three-dimensional system state, $[x\ y\ z] \equiv [v_G\ v_D\ i_L]$, of our example circuit. (b) Projection of the 3D state-space trajectory onto the $[v_G\ v_D]$ plane. (c) Projection of the 3D state-space trajectory onto the $[v_G\ i_L]$ plane. (d) Projection of the 3D state-space trajectory onto the $[v_D\ i_L]$ plane.

$$\frac{\partial v(x,t)}{\partial x} = -L\frac{\partial i(x,t)}{\partial t} - Ri(x,t)$$

$$\frac{\partial i(x,t)}{\partial x} = -C\frac{\partial v(x,t)}{\partial t} - Gv(x,t)$$

(2.8)

where $i(x,t)$ and $v(x,t)$ are the distributed current and voltage along the line and $R, L, C,$ and $G,$ stand for the usual line's distributed resistance, inductance, capacitance, and conductance per unit length, respectively.

Please note that, because all time-domain circuit simulators are programmed to solve ODEs, they require that the circuit be represented by the state- and output-equations of (2.1) and (2.2). Consequently, a lumped representation is needed for distributed components. Such lumped representation is usually given in the form of an approximate equivalent circuit, obtained from the frequency-domain behavior observed at the device

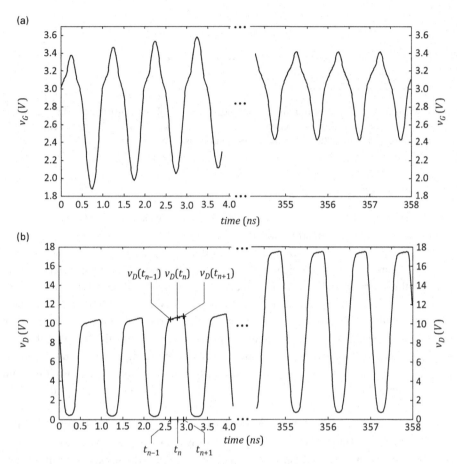

Figure 2.4 Time evolution of two of the system's state-variables over time. (a) $s_1(t) = v_G(t)$ and (b) $s_2(t) = v_D(t)$.

ports, which mimics the operation of the original multiport [1]. However, alternative behavioral representations, such as the one based on the impulse response function and subsequent convolution with the component's excitation, are also possible [2].

It should be emphasized that these are behavioral representations, not real equivalent circuits, as distributed elements do not have a true lumped representation. For example, even our uniform transmission line of Figure 2.5 would need an infinite number of elementary lumped RLCG sections. Therefore, if one is willing to simulate such a circuit in the time domain, he or she should be aware that he is not simulating the actual circuit but another one that, hopefully, produces a similar result. For example, if the extracted "equivalent" circuit is not guaranteed to obey certain properties, such as passivity [3], one may end up with nonconvergence results or erroneous solutions such as a nonphysical (and so impossible) oscillatory response from an original circuit composed of only passive components.

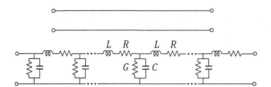

Figure 2.5 The uniform transmission line as an example of a distributed component. Such distributed and dispersive components are represented by partial differential equations in time and space, impeding the use of the state-equation as a general model.

2.1.2   Representation of Signals

As far as the signals are concerned, we could group them into baseband – or slowly evolving – information signals, fast-varying RF carriers, and modulated signals, in which the first two are combined. Furthermore, as shown in the spectra included in Figure 2.1, we could also note that, contrary to the RF carriers, which are periodic and thus narrowband, baseband signals are wideband (that span through many octaves or decades) and are necessarily aperiodic, as they carry information (nonrepeatable, by definition). Therefore, while the periodic RF carrier can be represented in either the time or frequency domains, baseband signals must be necessarily treated in the time domain since the discrete Fourier series assumes periodicity in both time and frequency. Actually, because we will be using a digital computer, i.e., a machine with a finite number of states, it will necessarily impose a discrete, or sampled and of confined nature, description of any time- or frequency-domain entities. This means that if you want to represent, in the frequency domain, a certain continuous-time signal, extending from, say, 0 to T seconds, what you will get is the frequency-domain representation of another discrete and periodic signal, of period T, which extends from $-\infty$ to $+\infty$, and whose sampling frequency, f_s, is determined by the computer time resolution. Such a signal has a discrete and periodic spectrum whose frequency resolution is $\Delta f = 1/T$ and whose period is f_s. Conversely, any continuous spectrum signal will be treated as one of discrete spectrum, which has a periodic time-domain representation. In addition, any Fourier conversion between these two domains will necessarily be executed via the discrete Fourier transform, DFT, or its fast version, the FFT, in case the number of time and frequency points is a multiple of two.

2.2   Time-Domain Circuit Analysis

This section deals with circuit analysis techniques that operate in the time domain, such as *SPICE* or SPICE-like transient simulators [4], [5], and the ones known for calculating the periodic steady-state in the time domain. These simulators assume that the excitation-vector is known for the period of time where the response is desired and that the circuit is represented by the state- and output-equations of (2.1) and (2.2). In addition, it is assumed that the response is desired as a waveform in the time-domain.

2.2.1   Time-Step Integration Engines

Time-step integration is used in all SPICE-like *transient simulators* [4], [5] but also in many system simulators [6–8]. Since it aims at the evolution of the circuit in time, it can be seen as the most "natural" circuit analysis technique because it closely follows the conceptual interpretation we make of the circuit operation. Actually, assuming one starts from a known excitation, $x(t)$, discretized in a number of appropriate time-points, $x(t_n)$, and an also known, initial state, $s(t_0)$, one can easily compute the response at all future time samples, $s(t_n)$, from the discretization of the state-equation. This discretization reflects the knowledge that the derivative of the state at the time, t_n, is the limit when the time-step length, h_n, tends to zero, of the ratio $[s(t_{n+1}) - s(t_n)]/h_n$. Therefore, the state-equation (2.1) can be approximated by the following difference equation

$$\frac{s(t_{n+1}) - s(t_n)}{h_n} = f[s(t_n), x(t_n)] \tag{2.9}$$

which allows the calculation of all $s(t_n)$ using the following recursive scheme known as the forward Euler method [9]

$$s(t_{n+1}) = s(t_n) + h_n f[s(t_n), x(t_n)] \tag{2.10}$$

in a time-step by time-step basis. It is this time-step integration of the state-equation that constitutes the root of the name of this family of methods.

Alternatively, we could have also approximated the state derivative at t_n by

$$\frac{s(t_{n+1}) - s(t_n)}{h_n} = f[s(t_{n+1}), x(t_{n+1})] \tag{2.11}$$

which would then lead to the alternative recursive scheme of

$$s(t_{n+1}) = s(t_n) + h_n f[s(t_{n+1}), x(t_{n+1})] \tag{2.12}$$

which is known as the backward Euler method.

Despite the obvious similarities between (2.10) and (2.12) (originated in their common genesis), they correspond to very distinct numerical methods. First of all, (2.10) gives the next state, $s(t_{n+1})$, in an explicit way, since everything in its right-hand side is a priori known. On the contrary, (2.12) has an implicit formulation since the next state $s(t_{n+1})$ depends on the known $s(t_n)$ and $x(t_{n+1})$, but also on itself. Hence, $s(t_{n+1})$ can be obtained through simple and fast algebraic operations from (2.10), while the solution of (2.12) demands for an iterative nonlinear solver, for which it is usually formulated as

$$s(t_{n+1}) - s(t_n) - h_n f[s(t_{n+1}), x(t_{n+1})] = 0 \tag{2.13}$$

Actually, (2.13) can be solved using, e.g., a gradient method like the *Newton-Raphson iteration* [10].

A Brief Note on the Newton-Raphson Iteration

The Newton-Raphson iteration plays such an important role in nonlinear circuit simulation that we will outline it here. Based on a first order Taylor series approximation, it finds the zero of a nonlinear function, or vector of functions, through the solution of a successively approximated linear system of equations. In the present case, the vector of functions whose zero is to be calculated is $\boldsymbol{\varphi}[\mathbf{s}(t_{n+1})] \equiv \mathbf{s}(t_{n+1}) - \mathbf{s}(t_n) - h_n\mathbf{f}[\mathbf{s}(t_{n+1}), \mathbf{x}(t_{n+1})]$ and so its first-order (linear) Taylor approximation in the vicinity of a certain predetermined estimate, $\mathbf{s}(t_{n+1})_i$, becomes

$$\boldsymbol{\varphi}[\mathbf{s}(t_{n+1})] \approx \boldsymbol{\varphi}\left[\mathbf{s}(t_{n+1})_i\right] + \mathbf{J}_\varphi\left[\mathbf{s}(t_{n+1})_i\right]\left[\mathbf{s}(t_{n+1})_{i+1} - \mathbf{s}(t_{n+1})_i\right] = 0 \qquad (2.14)$$

where $\mathbf{J}_\varphi\left[\mathbf{s}(t_{n+1})_i\right]$ is the Jacobian matrix of $\boldsymbol{\varphi}[\mathbf{s}(t_{n+1})]$ whose element of row m and column l is calculated by

$$\mathbf{J}_\varphi\left[\mathbf{s}(t_{n+1})_i\right]_{ml} \equiv \left.\frac{\partial \varphi_m[\mathbf{s}(t_{n+1})]}{\partial s_l(t_{n+1})}\right|_{\mathbf{s}(t_{n+1})_i} \qquad (2.15)$$

i.e., it is the derivative of the m'th component of the vector of functions $\boldsymbol{\varphi}[\mathbf{s}(t_{n+1})]$ with respect to the l'th component of the state-vector $\mathbf{s}(t_{n+1})$, evaluated at the previous estimate $\mathbf{s}(t_{n+1})_i$.

Since (2.14) is now a linear system in the unknown $\mathbf{s}(t_{n+1})_{i+1}$, we can get a (hopefully) closer estimate of the desired zero by the following iteration:

$$\mathbf{s}(t_{n+1})_{i+1} = \mathbf{s}(t_{n+1})_i - \left[\mathbf{J}_\varphi\left[\mathbf{s}(t_{n+1})_i\right]\right]^{-1}\boldsymbol{\varphi}\left[\mathbf{s}(t_{n+1})_i\right] \qquad (2.16)$$

This is graphically illustrated in Figure 2.6 (a) for the simplest one-dimensional case. Please note that, if $\boldsymbol{\varphi}[\mathbf{s}(t_{n+1})]$ were linear, its Jacobian matrix would be constant, i.e., independent of $\mathbf{s}(t_{n+1})_i$, and the Newton-Raphson algorithm would converge in only one iteration. In general, the Newton-Raphson method is faster in presence of mildly nonlinear problems and slower – i.e., it needs more iterations – when dealing with strongly nonlinear problems (higher curvature in the function of Figure 2.6). Furthermore, it requires an initial estimate of the solution, and converges if the function $\boldsymbol{\varphi}[\mathbf{s}(t_{n+1})]$ is well behaved, i.e., if its Jacobian matrix has no points of zero determinant. In fact, if $\mathbf{s}(t_{n+1})_i$ is a point in the state-space in which the determinant of the Jacobian is zero, or close to zero, the next estimate will suffer a big jump, possibly to a point very far from the desired solution $\boldsymbol{\varphi}[\mathbf{s}(t_{n+1})] = 0$, even if it were already close to it at $\mathbf{s}(t_{n+1})_i$. As illustrated now in Figure 2.6 (b), the Newton-Raphson iteration of (2.16) may diverge from the target solution even when it is already quite close to it. Alternatively, since in the vicinity of a zero determinant $\mathbf{J}_\varphi\left[\mathbf{s}(t_{n+1})_i\right]$ changes sign, the iteration may be stalled, endlessly jumping from two distinct points as depicted in Figure 2.6 (c). In this case, (2.16) does not diverge, but it cannot also converge to the solution.

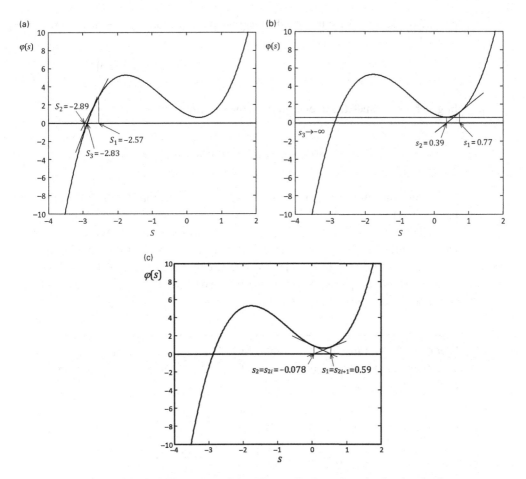

Figure 2.6 Graphical illustration of the Newton-Raphson iteration in the simplest one-dimensional case. (a) Normal convergence of the Newton-Raphson iteration. (b) When the function of which we want to calculate the zero, $\varphi(s)$, has a zero in its derivative (the Jacobian), the iteration may diverge from the desired solution. (c) Situation in which the iteration does not either diverge or converge but is kept wandering around the vicinity of a zero derivative point.

Unfortunately, the simplicity of (2.10) has a high price in numerical stability. Since it is a predictive method, its associated error tends to build up very rapidly, a problem that must be obviated by reducing the time-step, and thus increasing the overall simulation time. In practice, this increase in simulation time and the potential numerical instability are so significant, when compared to (2.12), that (2.10) is hardly ever used in practical circuit simulators.

Actually, (2.10) and (2.12) do not constitute the only way that we could have used to integrate the state-equation. They are just examples of a zero-order method in which the derivative is assumed constant during the time-step interval. For example, if a first-order integration (known as the trapezoidal integration rule) would have been used, we would obtain [8]

$$\mathbf{s}(t_{n+1}) = \mathbf{s}(t_n) + h_n \frac{\mathbf{f}[\mathbf{s}(t_{n+1}), \mathbf{x}(t_{n+1})] + \mathbf{f}[\mathbf{s}(t_n), \mathbf{x}(t_n)]}{2} \qquad (2.17)$$

In general, the time-step integration scheme is formulated as

$$\mathbf{s}(t_{n+1}) = \mathbf{s}(t_n) + \int_{t_n}^{t_{n+1}} \mathbf{f}[\mathbf{s}(t), \mathbf{x}(t)] \, dt \qquad (2.18)$$

from which the 4th-order Runge-Kutta method [8] is one of its most popular implementations (see Exercise 2.1).

Finally, it should be added that there is no need to keep the time-step, h_n, constant. In fact, all modern time-domain simulators use dynamic time-steps, measuring the variation in time of s(t). Larger time-steps are used when s(t) is slowly varying, while smaller ones are adopted in zones of fast change of state. This way, a good compromise between accuracy and computation time is obtained.

Example 2.2 The Time-Step Integration Formulation   As an example of time-step integration, the trapezoidal rule discretization of our circuit example described by (2.5) and (2.6) would lead to

$$\begin{cases} s_1(t_{n+1}) = s_1(t_n) + h_n \dfrac{f_1(t_{n+1}) + f_1(t_n)}{2} \\[2mm] s_2(t_{n+1}) = s_2(t_n) + h_n \dfrac{f_2(t_{n+1}) + f_2(t_n)}{2} \\[2mm] s_3(t_{n+1}) = s_3(t_n) + h_n \dfrac{f_3(t_{n+1}) + f_3(t_n)}{2} \end{cases} \qquad (2.19)$$

where the three $f(t_n)$ were defined in (2.6).

For each time step, t_n, (2.19) is a nonlinear system of three equations in three unknowns, $s_1(t_{n+1})$, $s_2(t_{n+1})$ and $s_3(t_{n+1})$, which can be solved by the Newton-Raphson iteration of (2.16) making

$$\varphi[\mathbf{s}(t_{n+1})] = \begin{bmatrix} s_1(t_{n+1}) - s_1(t_n) - \dfrac{h_n}{2} [f_1(t_{n+1}) + f_1(t_n)] \\[2mm] s_2(t_{n+1}) - s_2(t_n) - \dfrac{h_n}{2} [f_2(t_{n+1}) + f_2(t_n)] \\[2mm] s_3(t_{n+1}) - s_3(t_n) - \dfrac{h_n}{2} [f_3(t_{n+1}) + f_3(t_n)] \end{bmatrix} = 0 \qquad (2.20a)$$

or in a compacted form

$$\varphi[\mathbf{s}(t_{n+1})] = \mathbf{s}(t_{n+1}) - \mathbf{s}(t_n) - \frac{h_n}{2} [\mathbf{f}(t_{n+1}) + \mathbf{f}(t_n)] = 0 \qquad (2.20b)$$

and

$$\mathbf{J}_\varphi[\mathbf{s}(t_{n+1})] = \begin{bmatrix} 1 - \dfrac{h_n}{2}\dfrac{\partial f_1(t_{n+1})}{\partial s_1(t_{n+1})} & -\dfrac{h_n}{2}\dfrac{\partial f_1(t_{n+1})}{\partial s_2(t_{n+1})} & -\dfrac{h_n}{2}\dfrac{\partial f_1(t_{n+1})}{\partial s_3(t_{n+1})} \\[2ex] -\dfrac{h_n}{2}\dfrac{\partial f_2(t_{n+1})}{\partial s_1(t_{n+1})} & 1 - \dfrac{h_n}{2}\dfrac{\partial f_2(t_{n+1})}{\partial s_2(t_{n+1})} & -\dfrac{h_n}{2}\dfrac{\partial f_2(t_{n+1})}{\partial s_3(t_{n+1})} \\[2ex] -\dfrac{h_n}{2}\dfrac{\partial f_3(t_{n+1})}{\partial s_1(t_{n+1})} & -\dfrac{h_n}{2}\dfrac{\partial f_3(t_{n+1})}{\partial s_2(t_{n+1})} & 1 - \dfrac{h_n}{2}\dfrac{\partial f_3(t_{n+1})}{\partial s_3(t_{n+1})} \end{bmatrix}$$

$$= \begin{bmatrix} 1 + \dfrac{h_n}{2C_g}\left(\dfrac{1}{R_s} + g_m(t_{n+1})\right) & \dfrac{h_n}{2C_g}\left(\dfrac{1}{R_L} + g_{ds}(t_{n+1})\right) & \dfrac{h_n}{2C_g} \\[2ex] \dfrac{h_n}{2C_g}\left[\dfrac{1}{R_s} + \left(\dfrac{C_b+C_g}{C_b}\right)g_m(t_{n+1})\right] & 1 + \dfrac{h_n(C_b+C_g)}{2C_gC_b}\left(\dfrac{1}{R_L} + g_{ds}(t_{n+1})\right) & \dfrac{h_n(C_b+C_g)}{2C_gC_b} \\[2ex] 0 & -\dfrac{h_n}{2L_d} & 1 \end{bmatrix}$$

$$\tag{2.21a}$$

or

$$\mathbf{J}_\varphi[\mathbf{s}(t_{n+1})] = \mathbf{I} - \dfrac{h_n}{2}\dfrac{\partial \mathbf{f}(t_{n+1})}{\partial \mathbf{s}(t_{n+1})} \tag{2.21b}$$

where $\mathbf{I}$ stands for the identity matrix, and $g_m(t)$ and $g_{ds}(t)$ are, respectively, the FET's transcoductance and output conductance, defined by $g_m(t) = \left.\dfrac{\partial i_D(v_G,v_D)}{\partial v_G}\right|_{v_G(t),v_D(t)}$ and $g_{ds}(t) = \left.\dfrac{\partial i_D(v_G,v_D)}{\partial v_D}\right|_{v_G(t),v_D(t)}$.

Figures 2.3 and 2.4 depict the results of the simulation of (2.16) for an excitation of $v_I(t) = V_{GG} + v_S(t) = 3 + 20\cos(2\pi f_c t)$, $f_c = 1$ GHz, and with $s_1(0) = 3.0$ V, $s_2(0) = 10.0$ V and $s_3(0) = -10/6.6$ A initial conditions. This time-step integration used the trapezoidal rule of (2.19) and a fixed time step of $h = 1/(100 f_c)$.

If the frequency-domain response representation is desired, as is quite common in RF and microwave circuit analysis and design, then a DFT must be performed on the time-domain response, which requires that the signal is uniformly sampled and periodic.

Unfortunately, and as we have seen, simulation speed determines the use of dynamic time-steps, which produces a non-uniformly sampled response. So, either one forces the simulator to use a fixed time-step, e.g., using the simulator's time-step ceiling option – turning the time-domain simulation highly inefficient – or relies on the hidden interpolation and resampling functions operated before the DFT is applied, which are not error free – turning the frequency-domain representation potentially inaccurate.

If the signal is not periodic, then the DFT will make it periodic, with a period equal to the simulated time, producing an error that is visible as *spectral leakage* [10], i.e.,

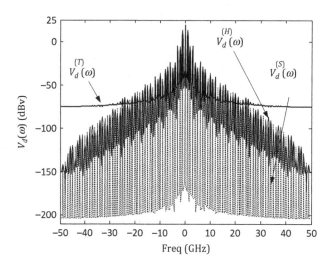

Figure 2.7 Illustrative example of the spectral leakage generated when an aperiodic time-domain function is (discrete) Fourier transformed to the frequency domain. Please note the spectral leakage manifested as erroneous spectral power around the desired signal and its mitigation by windowing. In this example, $V_d^{(T)}(\omega)$ is the frequency-domain representation of five periods of the result of our transient simulation, $v_D(t)$, collected between 50 ns and 55 ns, while $V_d^{(H)}(\omega)$ is the corresponding signal passed through a Hanning window [10]. The dotted spectrum, $V_d^{(S)}(\omega)$, is the same five periods but now collected after the steady-state was reached, i.e., collected between 380 ns and 385 ns.

in which the energy of individual spectral lines is distributed onto (leaks to) their vicinity. This error is caused by the time-domain signal discontinuities that may exist in the beginning and end points of the constructed periodic signal and is illustrated in Figure 2.7. It is particularly harmful in distortion analysis where the distortion components may be totally immersed in the spectral leakage error. Hence, it is quite common to adopt measures to mitigate spectral leakage, increasing the simulation time (in order to reduce the relative weight of these discontinuities in the overall signal) or to simply eliminate these discontinuities, forcing the output signal to start and end in the same zero value – $\mathbf{y}(0) = \mathbf{y}(T) = 0$ – multiplying $\mathbf{y}(t)$ by an appropriate window function [10].

This lack of periodicity of the output signal can be due to the nature of the problem at hand. For example, a two-tone signal of f_1 and f_2 is only periodic if its central frequency $f_c = (f_1 + f_2)/2$ is a multiple of its frequency separation $\Delta f = f_2 - f_1$. But it may also be due to the presence of the transient response. In such a case, it is necessary to truncate the signal to its last few periods. This process is usually quite frustrating when simulating RF circuits since the length of this annoying transient response can be many orders of magnitude higher than the length of the desired period, or of a few periods. Indeed, the length of these long transients is usually determined by the time-constants associated to the large dc decoupling capacitors and RF chokes of bias networks.

Therefore, this transient time-domain simulation of microwave and RF circuits tends to be either inefficient, inaccurate, or both, which motivated the development of analysis techniques specifically devoted to calculate the periodic steady-state.

2.2.2   Periodic Steady-State Analysis in the Time-Domain

Although one might think that the integration of the transient response is unavoidable, as it corresponds to the way the real circuit behaves, that is not necessarily the case. To understand why, let us recall the previously discussed problem that consists in determining the response of a forced system to a periodic excitation of period T. To guarantee that this problem is solvable, let us also consider that the steady-state response is, indeed, periodic, of the same period T of the excitation.

Obviously, the complete response starts with the aperiodic transient, which is then followed by the periodic steady-state. This transient is naturally governed by the system, the excitation and the assumed initial conditions, or initial state, while the periodic steady-state is only dependent on the system and the excitation (in fact, its period is solely determined by the excitation). Therefore, different initial states lead to distinct transient responses, but always to the same steady-state. Actually, a close observation of Figures 2.3 and 2.4 indicates that different initial states may lead to transients of very different length, depending on whether we start from an initial state that is close, or very far, from the final cyclic orbit that describes the periodic steady-state. In fact, the notion of state tells us that all states of that cyclic orbit keep the system in the same orbit, i.e., lead to a transient response of zero length or, in other words, a response without transient.

Example 2.3 A Time-Step Integration Simulation that Does Not Exhibit a Transient

Although the mere existence of a transient simulation without a transient may sound absurd, you can convince yourself that this can indeed be true by running the following example. Pick up any lumped-element circuit you know, which, being excited by a periodic stimulus, produces a periodic steady-state response. Perform a normal transient simulation using a SPICE-like circuit simulator for a time that is sufficiently long to let the circuit reach its periodic steady-state. You will naturally see a transient response followed by the periodic steady-state. Analyze this steady-state response, observing and recording the values of all state-variables, for example all capacitor voltages and inductor currents, at the beginning (arbitrary) of one period of the excitation. Now, run the same transient simulation but use the simulator option that allows you to set the initial values of all energy-storage elements to the values previously recorded (use all precision digits the simulator provides). You can run the transient simulator for a much smaller time-span as you will see that the response is completely periodic, i.e., that there is no transient response.

We actually performed this experiment in our circuit example, starting with the same initial conditions and excitation used for our previous transient simulation, i.e. $v_G(0) = 3.0\,\text{V}$, $v_D(0) = 10.0\,\text{V}$, $i_L(0) = -10/6.6\,\text{A}$ and $v_I(t) = 3 + 20\cos\left(2\pi 10^9 t\right)$, measured these variables at $t = 384$ ns, and then used these values $[v_G(384 \text{ ns}) = 3.118026263640600$ V, $v_D(0) = 11.183192073318860$ V and $i_L(0) = -2.625494710083924$ A] as the new initial conditions. As expected, the simulator response went directly to the steady-state response of Figures 2.3 and 2.4 without any transient.

By the way, this example leads us to this very curious question: which were, after all, the energy-storage initial conditions used by default by the simulator in the first run? And how did it select them? You will find a response to these questions in the next section.

Now that we are convinced that our problem of bypassing the transient response may have a solution, it becomes restricted to find a way to determine these appropriate initial states (see Exercise 2.2). There are, at least, three known techniques to do that.

The first method we can imagine to determine the proper initial states (i.e., the ones that are already contained in the cyclic orbit) was already discussed. Indeed, although the time-step integration starts from an initial state that may be very far from the desired one, the state it reaches after one period, is closer. Then, this new updated state is inherently used as the initial state for the time-step integration of the second period, from which a new update is found, in a process that we know is convergent to the desired proper initial state. Our problem is that this process is usually too slow due to the large dc decoupling bias network capacitors and RF chokes. In an attempt to speed up the transient calculation, a simple dc analysis (in which all capacitors and inductors are treated as open or short circuits, respectively) is performed prior to any transient simulation. Then, the dc open circuit voltages across the capacitors and the short-circuit currents across the inductors are used as the initial conditions of the corresponding voltages and currents in the actual transient simulation, unless other initial conditions are imposed by the user. These dc values constitute, therefore, the default initial state that the simulator uses. When these dc quiescent values are already very close to the actual dc bias values, as is the case of all linear or quasi-linear circuits, such as small-signal or class A power amplifiers, the transient response is short and this method of finding the default initial conditions can be very effective. Otherwise, i.e., in case strongly nonlinear circuits such as class C amplifiers or rectifiers are to be simulated, these default values are still very far from the actual bias point and a large transient response subsists.

The second and third methods find the proper initial condition using a completely different strategy: they change the nature of the numerical problem.

2.2.2.1 Initial-Value and Boundary-Value Problems

Up to now, we have considered that our circuit analysis could be formulated as an *initial-value problem*, i.e., a problem defined by setting initial boundary conditions (the initial state) and a rule for the time evolution of the other internal time-steps (the state-equation). No boundary conditions are defined at the end of the simulation time; as we know, this simulation time can be arbitrarily set by the user. When we are willing to determine the periodic steady-state, and we know a priori that the steady-state period is the period of the excitation, we can also define a final boundary condition and reformulate the analysis as a *boundary-value problem*. In this case the initial state, the internal rule, and the final state are all a priori defined. Well, rigorously speaking, what is defined is the internal rule and an equality constraint for the initial and the final states: $s(t_0 + T) = s(t_0)$.

2.2.2.2   Finite-Differences in Time-Domain

Having this boundary-value problem in mind, the second method is known as the finite-differences in time-domain and is directly formulated from the discretization of the state-equation and the periodic boundary condition $s(t_0 + T) = s(t_0)$. Using again a zero-order backward Euler discretization in $N + 1$ time-samples, we can define the following (incomplete) system (for each state-variable or each circuit node) of N equations in $N + 1$ unknowns

$$\begin{cases} \mathbf{s}(t_1) = \mathbf{s}(t_0) + h_0 \mathbf{f}[\mathbf{s}(t_1), \mathbf{x}(t_1)] \\ \quad \vdots \\ \mathbf{s}(t_{n+1}) = \mathbf{s}(t_n) + h_n \mathbf{f}[\mathbf{s}(t_{n+1}), \mathbf{x}(t_{n+1})] \\ \quad \vdots \\ \mathbf{s}(t_N) = \mathbf{s}(t_{N-1}) + h_N \mathbf{f}[\mathbf{s}(t_N), \mathbf{x}(t_N)] \end{cases} \tag{2.22}$$

which, augmented by the boundary equation $s(t_N) = s(t_0)$, constitutes a (now complete) system of $N + 1$ equations in $N + 1$ unknowns that can be readily solved through a $N + 1$-dimensional Newton-Raphson iteration scheme.

2.2.2.3   Shooting Method

The third alternative to find the proper initial condition that leads to the desired steady-state is known as the *shooting* or *periodic steady-state*, *PSS*, method. Conceptually, it can be described as a trial-and-error process in which we start by simulating the circuit for one period with some initial state guess; for example, the one given by an a priori dc analysis, $\mathbf{s}(t_0)_1$. In our circuit example, this attempt would lead to the first set of trajectories depicted in the phase-space plots of Figure 2.8, in which $\mathbf{s}(t_N)_1 \neq \mathbf{s}(t_0)_1$, indicating, as expected, that we failed the periodic steady-state. By slightly changing the tested initial condition by some $\Delta \mathbf{s}(t_0)$, say $\mathbf{s}(t_0)_2 = \mathbf{s}(t_0)_1 + \Delta \mathbf{s}(t_0)$, we can try the alternative attempt identified as the second set of trajectories of Figure 2.8. Although this one led to another failure, since $\mathbf{s}(t_N)_2$ is still different from $\mathbf{s}(t_0)_2$, it may provide us an idea of the sensitivity of the final condition to the initial condition, $\Delta \mathbf{s}(t_N) / \Delta \mathbf{s}(t_0)$, which we then may use to direct our search for better trial $\mathbf{s}(t_0)_3$, $\mathbf{s}(t_0)_3$, ..., $\mathbf{s}(t_0)_K$, until the desired steady-state is achieved, i.e., in which $\mathbf{s}(t_N)_K = s(t_0)_K$.

Actually, using a first order Taylor series approximation of the desired null error, $\mathbf{s}(t_N) - \mathbf{s}(t_0) = 0$, we may write [11], [12]:

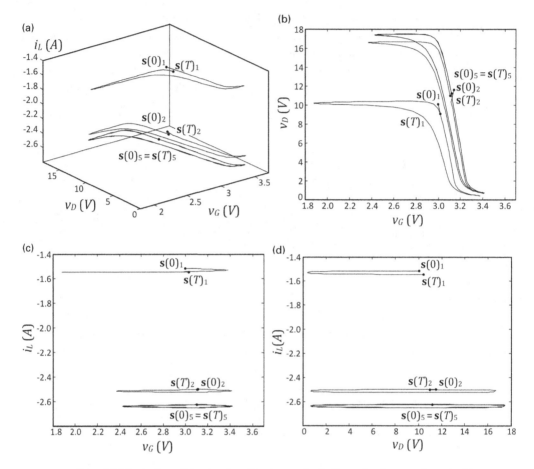

Figure 2.8 Illustration of the shooting-Newton method of our circuit example excited by $v_I(t) = 3 + 20\cos\left(2\pi 10^9 t\right)$, where all the required five Newton iterations are shown. As done for the previous transient analysis, we started from the initial conditions derived from dc analysis, $v_G(0) = 3.0$ V, $v_D(0) = 10.0$ V and $i_L(0) = -10/6.6$ A, and ended up on $v_G(0) = 3.118026121336410$ V, $v_D(0) = 11.183189430567937$ V and $i_L(0) = -2.625492967125298$ A.

$$s(t_N)_{i+1} - s(t_0)_{i+1} \approx s(t_N)_i - s(t_0)_i + \left[\frac{\Delta s(t_N)}{\Delta s(t_0)} - \mathbf{I}\right]\left[s(t_0)_{i+1} - s(t_0)_i\right] = 0 \quad (2.23)$$

which leads to the following Newton-Raphson iteration:

$$s(t_0)_{i+1} = s(t_0)_i - \left[\frac{\Delta s(t_N)}{\Delta s(t_0)} - \mathbf{I}\right]^{-1}\left[s(t_N)_i - s(t_0)_i\right] \quad (2.24)$$

known as the *shooting-Newton PSS* method. It should be noted that this sensitivity matrix, $\frac{\Delta s(t_N)}{\Delta s(t_0)}$, can be evaluated along the time-step integration of the period, using the chain differentiation rule as

$$\frac{\Delta \mathbf{s}(t_N)}{\Delta \mathbf{s}(t_0)} \equiv \frac{\partial \mathbf{s}(t_N)}{\partial \mathbf{s}(t_0)} = \frac{\partial \mathbf{s}(t_N)}{\partial \mathbf{s}(t_{N-1})} \frac{\partial \mathbf{s}(t_{N-1})}{\partial \mathbf{s}(t_{N-2})} \cdots \frac{\partial \mathbf{s}(t_{n+1})}{\partial \mathbf{s}(t_0)} \tag{2.25a}$$

where

$$\frac{\partial \mathbf{s}(t_{n+1})}{\partial \mathbf{s}(t_0)} = \frac{\partial \mathbf{s}(t_{n+1})}{\partial \mathbf{s}(t_n)} \frac{\partial \mathbf{s}(t_n)}{\partial \mathbf{s}(t_{n-1})} \cdots \frac{\partial \mathbf{s}(t_1)}{\partial \mathbf{s}(t_0)} \tag{2.25b}$$

directly from the time-step integration formula of (2.18).

Example 2.4 The Shooting-Newton Formulation   In our circuit example, where we used the trapezoidal integration rule, this recurrent formula can be derived from (2.20) and (2.21) by noting that, if $\boldsymbol{\varphi}[\mathbf{s}(t_{n+1})] = 0$, then

$$\frac{\partial \boldsymbol{\varphi}[\mathbf{s}(t_N)]}{\partial \mathbf{s}(t_0)} = 0 \tag{2.26}$$

and so

$$\frac{\partial \mathbf{s}(t_{n+1})}{\partial \mathbf{s}(t_0)} - \frac{\partial \mathbf{s}(t_n)}{\partial \mathbf{s}(t_0)} - \frac{h_n}{2} \left[\frac{\partial \mathbf{f}(t_{n+1})}{\partial \mathbf{s}(t_0)} + \frac{\partial \mathbf{f}(t_n)}{\partial \mathbf{s}(t_0)} \right]$$

$$= \frac{\partial \mathbf{s}(t_{n+1})}{\partial \mathbf{s}(t_0)} - \frac{\partial \mathbf{s}(t_n)}{\partial \mathbf{s}(t_0)} - \frac{h_n}{2} \left[\frac{\partial \mathbf{f}(t_{n+1})}{\partial \mathbf{s}(t_{n+1})} \frac{\partial \mathbf{s}(t_{n+1})}{\partial \mathbf{s}(t_0)} + \frac{\partial \mathbf{f}(t_n)}{\partial \mathbf{s}(t_n)} \frac{\partial \mathbf{s}(t_n)}{\partial \mathbf{s}(t_0)} \right]$$

$$= \left[\mathbf{I} - \frac{h_n}{2} \frac{\partial \mathbf{f}(t_{n+1})}{\partial \mathbf{s}(t_{n+1})} \right] \frac{\partial \mathbf{s}(t_{n+1})}{\partial \mathbf{s}(t_0)} - \left[\mathbf{I} + \frac{h_n}{2} \frac{\partial \mathbf{f}(t_n)}{\partial \mathbf{s}(t_n)} \right] \frac{\partial \mathbf{s}(t_n)}{\partial \mathbf{s}(t_0)}$$

$$= \mathbf{J}_{\varphi}[\mathbf{s}(t_{n+1})] \frac{\partial \mathbf{s}(t_{n+1})}{\partial \mathbf{s}(t_0)} - [2\mathbf{I} - \mathbf{J}_{\varphi}[\mathbf{s}(t_n)]] \frac{\partial \mathbf{s}(t_n)}{\partial \mathbf{s}(t_0)} = 0$$

and finally

$$\frac{\partial \mathbf{s}(t_{n+1})}{\partial \mathbf{s}(t_0)} = \mathbf{J}_{\varphi}[\mathbf{s}(t_{n+1})]^{-1} [2\mathbf{I} - \mathbf{J}_{\varphi}[\mathbf{s}(t_n)]] \frac{\partial \mathbf{s}(t_n)}{\partial \mathbf{s}(t_0)} \tag{2.27}$$

Although the shooting-Newton PSS is based on an iterative process to find the appropriate initial state, it is, by far, much more efficient than the other two preceding methods for at least three reasons [12]. The first one is that, while (2.22) attempts to solve a $N+1$-dimensional problem at once, (2.24) solves it resolving $N+1$ one-dimensional problems, one at a time. The second is that the sensitivity, or the Jacobian, of (2.24) can be computed along with the transient simulation. Finally, the third is that although the circuit may be strongly nonlinear, the dependence of the final condition on the initial condition seems to be only mildly nonlinear in most practical problems, so that the Newton-Raphson solver converges in just a few iteration steps. For example, as is illustrated in the phase-state plots of Figure 2.8, the application of the shooting-Newton algorithm to our circuit example needed only five iterations to reach the desired periodic steady-state, even if we started from the same initial condition used for the transient analysis.

These properties make the PSS the technique of choice for efficiently computing the periodic steady-state regimes in time-domain of RF and microwave circuits. However, the inexistence of time-domain models for many RF and microwave components (mostly passive components such as filters and distributed elements, which are usually described by scattering parameter matrices) and the fact that, many times, RF/microwave designers operate on stimuli and responses represented as spectra, make frequency-domain methods, such as the harmonic-balance, the most widely used simulation techniques in almost all RF/microwave circuit designs.

2.3   Frequency-Domain Circuit Analysis

As you may still recall from basic linear analysis, and from what was discussed in Chapter 1, frequency-domain methods are particularly useful to study the periodic steady-state for two different reasons.

The first of these reasons concerns the computational efficiency, because a sinusoid, which should be represented in the time domain by several time samples, requires only two real values (or a single complex value) when represented in the frequency domain. Indeed, knowing that the response of a linear system to a sinusoid is also a sinusoid of the same frequency, the sinusoidal response becomes completely described by its amplitude and phase. Furthermore, if the system's excitation and response are not pure sines but a superposition of sines that are harmonically related, then we still need two real values for each represented harmonic. Actually, the harmonic zero, i.e., dc, requires only one value, as its phase is undefined.

The second reason for the paramount role played by frequency-domain techniques, is a consequence of the fact that the time-derivative of a complex exponential of time (or of a sinusoid, as they are related by the Euler's formula) is still the same complex exponential of time multiplied by a constant:

$$\frac{de^{j\omega_0 t}}{dt} = j\omega_0 e^{j\omega_0 t} \tag{2.28}$$

This converts an ordinary linear differential equation into a much simpler algebraic equation. In fact, assuming that both the excitation and the state are periodic of period $T = 2\pi/\omega_0$, then, $x(t)$ and $s(t)$ can be represented by the following Fourier series

$$\mathbf{x}(t) = \sum_{k=-K}^{K} \mathbf{X}_k e^{jk\omega_0 t} \tag{2.29}$$

$$\mathbf{s}(t) = \sum_{k=-K}^{K} \mathbf{S}_k e^{jk\omega_0 t} \tag{2.30}$$

which, substituted in the state-equation of (2.1)

$$\frac{d\mathbf{s}(t)}{dt} = \mathbf{f}[\mathbf{s}(t), \mathbf{x}(t)] = \mathbf{A}_1 \mathbf{s}(t) + \mathbf{A}_2 \mathbf{x}(t) \tag{2.31}$$

in which $\mathbf{A}_1$ and $\mathbf{A}_2$ are the two matrices that uniquely identify our linear system, results in

$$\sum_{k=-K}^{K} jk\omega_0 \mathbf{S}_k e^{jk\omega_0 t} = \mathbf{A}_1 \sum_{k=-K}^{K} \mathbf{S}_k e^{jk\omega_0 t} + \mathbf{A}_2 \sum_{k=-K}^{K} \mathbf{X}_k e^{jk\omega_0 t} \tag{2.32}$$

or

$$jk\omega_0 \mathbf{S}_k = \mathbf{A}_1 \mathbf{S}_k + \mathbf{A}_2 \mathbf{X}_k \tag{2.33}$$

an algebraic linear system of equations in the vector $\mathbf{S}_k$, the Fourier components of $\mathbf{s}(t)$. Exercise 2.3 deals with an example of this linear formulation.

2.3.1   Harmonic-Balance Engines

The application of the Fourier series representation to the analysis of nonlinear systems assumes that, for a broad set of systems (namely the ones that are stable and of fading memory [13]), their steady-state response to a sinusoid, or to a periodic excitation, of period T, is still a periodic function of the same period. However, as we already know from the harmonic generation property of nonlinear systems, this does not mean that the frequency content of the response is the same of the input. It simply means that it is located on the same frequency grid, $\omega_k = k\omega_0$. In general, the harmonic content of the output is richer than the one of the input, although there is, at least, one counter-intuitive example of this: linearization by pre- or post-distortion, in which a nonlinear function and its inverse are cascaded. Please note that this assumption of periodicity of the steady-state response of our system to a periodic excitation (in the same frequency grid) is a very strong one. If it fails (for example, if the system is chaotic or is unstable, oscillating in a frequency that falls outside the original frequency grid) frequency-domain techniques may not converge to any solution or, worse, may converge to wrong solutions without any warning.

The extension of the frequency-domain analysis of linear systems to nonlinear ones is known as the *harmonic-balance*, HB, since it tries to find an equilibrium between the Fourier components that describe the currents and voltages in all circuit nodes or branches. It is quite easy to formulate, although, as we will see, not so easy to solve. To formulate it, we simply state again the Fourier expansions of (2.29) and (2.30) and substitute them in the state-equation (2.1), to obtain

$$\sum_{k=-K}^{K} jk\omega_0 \mathbf{S}_k e^{jk\omega_0 t} = \mathbf{f}\left[\sum_{k=-K}^{K} \mathbf{S}_k e^{jk\omega_0 t}, \sum_{k=-K}^{K} \mathbf{X}_k e^{jk\omega_0 t}\right] \tag{2.34}$$

Within the stable and fading memory system restriction, we do know that the right-hand side of (2.34) will be periodic of fundamental frequency ω_0, and thus admits the following Fourier expansion:

$$\mathbf{f}\left[\sum_{k=-K}^{K} \mathbf{S}_k e^{jk\omega_0 t}, \sum_{k=-K}^{K} \mathbf{X}_k e^{jk\omega_0 t}\right] = \sum_{k=-K}^{K} \mathbf{F}_k[\mathbf{S}(\omega), \mathbf{X}(\omega)] e^{jk\omega_0 t} \tag{2.35}$$

in which $\mathbf{S}(\omega)$ and $\mathbf{X}(\omega)$ stand for the Fourier components vectors of $\mathbf{s}(t)$ and $\mathbf{x}(t)$, respectively. Substituting (2.35) in (2.34) results in

$$\sum_{k=-K}^{K} jk\omega_0 \mathbf{S}_k e^{jk\omega_0 t} = \sum_{k=-K}^{K} \mathbf{F}_k[\mathbf{S}(\omega), \mathbf{X}(\omega)] e^{jk\omega_0 t} \tag{2.36}$$

which would not be of much use if it weren't for the orthogonality property of the Fourier series. In fact, since the complex exponentials of the summations of (2.36) are orthogonal – which means that no component of $e^{jk_1\omega_0 t}$ can contribute to $e^{jk_2\omega_0 t}$ unless $k_1 = k_2$ – (2.36) can actually be unfolded as a (nonlinear) system of $2K+1$ equations (for each state-variable or circuit node), of the form

$$jk\omega_0 \mathbf{S}_k = \mathbf{F}_k[\mathbf{S}(\omega), \mathbf{X}(\omega)] \tag{2.37}$$

in $2K+1$ unknowns (the $\mathbf{S}_k$). Written in vector-matrix form, (2.37) can be expressed as

$$\mathbf{j}\boldsymbol{\Omega}\mathbf{S}(\omega) = \mathbf{F}[\mathbf{S}(\omega), \mathbf{X}(\omega)] \tag{2.38}$$

in which $\mathbf{j}\boldsymbol{\Omega}$ is a diagonal matrix whose diagonal values are the $jk\omega_0$.

Please note that since there is no known method to compute the nonlinear mapping $[\mathbf{S}(\omega), \mathbf{X}(\omega)] \to \mathbf{F}(\omega)$ for any general function $\mathbf{f}[\mathbf{s}(t), \mathbf{x}(t)]$ or $\mathbf{F}[\mathbf{S}(\omega), \mathbf{X}(\omega)]$ directly in the frequency domain, one has to rely on the time-domain calculation, $\mathbf{f}[\mathbf{s}(t), \mathbf{x}(t)]$, converting first $\mathbf{S}(\omega)$ and $\mathbf{X}(\omega)$ to the time domain, and then converting the resulting $\mathbf{f}(t) = \mathbf{f}[\mathbf{s}(t), \mathbf{x}(t)]$ back to the frequency domain. These frequency-time-frequency domain transforms constitute a heavy burden to the HB algorithm, which has motivated the quest for a pure frequency-domain alternative, nowadays known as spectral-balance [14].

The idea is to express $\mathbf{f}[\mathbf{s}(t), \mathbf{x}(t)]$ with only arithmetic operations, for which the frequency-domain mapping is known. For example, if $\mathbf{f}[\mathbf{s}(t), \mathbf{x}(t)]$ is expressed as a polynomial – which is nothing else but a summation of products – the time-domain addition corresponds to a frequency-domain addition, while the time-domain product corresponds to a frequency-domain spectral convolution (which can be efficiently implemented as the product of a Toplitz matrix by a vector [15]). Actually, this same idea can be extended to more powerful rational-function approximators (i.e., ratios of polynomials) since the time-domain division corresponds to a frequency-domain deconvolution, which is implemented as the product of the inverse of the Toplitz matrix just referred by a vector [15]. Although the spectral-balance has not seen the same widespread implementation of the traditional harmonic-balance, it is still interesting as a pure frequency-domain nonlinear simulation technique.

To solve the nonlinear system of equations of (2.38), we can again make use of the multidimensional Newton-Raphson algorithm, for which we need to rewrite (2.38) in its canonical form as

$$\mathbf{j\Omega S}(\omega) - \mathbf{F}[\mathbf{S}(\omega), \mathbf{X}(\omega)] = 0 \tag{2.39}$$

Then, the Newton-Raphson iteration would be

$$\mathbf{S}_{i+1}(\omega) = \mathbf{S}_i(\omega) - [\mathbf{j\Omega} - \mathbf{J}_F[\mathbf{S}_i(\omega), \mathbf{X}(\omega)]]^{-1}[\mathbf{j\Omega S}_i(\omega) - \mathbf{F}[\mathbf{S}_i(\omega), \mathbf{X}(\omega)]] \tag{2.40}$$

in which $\mathbf{J}_F[\mathbf{S}(\omega), \mathbf{X}(\omega)]$ stands for the Jacobian matrix of the frequency-domain function $\mathbf{F}[\mathbf{S}(\omega), \mathbf{X}(\omega)]$.

Unfortunately, the previous phrase may be misunderstood because of its apparent simplicity. In fact, if we said that we did not know any direct way to compute $\mathbf{F}[\mathbf{S}(\omega), \mathbf{X}(\omega)]$ in the frequency domain, we certainly do not know of any way to compute its Jacobian either. However, this Jacobian plays such an important role in nonlinear RF circuit analysis that we cannot avoid digging into its calculation. Therefore, to retain the essential, without being distracted by accessory numerical details, let us imagine we have a simple one-dimensional function of a one-dimensional state-variable: $f(t) = f[s(t)]$ or $F(\omega) = F[S(\omega)]$:

$$F_k[S(\omega)] = \frac{1}{T} \int_{-T/2}^{T/2} f[s(t)] e^{-jk\omega_0 t} dt \tag{2.41}$$

The row m and column l of the wanted Jacobian matrix, $J_F[S(\omega)]_{ml}$, is thus the partial derivative of the m'th Fourier component of $F(\omega)$, F_m, with respect to the l'th Fourier component of $S(\omega)$, S_l:

$$J_F[S(\omega)]_{ml} \equiv \frac{\partial F_m}{\partial S_l} = \frac{\partial}{\partial S_l} \frac{1}{T} \int_{-T/2}^{T/2} f \left[\sum_{k=-K}^{K} S_k e^{jk\omega_0 t} \right] e^{-jm\omega_0 t} dt \tag{2.42}$$

Now, as the order of the derivative and the integral can be interchanged, using the derivative chain rule that allows us to write

$$\frac{\partial}{\partial S_l} f[s(t)] = \frac{df(s)}{ds} \frac{\partial s}{\partial S_l} = g(s) \frac{\partial}{\partial S_l} \left[\sum_{k=-K}^{K} S_k e^{jk\omega_0 t} \right] = g(s) e^{jl\omega_0 t} \tag{2.43}$$

and thus

$$J_F[S(\omega)]_{ml} \equiv \frac{\partial F_m}{\partial S_l} = \frac{1}{T} \int_{-T/2}^{T/2} g[s(t)] e^{-j(m-l)\omega_0 t} dt \tag{2.44}$$

which is the $(m - l)$'th Fourier component of the derivative of $f(s)$ with respect to s, $g(s)$.

A physical interpretation of this result can be given if we consider that, for example, $f(s)$ represents a FET's drain-source current dependence on v_{GS}, $i_{DS}[v_{GS}(t)]$. In this case, $g[s(t)]$ is the FET's transconductance and so (2.44) stands for the $(m-l)$'th Fourier component of the time-variant transconductance, $g_m(t) = g_m[v_{GS}(t)]$, in the *large-signal operating point, LSOP*, $[v_{GS}(t), i_{DS}(t)]$. It constitutes, thus, the frequency-domain representation of the device operation linearized in the vicinity of its LSOP, an extension of the linear gain, or transfer function, when the device is no longer biased with a dc quiescent point, but with one that is time-variant and periodic.

Therefore, this Jacobian is the frequency-domain time-variant transfer function of a mixer whose LSOP is determined by the local-oscillator (maybe with also a dc bias), known in the mixer design context as the *conversion matrix* [16], [17]. Beyond this, the position of the poles of the Jacobian's determinant provide valuable information on the *large-signal stability* of the circuit, in much the same way as the poles of the frequency-domain gain of a time-invariant linear amplifier – biased at a dc quiescent point – determine its small-signal stability [18].

But, this Jacobian can represent much more than this if a convenient linear transformation of variables is performed in the circuits' currents and voltages to build incident and scattered waves. If $\mathbf{F}[\mathbf{S}(\omega)]$ is defined so that it maps frequency-domain incident waves, $\mathbf{A}(\omega) = \mathbf{S}(\omega)$, into their corresponding scattered waves, $\mathbf{B}(\omega) = \mathbf{F}[\mathbf{A}(\omega)]$, the Jacobian would correspond to a form of large-signal S-parameters (available in some commercial simulators), or, more rigorously, to the S and T parameters of the *poly-harmonic distortion, PHD, or X-Parameters model*, [19], [20] (as will be discussed in Chapter 4).

As always, an important issue that arises when one is trying to find the solution of a nonlinear equation like (2.39) with the Newton-Raphson iteration of (2.40), is the selection of an appropriate initial condition. From the various possible alternatives we might think of, there is one that usually gives good results. It assumes that most practical systems are mildly nonlinear, i.e., that their actual response is not substantially distinct from the one obtained if the circuit is considered linear in the vicinity of its quiescent point. Therefore, the simulator starts by determining this dc quiescent point, setting to zero all ac excitations, and then estimates an initial condition from the response of this linearized circuit.

If this does not work, i.e., if it is found that the Newton-Raphson iteration cannot converge to a solution, then the simulator reduces the excitation level down to a point where this linear or, quasi-linear, approximation holds. Having reached convergence for this reduced stimulus, it uses this response as the initial condition for another simulation of the same circuit, but now with a slightly higher excitation level. Then, this *source-stepping* process is repeated up to the excitation level originally desired.

Example 2.5 The Harmonic-Balance Formulation   Continuing to use our circuit example of Figure 2.2, with the same periodic stimulus $v_I(t) = 3 + 20 \cos\left(2\pi 10^9 t\right)$, we will now solve it for its periodic steady-state using the harmonic-balance formulation. For that, we will use the time-domain state equations of (2.5) and (2.6) to identify $\mathbf{F}[\mathbf{S}(\omega), \mathbf{X}(\omega)]$ in (2.39) and (2.40) as

$$
\mathbf{F}[\mathbf{S}(\omega), \mathbf{X}(\omega)] = \left\{
\begin{array}{l}
-\dfrac{1}{C_g}\left[\dfrac{S_1(\omega) - X_1(\omega)}{R_s} + I_d(S_1, S_2) + S_3(\omega) + \dfrac{S_2(\omega)}{R_L}\right] \\[2ex]
\dfrac{X_1(\omega) - S_1(\omega)}{C_g R_s} - \dfrac{C_b + C_g}{C_b C_g}\left[I_d(S_1, S_2) + S_3(\omega) + \dfrac{S_2(\omega)}{R_L}\right] \\[2ex]
\dfrac{S_2(\omega) - X_2(\omega)}{L_b}
\end{array}
\right.
$$

$$(2.45)$$

The application of the Newton-Raphson algorithm to (2.39) requires the construction of the Jacobian matrix $\mathbf{J}_F[\mathbf{S}(\omega), \mathbf{X}(\omega)]$. In this example of three state-variables, the Jacobian matrix will be a 3×3 block matrix whose elements are given by:

$$\mathbf{J}_F[\mathbf{S}(\omega), \mathbf{X}(\omega)]_{ml}$$

$$
= \begin{bmatrix}
-\dfrac{1}{C_g}\left[\dfrac{1}{R_s} + G_m(\omega_{m-l})\right] & -\dfrac{1}{C_g}\left[G_{ds}(\omega_{m-l}) + \dfrac{1}{R_L}\right] & -\dfrac{1}{C_g} \\[2ex]
-\dfrac{1}{C_g R_s} - \dfrac{C_b + C_g}{C_g C_b}G_m(\omega_{m-l}) & -\dfrac{C_b + C_g}{C_g C_b}\left[G_{ds}(\omega_{m-l}) + \dfrac{1}{R_L}\right] & -\dfrac{C_b + C_g}{C_g C_b} \\[2ex]
0 & \dfrac{1}{L_b} & 0
\end{bmatrix}
$$

$$(2.46)$$

in which $G_m(\omega_{m-l})$ and $G_{ds}(\omega_{m-l})$ are the $(m - l)$ frequency component of the device's time-varying transconductance, $g_m(t)$, and output conductance, $g_{ds}(t)$, respectively.

Figure 2.9 illustrates the evolution of the harmonic content of the output voltage of our example circuit, $|V_d(k\omega_0)| = |S_3(k\omega_0)|$, during the Newton-Raphson solution of (2.40), while Figure 2.10 provides this same result but now seen in the time-domain, $v_D(t) = s_3(t)$. Although the 3-D plot of Figure 2.9 presents only the first 15 harmonics plus dc, 50 harmonics were actually used in the simulation.

2.3.2   Piece-Wise Harmonic-Balance

In this section we introduce what became known as the piece-wise harmonic-balance method, to distinguish it from the nodal-based harmonic-balance analysis technique just described. In fact, in piece-wise HB we do not start by a nodal analysis – from which we derived the state-equation – but begin by separating all of the circuit's current, charge, or flux nonlinear sources from the remainder of the linear subcircuit network, as is illustrated in Figure 2.11. Then, we only analyze the exposed circuit nodes, i.e., the ones to which nonlinear voltage-dependent charge or current sources are connected. Let us, for the sake of language simplicity, designate these exposed

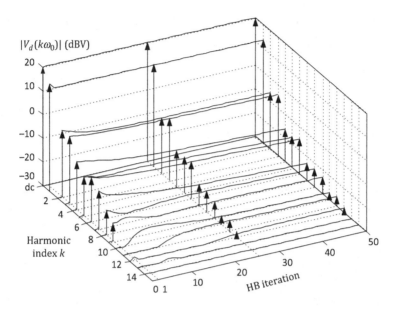

Figure 2.9 Illustration of the evolution of the first 15 harmonics, plus dc, of the amplitude spectrum of the output voltage, $|V_d(k\omega_0)| = |S_3(k\omega_0)|$, during the first 50 Newton-Raphson iterations of the HB simulation of our example circuit. Note how the simulation starts by assuming that the output voltage is a sinusoid (at the fundamental frequency, $k=1$) plus a dc value ($k=0$), and then evolves to build its full harmonic content.

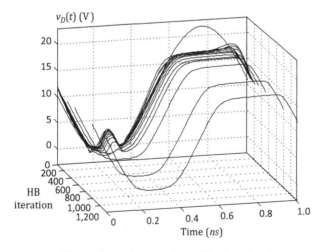

Figure 2.10 Illustration of the evolution of the time-domain steady-state harmonic-balance solution of $v_D(t) = s_3(t)$, during the whole Newton-Raphson iterations. Note how the simulation starts by assuming that the output voltage is a sinusoid plus a dc value, and then evolves to build the correct waveform shape.

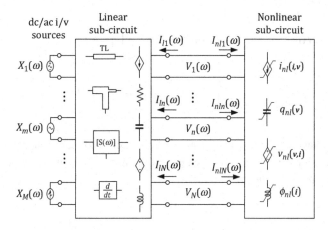

Figure 2.11 Decomposition of a nonlinear microwave circuit into a subcircuit linear block to which the nonlinear voltage dependent current or charge sources are connected, as needed for the application of the piece-wise harmonic-balance technique.

nodes as *nonlinear nodes* and *linear nodes* the ones contained in the linear subcircuit network. (Please note that, despite not mentioning them here, nonlinear current-dependent flux sources could have also been considered, although requiring a slightly different formulation).

In piece-wise HB, the variables we will determine are no longer the circuit's state-variables, but only the voltages of these exposed nodes, $\mathbf{V}(\omega)$. Because most microwave circuits have typically many more linear nodes than nonlinear ones, the dimension of $\mathbf{V}(\omega)$ is usually much smaller than the one of $\mathbf{S}(\omega)$, which constitutes an enormous computational advantage of the piece-wise HB with respect to its nodal-based counter-part. Now, we apply frequency-domain nodal analysis to these nonlinear nodes obtaining the following desired HB equation:

$$\mathbf{I}_{nl}[\mathbf{V}(\omega)] + \mathbf{I}_l[\mathbf{V}(\omega)] + \mathbf{I}_x[\mathbf{X}(\omega)] = 0 \qquad (2.47)$$

In (2.47), $\mathbf{I}_{nl}[\mathbf{V}(\omega)]$ is the frequency-domain current vector representing the currents entering the nonlinear current sources, which, as before, must be evaluated in the time-domain as

$$
\begin{aligned}
\mathbf{I}_{nl_k}[\mathbf{V}(\omega)] &= \frac{1}{T} \int_{-\frac{T}{2}}^{\frac{T}{2}} \mathbf{i}_{nl}[\mathbf{v}(t)] e^{-jk\omega_0 t} \, dt \\
&= \frac{1}{T} \int_{-T/2}^{T/2} \mathbf{i}_{nl}\left[\sum_{m=-K}^{K} \mathbf{V}_m e^{jm\omega_0 t} \right] e^{-jk\omega_0 t} \, dt
\end{aligned}
\qquad (2.48)
$$

in the case of voltage-dependent nonlinear current sources, or as

$$\mathbf{I}_{nl_k}[\mathbf{V}(\omega)] = jk\omega_0 \mathbf{Q}_{nl,k}[\mathbf{V}(\omega)] = jk\omega_0 \frac{1}{T} \int\limits_{-\frac{T}{2}}^{\frac{T}{2}} \mathbf{q}_{nl}[\mathbf{v}(t)]e^{-jk\omega_0 t} dt$$

$$= jk\omega_0 \frac{1}{T} \int\limits_{-T/2}^{T/2} \mathbf{q}_{nl}\left[\sum_{m=-K}^{K} \mathbf{V}_m e^{jm\omega_0 t}\right] e^{-jk\omega_0 t} dt$$

(2.49)

in the case of voltage-dependent nonlinear charge sources (nonlinear capacitances).

$\mathbf{I}_l[\mathbf{V}(\omega)]$ stands for the current vector entering the linear subcircuit network, which is easily computed in the frequency domain by

$$\mathbf{I}_{l,k}[\mathbf{V}(\omega)] = \mathbf{Y}(k\omega_0)\mathbf{V}_k(\omega)$$

(2.50)

in which $\mathbf{Y}(\omega)$ is the linear subcircuit network admittance matrix.

Finally, $\mathbf{I}_x[\mathbf{X}(\omega)]$ are the excitation currents that can be computed as

$$\mathbf{I}_x[\mathbf{X}(\omega)] = \mathbf{A}_x(\omega)\mathbf{X}(\omega)$$

(2.51)

where the transfer matrix $\mathbf{A}_x(\omega)$ can have either current gain or transadmittance gain entries, whether they refer to actual independent current source or independent voltage source excitations, respectively.

Please note that, while the nonlinear current sources, or the nonlinear subcircuit network, which is composed of only quasi-static elements, is treated in the time domain, the linear subcircuit, which contains all dynamic components (even the time derivatives of nonlinear charges), is directly treated in the frequency-domain.

This has an important advantage in terms of computational efficiency. Because static nonlinearities are treated in the time domain, they tend to require fast product relations, while they would need much more expensive convolution relations if dealt with in the frequency domain (as was seen when we discussed spectral-balance). On the other hand, linear dynamic elements necessitate convolution in the time domain, but only require product relations when treated in the frequency domain.

Finally, we should say that the piece-wise HB equation of (2.47) can be solved for the desired $\mathbf{V}(\omega)$ vectors through the following Newton-Raphson iteration:

$$\mathbf{V}_{i+1}(\omega) = \mathbf{V}_i(\omega) - [\mathbf{J}_{\mathbf{I}_{nl}}[\mathbf{V}_i(\omega)] + \mathbf{Y}(\omega)]^{-1}$$
$$\cdot[\mathbf{I}_{nl}[\mathbf{V}_i(\omega)] + \mathbf{Y}(\omega)\mathbf{V}_i(\omega) + \mathbf{A}_x(\omega)\mathbf{X}(\omega)]$$

(2.52)

For an application example of the piece-wise harmonic-balance method, see Exercise 2.4.

To summarize this description of the basics of the harmonic-balance – formulated in both its nodal and piece-wise form – Figure 2.12 depicts its algorithm in the form of a flowchart.

2.3.3   Harmonic-Balance in Oscillator Analysis

Because oscillators are autonomous systems, i.e., systems in which there is no excitation, we would not think of them in the context of harmonic-balance. Indeed, the whole

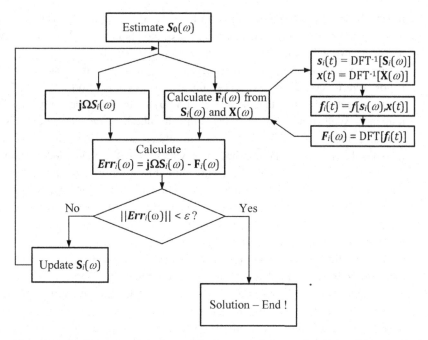

Figure 2.12 An illustrative flowchart of the harmonic-balance algorithm. Although this flowchart was formulated for HB applied to each node or state-variable, in its nodal form, it could be easily extended to be applied to only the nonlinear nodes, as is required for the piece-wise HB formulation.

idea of HB is based on the a priori knowledge of the state-vector frequency set, something we are looking for in the oscillator analysis. However, a clever modification of HB allows it to be used for the determination of the steady-state regime of this kind of autonomous systems.

To understand how it works, let us consider the piece-wise HB formulation of (2.47), in which all but the dc components of the excitation vector, $I_x[X(\omega)]$, are zero.

$$I_{nl}[V(\omega)] + I_l[V(\omega)] + I_x[X(0)] = 0 \tag{2.53}$$

Please note that, dividing both sides of (2.53) evaluated at the fundamental by the corresponding (non-null) node voltage vector, would lead to

$$\frac{I_{nl}[V(\omega_0)]}{V(\omega_0)} + \frac{I_l[V(\omega_0)]}{V(\omega_0)} = Y_{nl}(\omega_0) + Y_l(\omega_0) = 0 \tag{2.54}$$

which can be recognized as the Kurokawa oscillator condition [21]. So, using HB for oscillator analysis sums up to solve (2.53) for $V(\omega)$. The problem is that not only (2.53) is insufficient since it involves, at least, one more unknown than the available number of equations, as we already know one possible solution of it, that we do not want. In fact, if (2.47) was a determined system, (2.53) can no longer be because the oscillation base frequency ω_0 is now also unknown. In addition, because $I_x[X(\omega)]$ is now zero, except for $\omega = k\omega_0 = 0$, one possible solution of (2.53) is the so-called trivial, or degenerate, nonoscillatory condition where all but the dc components of $\mathbf{V}(\omega)$ are zero. In the following, we will deal with these two problems separately.

The first of these problems can be solved noting that if it is true that because $\mathbf{I}_x[\mathbf{X}(\omega)]$ is now zero (except for $\omega = 0$), and so there is no way to a priori know the basis of the frequency set, it also true that the phases of $\mathbf{V}(\omega)$ are undetermined, as we lost the phase reference. Well, rigorously speaking, only the phase of the fundamental is undetermined, as the phases of the harmonics are referenced to it. Actually, what is undetermined is the (arbitrary) starting time, t_0, at which we define the periodic steady-state. In fact, since the system is time-invariant, it should be insensitive to any time shift, which means that if $\mathbf{v}(t)$ is a periodic solution of (2.53), than $\mathbf{v}(t - t_0)$ should also be an identical solution of that same equation. In the frequency domain, this means that the phase of the fundamental frequency $\phi_1 = -\omega_0 t_0$ can be made arbitrary, while the phase of all the other higher-order harmonics are related to it as $\phi_k = -k\omega_0 t_0 = -k\phi_1$. So, (2.53) is still a system of $(2K + 1)$ equations in $2K + 1$ unknown vectors: the vector of dc values $\mathbf{V}(0)$, K vectors of amplitudes of the fundamental and its harmonics, $(K - 1)$ vectors of their phases $(\phi_2 \ldots \phi_K)$, plus the unknown frequency basis ω_0. Therefore, it can be solved in much the same way as before, using, for example, the Newton-Raphson iteration, except that now there is a column vector in the Jacobian that is no longer calculated as $\partial F_m / \partial S_l$ but as $\partial F_m / \partial \omega_0$.

The second problem consists in preventing the Newton-Raphson iteration on (2.53) to converge to the trivial, or degenerate, solution. As explained, this solution is the dc solution in which all but the 0'th harmonic components of $\mathbf{V}(\omega)$ are zero. One way to prevent this is to maximize the error for this particular trivial solution. This can be done normalizing the error of (2.53) by the root-mean-square voltage magnitude of non-dc $\mathbf{V}(\omega)$:

$$\frac{\mathbf{I}_{nl}[\mathbf{V}(\omega)] + \mathbf{I}_l[\mathbf{V}(\omega)]}{\sqrt{\sum_{k=1}^{K} |V_k|^2}} = 0 \qquad (2.55)$$

In commercial HB simulators, the adopted approach uses the so-called *oscillator-probe*. Its underlying idea is to analyze the oscillator, not as an autonomous system, but as a forced one. Hence, the user should select a particular node in which it is known that a small perturbation can significantly disturb the oscillator and apply a special source (the oscillator-probe) to it, as is represented in Figure 2.13.

Such an appropriate node is one that is in the oscillator feedback loop or close to the resonator. The oscillator-probe is a voltage source whose frequency ω_0 is left as an optimization variable, and whose source impedance is infinite at all harmonics but the fundamental at which it is zero. So, this $Z_S(\omega)$ behaves as a filter which is a short-circuit to the fundamental frequency, ω_0, and a nonperturbing open-circuit to all the other harmonics $k\omega_0$, where $k \neq 1$. Then, the oscillator analysis is based on finding the so-called nonperturbation condition, i.e., to determine ω_0 and $|V_S|$ that make $I_S(\omega_0) = 0$. Under this condition, the circuit behavior is the same whether it has the oscillator-probe attached to it or not. So, although this solution was found as the solution of a

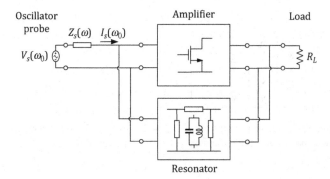

Figure 2.13 Oscillator analysis via harmonic-balance of a nonautonomous system using the oscillator-probe.

nonautonomous problem, it coincides with the solution of the original autonomous circuit.

2.3.4   Multitone Harmonic-Balance

The frequency-domain analysis technique presented above assumes a periodic regime for both the excitation and the state-vector, so that these quantities can be represented using the discrete Fourier transform. Unfortunately, there are a significant set of steady-state simulation problems that do not obey this restriction. Some of them, as chaotic systems, are so obviously aperiodic that we would not even dare to submit them to any frequency-domain analysis. Their spectra are continuous and so there is no point in trying to find the amplitude and phase of a set of finite discrete spectral lines. There are, however, other situations that are still aperiodic and thus cannot be treated using the DFT, but that can still be represented with a finite set of discrete spectral lines. They are the so-called multitone or *almost-periodic*, regimes. For example, consider a signal composed by the addition of two sinusoidal functions:

$$x(t) = A_1 \cos(\omega_1 t) + A_2 \cos(\omega_2 t) \tag{2.56}$$

This excitation can be either periodic or aperiodic, depending on whether or not ω_1 and ω_2 are two harmonics of the same fundamental frequency. That is, if $\omega_1 = k_1 \omega_0$ and $\omega_2 = k_2 \omega_0$, then $x(t)$ is periodic, and it is aperiodic if there is no ω_0 that verifies both of these conditions. An equivalent way of stating $x(t)$ periodicity is to stipulate that $\omega_2/\omega_1 = k_2/k_1$ is a rational number. Conversely, stating that $x(t)$ is aperiodic is equivalent to stipulating that there are no two nonzero integer numbers, m_1 and m_2, that satisfy $m_1 \omega_1 + m_2 \omega_2 = 0$. Therefore, periodicity requires that ω_1 and ω_2 are *commensurate* (ω_2/ω_1 is a rational number), while we say that the aperiodic two-tone excitation is composed of two *incommensurate frequencies* (ω_2/ω_1 is an irrational number). In general, an excitation composed by a set of frequencies, $\omega_1, \ldots, \omega_K$ is aperiodic if, and only if, there is no set of nonzero integers that verify

$k_1 \backslash k_2$	-3	-2	-1	0	+1	+2	+3
-2	X	X	X	X	X	X	X
-1	X	X	X	X	X	X	X
0	X	X	X	X	X	X	X
+1	X	X	X	X	X	X	X
+2	X	X	X	X	X	X	X

Figure 2.14 Box truncation frequency set for $|k_1| \leq 2$ and $|k_2| \leq 3$ (35 mixing products). "X" indicates a used position of the bidimensional spectrum truncation.

$$m_1 \omega_1 + \ldots + m_K \omega_K = 0 \tag{2.57}$$

In this context, we will restrict our multitone excitation to the case of incommensurate tones, as the other one can always be treated with the above described harmonic-balance algorithm for periodic regimes.

Let us then consider the two-tone excitation of (2.56), in which ω_1 and ω_2 are incommensurate. If it is true that neither $\mathbf{x}(t)$ nor $\mathbf{s}(t)$ can be expressed as a Fourier series, it is as well true that we may write $\mathbf{x}(t)$ as

$$\mathbf{x}(t) = \sum_{k=-K}^{K} \mathbf{X}_k e^{j\omega_k t} = \sum_{k_1=-K}^{K} \sum_{k_2=-K}^{K} \mathbf{X}_{k_1,k_2} e^{j(k_1\omega_1+k_2\omega_2)t} \tag{2.58}$$

in which, $\mathbf{X}_{k_1,k_2} = 0$ when $|k_1| + |k_2| > 1$, for a pure two-tone excitation, and

$$\mathbf{s}(t) = \sum_{k=-K}^{K} \mathbf{S}_k e^{j\omega_k t} = \sum_{k_1=-K}^{K} \sum_{k_2=-K}^{K} \mathbf{S}_{k_1,k_2} e^{j(k_1\omega_1+k_2\omega_2)t} \tag{2.59}$$

for $\mathbf{s}(t)$. In fact, experience tells us that the vast majority of nonlinear systems responds to a set of discrete spectral lines at ω_1 and ω_2, with a set of discrete lines located at all harmonics of ω_1, all harmonics of ω_2 and all possible mixing products of ω_1 and ω_2. So, although we cannot affirm that $\omega_k = k\omega_0$, because the response is aperiodic, and so neither (2.58) nor (2.59) are Fourier series, we can state that these signals can still be expressed as a sum of discrete spectral lines whose location is a priori known.

Naturally, as it happened with periodic harmonic-balance, a certain spectrum truncation must be established. However, while in the periodic regime this was easily set in a single axis of frequencies as $|k|\omega_0 \leq K\omega_0$, in the almost-periodic case, since we have two harmonic limits, K_1 and K_2, there is no unique obvious choice. On the contrary, there are at least two possibilities often used.

The first one establishes that $|k_1| \leq K_1$ and that $|k_2| \leq K_2$, in which K_1 and K_2 need not to be equal. For example, in mixer analysis, the highest harmonic order for the large-signal local oscillator must be significantly higher than the one used for the small RF signal. If the mixing products, $k_1\omega_1 + k_2\omega_2$, are represented as points in a plane defined by the orthogonal axes $k_1\omega_1$ and $k_2\omega_2$, then these conditions define a rectangular figure, or box, limited by $K_1\omega_1$ and $K_2\omega_2$. This is illustrated in Figure 2.14 for $K_1 = 2$ and $K_2 = 3$. That is the reason why this spectrum truncation scheme is known as *box truncation*.

$k_1 \backslash k_2$	-3	-2	-1	0	+1	+2	+3
-3				X			
-2			X	X	X		
-1		X	X	X	X	X	
0	X	X	X	X	X	X	X
+1		X	X	X	X	X	
+2			X	X	X		
+3				X			

Figure 2.15 Diamond truncation frequency set for $|k_1| + |k_2| \leq 3$ (25 mixing products). "X" indicates a used position of the bidimensional spectrum truncation.

The second truncation scheme derives from the practical knowledge that typical spectra tend to have smaller and smaller amplitudes when the order of nonlinearity is increased. This means that, if it can be accepted that the highest harmonic order for ω_1 is, say, K_1, and the maximum order for ω_2 is K_2, then a mixing product at $K_1\omega_1 + K_2\omega_2$ will certainly have negligible amplitude and box truncation may waste precious computational resources. So, the alternative is to consider a maximum harmonic order K and then state that $|k_1| + |k_2| \leq K$. As illustrated in Figure 2.15, this condition leads to a diamond-shaped coverage of the frequency plane above described, which is the reason why it is known as the ***diamond truncation***.

After stating that the response to be found is given by (2.59), under an appropriate spectrum truncation scheme, multitone HB determines the coefficients $\mathbf{S}_k$, or $\mathbf{S}_{k_1,k_2}$, knowing that they can be related by

$$
\sum_{k_1=-K}^{K} \sum_{k_2=-K}^{K} j(k_1\omega_1 + k\omega_2)\mathbf{S}_{k_1,k_2} e^{j(k_1\omega_1+k_2\omega_2)t}
$$

$$
= \mathbf{f}\left[\sum_{k_1=-K}^{K} \sum_{k_2=-K}^{K} \mathbf{S}_{k_1,k_2} e^{j(k_1\omega_1+k_2\omega_2)t}, \sum_{k_1=-K}^{K} \sum_{k_2=-K}^{K} \mathbf{X}_{k_1,k_2} e^{j(k_1\omega_1+k_2\omega_2)t} \right] \qquad (2.60)
$$

In fact, there is no substantial difference between the multitone HB algorithm of (2.60) and its original periodic HB except that, now, we are impeded to evaluate the right-hand side of (2.60) in the time domain as we lack a mathematical tool to jump between time and frequency domains. Hence, what we will outline in the following are three possible alternatives to overcome this limitation.

2.3.4.1   Almost-Periodic Fourier Transform

The first multitone HB technique we are going to address takes advantage of the fact that, although neither $\mathbf{x}(t)$ nor $\mathbf{s}(t)$ are periodic, they can still be represented by a finite set of discrete spectral lines. They are, thus, named almost-periodic and its frequency-domain coefficients can be determined in a way that is very similar to the DFT and is named the almost-periodic Fourier transform, APFT [22]. To understand how, let us assume that, for simplicity, $\mathbf{s}(t)$ is a one-dimensional vector, $s(t)$, defined in a discrete time grid of $t = nT_s$. If it can be expressed as (2.59) we may write

$$
\begin{bmatrix} s(1) \\ \vdots \\ s(n) \\ \vdots \\ s(N) \end{bmatrix} = \begin{bmatrix} e^{-j\omega_K T_s} & \cdots & 1 & \cdots & e^{j\omega_k T_s} & \cdots & e^{j\omega_K T_s} \\ \vdots & & \vdots & & \vdots & & \vdots \\ e^{-j\omega_K n T_s} & \cdots & 1 & \cdots & e^{j\omega_k n T_s} & \cdots & e^{j\omega_K n T_s} \\ \vdots & & \vdots & & \vdots & & \vdots \\ e^{-j\omega_K N T_s} & \cdots & 1 & \cdots & e^{j\omega_k N T_s} & \cdots & e^{j\omega_K N T_s} \end{bmatrix} \cdot \begin{bmatrix} S_{-K} \\ \vdots \\ S_0 \\ \vdots \\ S_k \\ \vdots \\ S_K \end{bmatrix} \tag{2.61a}
$$

or

$$
\mathbf{s}_n = \Gamma \mathbf{S}_k \tag{2.61b}
$$

In the periodic regime, NT_s can be made one entire period so that (2.61) corresponds to a square system of full rank where $N = 2K + 1$. In that case, the coefficients $\mathbf{S}_k$ can be determined by

$$
\mathbf{S}_k = \Gamma^{-1} \mathbf{s}_n \tag{2.62}
$$

which corresponds to the orthogonal DFT.

In the almost-periodic regime, the best we can do is to get an estimation of the $\mathbf{S}_k$. As always, the estimation is improved whenever we provide it more information, which, in this case, corresponds to increasing the sampled time, NT_s, and thus the size of $\mathbf{s}_n$. Actually, if we would make a test with only periodic regimes, but progressively longer periods (i.e., gradually tending to the almost-periodic regime), $\mathbf{S}_k$ would smoothly increase in size (because the frequency resolution is $\varDelta f = 1/T$) and the result of (2.62) would always be exact. Unfortunately, we cannot afford to do this and have to limit the size of $\mathbf{S}_k$. This leads to a nonsquare system in which $N >> 2K + 1$, and the $\mathbf{S}_k$ can be estimated in the least-squares sense as

$$
\mathbf{S}_k = \left(\Gamma^\dagger \Gamma \right)^{-1} \Gamma^\dagger \mathbf{s}_n \tag{2.63}
$$

in which $\Gamma^\dagger$ stands for the conjugate transpose or Hermitian transpose of Γ. Actually, because the DFT matrix is a unitary matrix, i.e., $\Gamma^\dagger = \Gamma^{-1}$, the APFT of (2.63) reduces to the DFT of (2.62), under periodic regimes.

Because this almost-periodic Fourier transform constitutes an approximate way to handle the aperiodic regime, as if it were periodic, it is prone to errors. Therefore, many times, the following two alternative methods are preferred.

2.3.4.2 Multidimensional Fourier Transform

To treat the two-tone problem exactly, we need to start by accepting that it cannot be described by any conventional Fourier series. Instead, it must be represented in a frequency set supported by a bidimensional basis defined by the incommensurate frequencies ω_1 and ω_2. In fact, stating the incommensurate nature of ω_1 and ω_2 is similar to stating their independence, in the sense that they do not share any common fundamental. But, if ω_1 and ω_2 are independent, then $\theta_1 = \omega_1 t$ and $\theta_2 = \omega_2 t$ are also two independent variables, which means that (2.59) could be rewritten as a bidimensional Fourier series expansion [23].

Actually,

$$s_{n_1,n_2} = \sum_{k_1=-K_1}^{K_1} \sum_{k_2=-K_2}^{K_2} S_{k_1,k_2} e^{jk_1\omega_1 n_1 T_{s_1}} e^{jk_2\omega_2 n_2 T_{s_2}} \tag{2.64}$$

and

$$S_{k_1,k_2} = \frac{1}{(2K_1+1)(2K_2+1)} \sum_{n_1=0}^{2K_1+1} \sum_{n_2=0}^{2K_2+1} s_{n_1,n_2} e^{jk_1\omega_1 n_1 T_{s_1}} e^{jk_2\omega_2 n_2 T_{s_2}} \tag{2.65}$$

define a bidimensional DFT pair, in which the time-domain data is sampled according to

$$T_{s_1} = \frac{2\pi}{(2K_1+1)\omega_1} \quad \text{and} \quad T_{s_2} = \frac{2\pi}{(2K_2+1)\omega_2} \tag{2.66}$$

Since s_{n_1,n_2} is periodic in both $2\pi/\omega_1$ and $2\pi/\omega_2$, the bidimensional DFT coefficients S_{k_1,k_2} can be computed as $(2K_2+1)$ one-dimensional DFTs followed by another set of $(2K_1+1)$ DFTs (actually, much more computationally efficient FFTs if $2K_1+1$ and $2K_2+1$ are made integer multiples of 2).

Since this bidimensional 2D - DFT (or, in general, the multidimensional DFT, MDFT) is benefiting from the periodicity of the waveforms in the two bases ω_1 and ω_2, it does not suffer from the approximation error of the almost-periodic Fourier transform. However, it is much more computationally demanding which limits its use to a small number of incommensurate tones. Moreover, it requires that the ω_1 and ω_2 are, indeed, two incommensurate frequencies. If they are not, then there will be certain mixing products, at $\omega_{k1} = m_1\omega_1 + m_2\omega_2$ and, say, $\omega_{k2} = m_3\omega_1 + m_4\omega_2$, that will coincide in frequency, and so whose addition should be made in phase, while the MDFT based harmonic-balance will add them in power, as if they were uncorrelated. This may naturally introduce simulation errors, whose significance depends on the amplitudes, and thus on the orders of the referred mixing products (higher order products, i.e., with higher $|m_1| + |m_2|$, usually have smaller amplitudes).

2.3.4.3   Artificial Frequency Mapping

To understand the underlying idea of artificial frequency mapping techniques, we first need to recognize two important harmonic-balance features.

The first one is that the role played by the Fourier transform in HB is restricted to solve the problem of calculating the spectral mapping imposed by a certain static nonlinearity, $Y(\omega) = F[X(\omega)]$. So, as long as this spectral mapping is correctly obtained, we do not care about what numerical technique, or trick, was used to get it. Actually, the difference between the traditional DFT-based HB and the spectral-balance is the most obvious proof of this.

The second feature worth discussing is less obvious. It refers to the fact that, since $f[x(t)]$ is assumed to be static, the actual positions of the spectral lines of $X(\omega)$ are irrelevant for

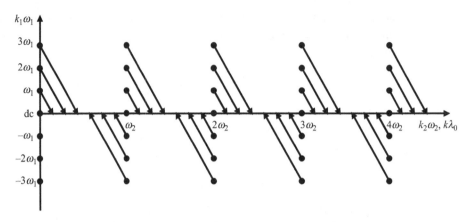

Figure 2.16 Artificial frequency mapping technique for box truncated spectra defined by $|k_1| \leq 2$ and $|k_2| \leq 3$.

the calculation of $Y(\omega)$; only their relative positions are important. In fact, note that, if you have no doubt that the amplitudes of the output spectral lines of, e.g., a squarer, $a_0 + a_1 x(t) + a_2 x^2(t)$, subject to a tone at ω_1 of amplitude A_1, and another one at ω_2 of amplitude A_2, are $a_0 + (1/2)a_2 A_1^2 + (1/2)a_2 A_2^2$ at dc, $a_1 A_1$ and $a_1 A_2$ at ω_1 and ω_2, $(1/2)a_2 A_1^2$ and $(1/2)a_2 A_2^2$ at $2\omega_1$ and $2\omega_2$, $a_2 A_1 A_2$ at $\omega_1 - \omega_2$ and $\omega_1 + \omega_2$, it is because you know that the response shape (the amplitude of each frequency component) is independent of the frequencies ω_1 and ω_2. Otherwise, you would not have even dared to think of a result without first asking for the actual values of ω_1 and ω_2. Indeed, you know that you can calculate these amplitudes independently of whether the frequencies are, say, 13 Hz and 14 Hz, 13 GHz and 14 GHz, or even 13 GHz and $10\sqrt{2}$ GHz. However, there is a fundamental difference in HB when you consider 13 GHz and 14 GHz or 13 GHz and $10\sqrt{2}$ GHz as the former pair defines a periodic regime, allowing the use of the DFT, while the latter does not. Artificial frequency mapping techniques are algorithms that map an almost-periodic real time and frequency grid, $(n_1 T_{s1}, n_2 T_{s2})$ and $(k_1 \omega_1, k_2 \omega_2)$, onto one that is periodic, in the artificial time, $r T_s$, and artificial frequency, $s\lambda_0$, before applying the IDFT. Then, after the $f(r T_s,) = f[x(r T_s,)]$ and $Y(\lambda) = \text{DFT}[f(r T_s,)]$ are calculated, they apply the inverse mapping to $Y(\lambda)$ to obtain the desired $Y(\omega)$.

For example, if the original real frequency grid were defined over $f_1 = 13$ GHz and $f_2 = 10\sqrt{2}$ GHz, and we were assuming box truncation up to $K_1 = 2$ and $K_2 = 3$, then, the artificial frequency mapped spectrum could be set on a normalized one-dimensional grid of $\lambda_0 = 1$, where $k_1 f_1$ would be transformed onto $k_1 \lambda_0$ and $k_2 f_2$ onto $(1 + 2K_1)k_2 \lambda_0$.

This illustrative artificial frequency mapping for box truncation is depicted in Figure 2.16 but many other similar maps have been proposed [24], [25].

2.4   Envelope-Following Analysis Techniques

This section is devoted to simulation techniques that operate in both the time and frequency domains. As we explained, HB is a pure frequency-domain algorithm, in which the role played by time is merely circumstantial, as we never seek to determine amplitude time samples that describe the waveforms in the time domain. On the contrary, in the techniques addressed in this section we will calculate both time-sampled amplitudes and Fourier coefficients of the corresponding state-variables' descriptions in time and frequency.

Mixed-domain techniques were conceived to resolve a fundamental problem of microwave/RF circuits used in communications related to appropriately handling modulated signals. As already discussed in Section 2.1, typical modulated signals use a periodic RF carrier (which is usually a sinusoid in wireless RF front-ends, but that can be a square wave in switched-capacitor circuits or switched-mode amplifiers, or even a triangular wave as found in pulse-width modulators) and a baseband modulation. Because this modulation carries information, it must be, necessarily, aperiodic. This impedes the use of the DFT, limiting our algorithm choices to time-domain transient analysis. Unfortunately, the very disparate time rates of the RF carrier and the baseband modulation make time-step integration highly inefficient, since the simulation of any statistically representative length of the modulation signal would require the integration of millions of carrier periods, and thus many more time samples.

In an approximate way to get around this problem, we could assume that the modulation is periodic – although with a much longer period, compared to the carrier – and then represent it as a very large number of equally spaced tones that fill in the modulated signal's bandwidth. This way, we could approximately represent the modulated signal as a narrowband multitone signal of many tones but with only two incommensurate bases (the modulation and the carrier periods), which could already be treated with one of the above described two-tone HB engines [26]. Naturally, the number of considered tones can easily become very large if we want the modulation to be progressively more complex and thus decide to increase the bandwidth/frequency-resolution ratio. In the limit, when the modulation period tends to infinity (to make it more statistically representative of the actual information signal), the frequency resolution – i.e., the constant tone separation – tends to zero, and this multitone HB approximation becomes useless. This is when we have to turn our attention to the envelope-following engines addressed below.

2.4.1   Representation of Modulated Signals

The signals we are willing to address can be represented as an amplitude, $a_x(t)$, and phase, $\phi_x(t)$, modulated carrier of frequency ω_0, of the form

$$x(t) = a_x(t) \cos \left[\omega_0 t + \phi_x(t) \right] = \mathrm{Re} \left[a_x(t) e^{j\phi_x(t)} e^{j\omega_0 t} \right] \tag{2.67}$$

In communication systems, the entity $\tilde{x}(t) \equiv a_x(t) e^{j\phi_x(t)}$ is the so-called *low-pass equivalent complex envelope* because it is a generalization of the real amplitude envelope $a(t)$, but now including also the phase information, $\phi_x(t)$. Please note that, being a complex signal (i.e., whose imaginary part is not zero), it is not constrained to obey the complex conjugate symmetry of real signals. So, not only $\tilde{X}(\omega) \neq \tilde{X}^*(-\omega)$, as there may be no relation between $\tilde{X}(\omega)$ and $\tilde{X}(-\omega)$. However, as $x(t)$ is still a real signal $X(\omega) = X^*(-\omega)$.

Because the system is nonlinear, it will distort both the carrier and its modulation. Therefore, not only it will generate harmonics of ω_0, but also the envelope of each of these harmonics will carry distorted versions of the original amplitude, $a_x(t)$, and phase, $\phi_x(t)$. Hence, in general, we will have to assume that, even if the excitation could be restricted to (2.67), the state-variables (and thus any generalized modulated periodic excitation) will share the form of

$$s(t) = \mathbf{a}_0(t) + \sum_{\substack{k=-K \\ k \neq 0}}^{K} \mathbf{a}_k(t) e^{j[k\omega_0 t + \phi_k(t)]} \tag{2.68}$$

and the objective of the simulation is to obtain the corresponding complex envelope amplitudes, $\mathbf{a}_k(t)$, and phases, $\phi_k(t)$.

2.4.2   Envelope-Transient Harmonic-Balance

The underlying idea behind envelop-following techniques is to admit that, because of their distinct nature, the envelope and the carrier behave as if they were evolving in two independent, or orthogonal, time scales, τ_1 and τ_2. This way, any modulated excitation-vector, $\mathbf{x}(t)$, and state-vector, $\mathbf{s}(t)$, could be rewritten as

$$\mathbf{x}(\tau_1, \tau_2) = \mathbf{a}_{x0}(\tau_1) + \sum_{\substack{k=-K \\ k \neq 0}}^{K} \mathbf{a}_{x_k}(\tau_1) e^{j[k\omega_0 \tau_2 + \phi_{xk}(\tau_1)]} \tag{2.69}$$

and

$$\mathbf{s}(\tau_1, \tau_2) = \mathbf{a}_0(\tau_1) + \sum_{\substack{k=-K \\ k \neq 0}}^{K} \mathbf{a}_k(\tau_1) e^{j[k\omega_0 \tau_2 + \phi_k(\tau_1)]} \tag{2.70}$$

and the ordinary differential equation, ODE, in time that we used to model our system, the state-equation of (2.1), becomes the following multirate partial differential equation, MPDE, in τ_1 and τ_2 [27]

$$\frac{\partial \mathbf{s}(\tau_1, \tau_2)}{\partial \tau_1} + \frac{\partial \mathbf{s}(\tau_1, \tau_2)}{\partial \tau_2} = \mathbf{f}[\mathbf{s}(\tau_1, \tau_2), \mathbf{x}(\tau_1, \tau_2)] \tag{2.71}$$

The solution of this MPDE is a trajectory of $\mathbf{s}(\tau_1, \tau_2)$ whose physical significance is restricted to $\tau_1 = \tau_2 = t$.

At this point, the reader may be wondering where we are heading, since, to solve an already difficult problem, it seems we have complicated it even more. And we have to admit he is right. However, there is an important advantage hidden in (2.71): the dynamics with respect to the envelope and the carrier, i.e., with respect to τ_1 and τ_2, are decoupled so that we can now take advantage of the periodicity of $\mathbf{s}(\tau_1, \tau_2)$ in τ_2.

This means that we could try to solve (2.71) in the frequency domain for the periodic regime determined by the RF carrier and keep the time-domain representation for the aperiodic complex envelope. This way, we could rewrite (2.70) as

$$s(\tau_1, \tau_2) = \sum_{k=-K}^{K} \mathbf{S}_k(\tau_1) e^{jk\omega_0\tau_2} \tag{2.72}$$

which, substituted in the MPDE of (2.71), would lead to

$$\sum_{k=-K}^{K} \frac{d\mathbf{S}_k(\tau_1)}{d\tau_1} e^{jk\omega_0\tau_2} + \sum_{k=-K}^{K} jk\omega_0 \mathbf{S}_k(\tau_1) e^{jk\omega_0\tau_2}$$
$$= \mathbf{f}\left[\sum_{k=-K}^{K} \mathbf{S}_k(\tau_1) e^{jk\omega_0\tau_2}, \sum_{k=-K}^{K} \mathbf{X}_k(\tau_1) e^{jk\omega_0\tau_2} \right] \tag{2.73}$$

or

$$\frac{d\mathbf{S}(\tau_1, \omega)}{d\tau_1} + \mathbf{j}\Omega\mathbf{S}(\tau_1, \omega) = \mathbf{f}[\mathbf{S}(\tau_1, \omega), \mathbf{X}(\tau_1, \omega)] \tag{2.74}$$

(2.73) and (2.74) can be seen as a τ_1-varying HB equation, and also a $2K+1$ set of ODEs, which describe the evolution in the time τ_1 of the state-vector Fourier coefficients $\mathbf{S}(\omega)$ or $\mathbf{S}_k$. It constitutes, therefore, a time and frequency mixed-mode simulation technique.

Using, for example, a backward Euler discretization,

$$\mathbf{S}(\tau_{1_{n+1}}, \omega) - \mathbf{S}(\tau_{1_n}, \omega) + h_n\mathbf{j}\Omega\mathbf{S}(\tau_{1_{n+1}}, \omega) - h_n\mathbf{f}\left[\mathbf{S}(\tau_{1_{n+1}}, \omega), \mathbf{X}(\tau_{1_{n+1}}, \omega)\right] = 0 \quad (2.75)$$

the ODE in τ_1 can be solved as a succession of harmonic-balance problems, in which the Fourier coefficients to be determined are the $\mathbf{S}(\tau_{1_{n+1}}, \omega)$, while, because in each time-step the $\mathbf{S}(\tau_{1_n}, \omega)$ are known, they are treated as excitations.

This formulation – for obvious reasons known as the *envelope-transient harmonic-balance, ETHB* – can be interpreted as the solution of a set of periodic steady-state problems in which the excitation is evolving at the much slower pace of the envelope [28–30]. Actually, this process is what is illustrated in Figures 2.17 and 2.18. (As an example of application of ETHB to a circuit, see Exercise 2.6).

Because it is assumed that, at each envelope sample, the system is kept in a periodic steady-state, the envelope must be slowly varying in comparison to the carrier, so that it can be considered frozen during, at least, one RF period. Hence, ETHB requires that the modulated excitations and the state-variables can be expressed as narrowband signals whose bands, centered at dc and each of the carrier harmonics, do not overlap. Further-more, contrary to HB, which can handle devices with only a frequency-domain repre-sentation (such as a scattering matrix), this ETHB formulation requires a time-domain description for all system elements. If those time-domain models do not exist, they have to be created, which is prone to errors or even to numerical instabilities (namely, when the extracted time-domain models of passive blocks are not guaranteed to be passive). On the other hand, it correctly handles the nonlinearity and the system's dynamics imposed on both the carrier and the envelope. For example, it can cope with the transients

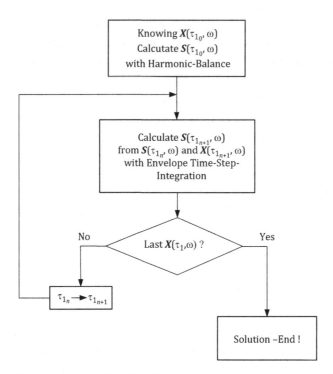

Figure 2.17 A simplified flowchart of the envelope-transient harmonic-balance algorithm.

observed in radar pulsed amplifiers or in any other RF pulsed systems, the initial transients of oscillators or even the capture or hold-in processes of phase-locked-loops, PLLs. In fact, ETHB should be the simulation option for all moderately strong nonlinear systems subject to multirate stimuli that only exhibit periodicity in the fast time-scale and are aperiodic in the slow one. This is the case of envelope-modulated RF carriers but could be also applied to phase-noise problems, electro-thermal transient behavior, etc.

2.4.3   Envelope-Transient over Shooting

In the same way we could apply either time-domain shooting or frequency-domain harmonic-balance to determine the system's RF periodic steady-state, we can now conceive a time-frequency envelope-transient harmonic-balance algorithm or a similar time-time *envelope-transient over shooting*, EToS [31]. This EToS takes the MPDE of (2.71) discretizes it in τ_1 and τ_2 (using, for example, a backward Euler scheme for both τ_1 and τ_2)

$$
\frac{\mathbf{s}\left(\tau_{1_{n_1+1}}, \tau_{2_{n_2+1}}\right) - \mathbf{s}\left(\tau_{1_{n_1}}, \tau_{2_{n_2+1}}\right)}{h_{n_1}} + \frac{\mathbf{s}\left(\tau_{1_{n_1+1}}, \tau_{2_{n_2+1}}\right) - \mathbf{s}\left(\tau_{1_{n_1+1}}, \tau_{2_{n_2}}\right)}{h_{n_2}}
$$
$$
= \mathbf{f}\left[\mathbf{s}\left(\tau_{1_{n_1+1}}, \tau_{2_{n_2+1}}\right), \mathbf{x}\left(\tau_{1_{n_1+1}}, \tau_{2_{n_2+1}}\right)\right] \tag{2.76}
$$

and carries a transient analysis on τ_1 over successive periodic steady-state analyses using shooting-Newton on τ_2. This envelope following PSS should be preferred to

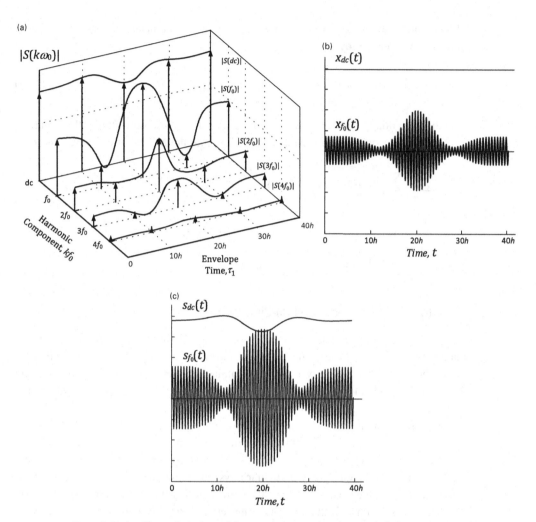

Figure 2.18 An illustrative view of the envelope-transient harmonic-balance process.
(a) Representation of the state-variable harmonic components in the carrier frequency and
envelope time domains. (b) and (c) depict the dc and fundamental components of the excitation
and the response, respectively.

ETHB whenever the periodic regime involves strong nonlinearities and so requires a
very large number of harmonics. This is the case, for example, of many new digital
assisted, or digital-like, RF circuits based on switching mode operation, such as
envelope tracking amplifiers or polar transmitters [32], RF pulse-width modulators
[33], and many other software defined radio circuits.

2.4.4   Another View of MDFT-Based Harmonic-Balance

For completeness, we could ask ourselves what would result from using the MPDE
formulation in presence of a periodic envelope, i.e., what would result from the MPDE

if both time scales were treated in the frequency domain. In that case, the solution would have to be expressed by

$$\mathbf{s}(\tau_1, \tau_2) - \sum_{k_1=-K_1}^{K_1} \sum_{k_2=-K_2}^{K_2} \mathbf{S}_{k_1,k_2} e^{jk_1\omega_1\tau_1} e^{jk_2\omega_2\tau_2} \qquad (2.77)$$

which is essentially equivalent to the MDFT of (2.64). Now, substituting this solution, and the corresponding bidimensional excitation, into the MPDE of (2.71) results in

$$\sum_{k_1=-K_1}^{K_1} \sum_{k_2=-K_2}^{K_2} j(k_1\omega_1 + k_2\omega_2) \mathbf{S}_{k_1,k_2} e^{j(k_1\omega_1\tau_1 + k_2\omega_2\tau_2)}$$
$$= \mathbf{f} \left[\sum_{k_1=-K_1}^{K_1} \sum_{k_2=-K_2}^{K_2} \mathbf{S}_{k_1,k_2} e^{j(k_1\omega_1\tau_1 + k_2\omega_2\tau_2)}, \sum_{k_1=-K_1}^{K_1} \sum_{k_2=-K_2}^{K_2} \mathbf{X}_{k_1,k_2} e^{j(k_1\omega_1\tau_1 + k_2\omega_2\tau_2)} \right] \qquad (2.78)$$

the MDFT-based harmonic-balance equation.

2.4.5 Circuit-Level and System-Level Co-Simulation

In wireless communications, we are also interested in simulating how the modulated data is treated by all blocks of the wireless system, conducting a so-called system-level simulation. Unfortunately, system-level simulators operate on modulated data and behavioral models of the RF or baseband circuits, and so do not support simulation at the circuit level. Envelope-following techniques are especially useful to bridge this gap, incorporating circuit-level simulation in these system-level simulators. While current and voltage waveforms, or spectra, are the desired results in circuit-level simulation, in system-level simulators we aim to predict how the information is processed, regardless of how it is represented by any particular physical magnitude. For that, we extract the information from the original real signal and then use it as the excitation of system-level models especially conceived to disregard the details of the physical magnitudes carrying that information. The process by which we extract the information from the real signal was already addressed in this text when, in Section 2.4.1, we introduced the concept of the complex envelope. The system-level models we are thinking of are low-pass equivalent behavioral models [34], whose excitation and response are their input and output complex envelopes, respectively.

System-level simulation is, therefore, much more computationally efficient than circuit-level simulation as it does not have to simulate each time point of the RF carrier but simply the time samples of the complex envelope, a much slower signal. Unfortunately, as in any other simulation environment, system-level simulators are limited in accuracy by their low-pass equivalent behavioral models. And if these are widely accepted as very accurate when used to represent linear blocks, as filters, they still constitute hot topics of research when they are supposed to represent nonlinear dynamic blocks such as RF power amplifiers [35] or I/Q modulators [36]. In these cases, we would like to simulate all the other blocks at the system-level, and keep, e.g., the RF power amplifier at the circuit-level. This form of system-level/circuit-level

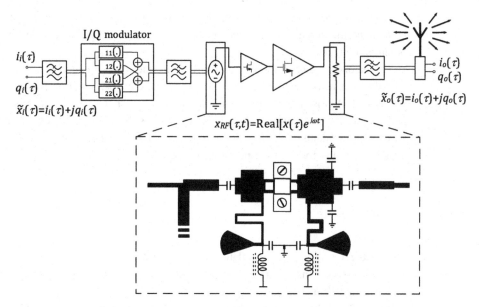

Figure 2.19 System-level/circuit-level co-simulation platform of a wireless transmitter in which all blocks except the RF power amplifier are represented as low-pass behavioral models for the complex envelope, whereas the power amplifier is represented at the circuit-level by its equivalent-circuit model for the real RF modulated signal.

co-simulation platform is depicted in Figure 2.19 and is enabled by our studied envelope-following techniques.

2.5   Summary

Because usual nonlinear lumped-element circuits and systems are represented by a state-equation, i.e., a set of coupled nonlinear differential equations in time, for which, in general, we do not have an analytical solution, we have to rely on numerical simulation. This means that, except for a few exceptional cases, of which linear systems constitute the best example, quantitative synthesis of nonlinear systems is impossible, and nonlinear system analysis is little more than the search for an approximate numerical solution of the state-equation for a particular excitation.

Similarly to what we learned in linear circuit analysis, where nodal analysis is general but other alternative techniques, such as mesh analysis or source transformation, could be more efficient for certain circuits, transient analysis is also general, but that does not mean it is the most efficient method for all applications. In fact, we explained how it is highly inefficient in determining the periodic-steady-state or the response to a narrow-band modulated signal. Therefore, in this chapter, we introduced other methods such as the time-domain shooting-Newton or the frequency-domain harmonic-balance, for the first problem, and the envelope-transient harmonic-balance or the envelope-transient-over-shooting, for the second one.

With that, we hope we could provide the RF engineer the knowledge necessary to understand the domain of applicability of each method and its capabilities, but also its potential dangers.

Simulators require a model for the circuit (the state-equation), a model for the stimulus and response signals, and models of the devices. The first two were already introduced and discussed in the beginning of this chapter. The device models are so vast and involve so much diverse knowledge (on both system identification and RF/microwave measurements) that they deserve several chapters specifically devoted to them. This is, thus, the focus of the next four chapters.

2.6   Exercises

Exercise 2.1   Write the formulation that would allow you to integrate the system of state-equations of the circuit of Figure 2.2, i.e., Eq. (2.4), using the 4th order Runge-Kutta method.

Exercise 2.2   Using the notions of state and state-equation, show that the periodic steady-state is indeed determined by the choice of appropriate initial conditions.

Exercise 2.3   Assuming a linear approximation for the FET channel current, i.e., $i_{ds}(v_{gs}, v_{ds}) \approx g_m v_{gs} + g_{ds} v_{ds}$, determine the matrices A_1 and A_2 of Eq. (2.33), that correspond to the circuit of Figure 2.2.

Exercise 2.4   Write the matrix equations that correspond to the analysis of the circuit example of Figure 2.2 via the piece-wise harmonic-balance method. Note how simplified this method becomes, in comparison to the nodal harmonic-balance, when the circuit contains many more linear elements than nonlinear ones.

Exercise 2.5   The resistance of conductors suffers an increase with the square-root of frequency due to the skin effect. Explain why it isn't possible to use a transient simulator as SPICE to deal with a circuit with such an element, having instead to use the impulse response and subsequent convolution or to use a harmonic-balance simulator?

Exercise 2.6   Write the envelope-transient harmonic-balance equations for the circuit example shown in Figure 2.2 and expressed by its state equations (2.4).

References

[1] L. T. Pillage and R. A. Rohrer, "Asymptotic waveform evaluation for timing analysis," *IEEE Trans. Comput.-Aided Design*, vol. CD-9, Apr. 1990.

[2] T. J. Brazil, "Causal-convolution – a new method for the transient analysis of linear systems at microwave frequencies," *IEEE Trans. Microw. Theory Techn.*, vol. MTT-43, Feb. 1995, pp. 315–323.

[3] A. Odabasioglu, M. Celik, and L. T. Pileggi, "PRIMA: passive reduced-order interconnect macromodeling algorithm," *IEEE Trans. Comput.-Aided Design Integr. Circuits Syst.*, vol. CAD-17, Aug. 1998, pp. 645–654.

[4] L. W. Nagel, *Spice2: A Computer Program to Simulate Semiconductor Circuits,* Electronic Research Laboratory, University of California-Berkeley, Memo ERL-M520, 1975.

[5] *HSPICE User Guide: Simulation and Analysis*, Mountain View, CA: Synopsys, Inc., 2008.

[6] *Matlab Simulink 7 User Manual*, Natick, MA: The Mathworks, Inc., 2010.

[7] *ADS Ptolemy Simulation*, Palo Alto, CA: Agilent Technologies, 2004.

[8] *Visual System Simulator Getting Started Guide – NI AWR Design Environment*, v12 ed., El Segundo, CA: National Instruments AWR, 2016.

[9] R. L. Burden and D. J. Faires, *Numerical Analysis*, 9th ed., Boston, MA: Brooks/Cole, 2011.

[10] C.-T. Chen, *Digital Signal Processing – Spectral Computation and Filter Design*, New York, NY: Oxford University Press, 2001.

[11] T. J. Aprille and T. N. Trick, "Steady-state analysis of nonlinear circuits with periodic inputs," *Proc. IEEE*, vol. 60, pp. 108–114, Jan. 1972.

[12] K. Kundert, J. White, and A. Sangiovanni-Vicentelli, *Steady-State Methods for Simulating Analog and Microwave Circuits*, Norwell, MA: Kluwer Academic Pub. 1990.

[13] M. Schetzen, *The Volterra and Wiener Theories of Nonlinear Systems*, New York, NY: John Wiley & Sons, 1980.

[14] G. W. Rhyne, M. B. Steer, and B. D. Bates, "Frequency-domain nonlinear circuit analysis using generalized power series," *IEEE Trans. Microw. Theory Techn.*, vol. MTT-36, Feb. 1988, pp. 379–387.

[15] C. R. Chang and M. B. Steer, "Frequency-domain nonlinear microwave circuit simulation using the arithmetic operator method," *IEEE Trans. Microw. Theory Techn.*, vol. MTT-38, Aug. 1990.

[16] S. A. Maas, *Microwave Mixers*, Norwood, MA: Artech House, 1986.

[17] S. A. Maas, *Nonlinear Microwave Circuits*, Norwood, MA: Artech House, 1988.

[18] A. Suárez, *Analysis and Design of Autonomous Microwave Circuits*, New York, NY: John Wiley & Sons, 2009.

[19] J. Verspecht, "Large signal network analysis," *IEEE Microw. Mag.*, vol. 6, Dec. 2005, pp. 82–92.

[20] D. Root, J. Verspecht, D. Sharrit, J. Wood, and A. Cognata, "Broad-band poly-harmonic distortion (PHD) behavioral models from fast automated simulations and large-signal vectorial network measurements," *IEEE Trans. Microw. Theory Techn.*, vol. MTT-53, Nov. 2005, pp. 3656–3664.

[21] K. Kurokawa, "Some basic characteristics of broadband negative resistance oscillator circuits," *The Bell System Tech. Jour.*, vol. 48, Jul.–Aug. 1969, pp. 1937–1955.

[22] A. Ushida and L. Chua, "Frequency-domain analysis of nonlinear circuits driven by multi-tone signals," *IEEE Trans. Circuits and Syst.*, vol. CAS-31, Sep. 1984, pp. 766–779.

[23] Rizzoli, V., Cecchetti, and C., Lipparini, A., "A general-purpose program for the analysis of nonlinear microwave circuits under multitone excitation by multidimensional Fourier transform," *17th European Microw. Conf. Proc*, pp. 635–640, 1987, pp. 766–779.

[24] N. B. Carvalho and J. C. Pedro, "Novel artificial frequency mapping techniques for multi-tone simulation of mixers," *2001 IEEE MTT-S Intern. Microw. Symp Dig.*, 2001, pp. 455–458.

[25] P. J. Rodrigues, *Computer Aided Analysis of Nonlinear Microwave Circuits*, Norwood, MA: Artech House, 1998.

[26] J. C. Pedro and N. B. Carvalho, "Efficient harmonic balance computation of microwave circuits' response to multitone spectra," *29th European Microw. Conf. Proc.*, 1999, pp. 103–106.

[27] H. Brachtendorf, G. Welsch, R. Laur, and A. Bunse-Gerstner, "Numerical steady-state analysis of electronic circuits driven by multitone signals," *Electrical Engineering*, vol. 79, Apr. 1996, pp. 103–112.

[28] E. Ngoya, and R. Larchevèque, "Envelop transient analysis: a new method for the transient and steady state analysis of microwave communication circuits and systems," *1996 IEEE MTT-S Int. Microw. Symp. Dig.* 1996, pp. 1365–1368.

[29] V. Rizzoli, A. Neri, and F. Mastri, "A Modulation-oriented Piecewise Harmonic Balance Technique Suitable for Transient Analysis and Digitally Modulated Analysis," *26th European Microw. Conf. Proc.*, Prague, Cheks Rep. 1996, pp. 546–550.

[30] D. Sharrit, "Method for simulating a circuit," US Patent No. 5588142, Dec. 24, 1996.

[31] J. Roychowdhury, "Analyzing circuits with widely separated time scales using numerical PDE methods," *IEEE Trans. Circuits and Syst.*, vol. CAS-48, May 2001, pp. 578–594.

[32] J. F. Oliveira and J. C. Pedro, "An efficient time-domain simulation method for multirate RF nonlinear circuits," *IEEE Trans. Microw. Theory Techn.*, vol. MTT-55, Nov. 2007, pp. 2384–2392.

[33] J. F. Oliveira and J. C. Pedro, "An innovative time-domain simulation technique for strongly nonlinear heterogeneous RF circuits operating in diverse time-scales," *3rd European Integrated Circuits Conf. Proc.*, Oct. 2008, pp. 530–533.

[34] S. Benedetto, E. Biglieri, and R. Daffara, "Modeling and performance evaluation of nonlinear satellite links - a Volterra series approach," *IEEE Trans. Aerospace and Electronic Syst.*, vol. AES-15, Apr. 1979, pp. 494–507.

[35] J. C. Pedro and S. A. Maas, "A comparative overview of microwave and wireless power-amplifier behavioral modeling approaches," *IEEE Trans. Microw. Theory Techn.*, vol. MTT-53, Apr. 2005, pp. 1150–1163.

[36] H. Cao, A. Soltani Tehrani, C. Fager, T. Eriksson, and H. Zirath, "I/Q imbalance compensation using a nonlinear modeling approach," *IEEE Trans. Microw. Theory Techn.*, vol. MTT-57, Mar. 2009, pp. 513–518.

3 Linear Behavioral Models in the Frequency Domain
S-parameters*

3.1 Introduction

This chapter presents a concise treatment of S-parameters, meant primarily as an introduction to the more general formalism of large-signal approaches of the next chapter. The concepts of *time invariance* and *spectral maps* are introduced at this stage to enable an easier generalization in the ensuing chapter. The interpretations of S-parameters as calibrated measurements, intrinsic properties of the device-under-test (DUT), intellectual property (IP)-secure component behavioral models, and composition rules for linear system design are presented. The cascade of two linear S-parameter components is considered as an example to be generalized to the nonlinear case later. The calculation of S-parameters for a transistor from a simple nonlinear device model is used as an example to introduce the concepts of (static) *operating point* and *small-signal conditions*, both of which must be generalized for the large-signal treatment.

3.2 S-parameters

Since the 1950s, S-parameters, or scattering parameters, have been among the most important of all the foundations of microwave theory and techniques.

S-parameters are easy to measure at high frequencies with a vector network analyzer (VNA). Well-calibrated S-parameter measurements represent properties of the DUT, independent of the VNA system used to characterize it. Calibration procedures [2] remove systematic measurement errors and enable a separation of the overall values into numbers attributable to the device, independent of the measurement system used to characterize it. In this context, we call such properties *intrinsic*.[1] These DUT properties (gain, loss, reflection coefficient, etc.) are familiar, intuitive, and important [3]. Another key property of S-parameters is that the S-parameters of a composite system are completely determined from knowledge of the S-parameters of the constituent components and their connectivity. S-parameters provide the complete specification of how a linear component responds to an arbitrary signal. Therefore, simulations of linear

---

* An earlier version of this chapter appeared in Root et al., *X-Parameters* (2013) [1].
[1] The word *intrinsic* is used elsewhere in the book to refer to behavior at internal points or planes of a device that may not be directly accessible to measurement. The meaning should be clear from the context.

systems that are designed by cascading S-parameter blocks are predictable with certainty. S-parameters define a complete behavioral description of the linear component at the external terminals, independent of the detailed physics or specifics of the realization of the component. S-parameters can be shared between component vendors and system integrators freely, without the possibility that the component implementation can be reverse engineered, protecting IP and promoting sharing and reuse. Indeed, one may ask the question, "are S-parameters measurements, or do they constitute a model?" The answer is really "both." They are numbers that can be measured accurately at RF and microwave frequencies, and these numbers are the coefficients of a behavioral model that expresses the scattered waves as linear contributions of the incident waves.

S-parameters need not come only from measurements. They can be calculated from physics by solving Maxwell's equations, by linearizing the semiconductor equations, or computed from matrix analysis of linear equivalent circuits. In this way, the many benefits of S-parameters can be realized starting from a more detailed representation of the component from first principles or from a complicated linear circuit model.

Graphical methods based around the Smith Chart were invented to visualize and interpret S-parameters and graphical design methodologies soon followed for circuit design [3], [4]. These days, electronic design automation (EDA) tools provide simulation components – S-parameter blocks – and design capabilities using the familiar S-parameter analysis mode.

One of the great utilities of S-parameters is the interoperability among the measurement, modeling, and design capabilities they provide. One can characterize the component with measured S-parameters, use them as a high-fidelity behavioral model of the component with complete IP protection, and design systems with them in the EDA environment.

3.3   Wave Variables

The term *scattering* refers to the relationship between incident and scattered (reflected and transmitted) travelling waves.

By convention, in this text the circuit behavior is described using generalized power waves [5], although there are alternative wave-definitions used in the industry.

The wave variables, A and B, corresponding to a specific port of a network, are defined as simple linear combinations of the voltage and current, V and I, at the same port, according to Figure 3.1 and equations (3.1).

$$A = \frac{V + Z_0 I}{2\sqrt{Z_0}}$$

$$B = \frac{V - Z_0 I}{2\sqrt{Z_0}}$$

$$(3.1)$$

The reference impedance for the port, Z_0, is, in general, a complex value. For the purpose of simplifying the concepts presented, the reference impedance is restricted to real values in this text.

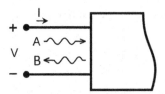

Figure 3.1 Wave definitions.

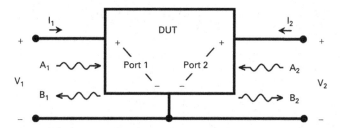

Figure 3.2 Incident and scattered waves of a two-port device. All ports should be referenced to the same pin for modeling purposes.

The currents and voltages can be recovered from the wave variables, according to equations (3.2).

$$V = \sqrt{Z_0} \cdot (A + B)$$

$$I = \frac{1}{\sqrt{Z_0}} \cdot (A - B)$$

(3.2)

Here A and B represent the incident, scattered waves V and I are the port voltage and current, respectively, and Z_0 is the reference impedance for the port. Z_0 can be different for each port, but we do not consider that further here. A typical value of Z_0 is 50 Ω by convention, but other choices may be more practical for some applications. A value for Z_0 closer to 1 Ω is more appropriate for S-parameter measurements of power transistors, for example, given that power transistors typically have very small output impedances.

The variables in equations (3.1) and (3.2) are complex numbers representing the RMS-phasor description of sinusoidal signals in the frequency domain. Later we will generalize to the envelope domain by letting these complex numbers vary in time.

A, B, V, and I can be considered RMS-vectors, the components of which indicate the values associated with sinusoidal signals at particular ports labeled by positive integers. Thus A_j is the incident wave RMS-phasor at port j and I_k is the current RMS-phasor at port k. For now, Z_0 is taken to be a fixed real constant, in particular, 50 Ω.

A graphical representation of the wave description is given in Figure 3.2 for the case of a system described by a two-port device. For definiteness, we assume a port description with a common reference pin, as indicated in the figure.

To retrieve the time-dependent sinusoidal voltage signal at the ith port, the complex value of the phasor and also the angular frequency, ω, to which the phasor corresponds, must be known. The voltage is then given by (3.3), and, similarly, for the other variables, where $V_i^{(pk)}$ are peak values.

$$v_i(t) = \text{Re}\left\{V_i^{(pk)}e^{j\omega t}\right\} \tag{3.3}$$

It is convenient to keep track of the frequency associated with a particular set of phasors by rewriting (3.1) according to (3.4), and (3.2) according to (3.5), where the port indexing notation is made explicit.

$$A_i(\omega) = \frac{1}{2\sqrt{Z_0}}\cdot(V_i(\omega) + Z_0 I_i(\omega))$$

$$\tag{3.4}$$

$$B_i(\omega) = \frac{1}{2\sqrt{Z_0}}\cdot(V_i(\omega) - Z_0 I_i(\omega))$$

$$V_i(\omega) = \sqrt{Z_0}\cdot(A_i(\omega) + B_i(\omega))$$

$$\tag{3.5}$$

$$I_i(\omega) = \frac{1}{\sqrt{Z_0}}(A_i(\omega) - B_i(\omega))$$

For each angular frequency, ω, Eq. (3.4) is a set of two equations defined at each port.

The assumption behind the S-parameter formalism is that the system being described is *linear* and therefore there must be a *linear relationship, implying superposition holds*, between the phasor representation of incident and scattered waves. This is expressed in (3.6) for an N-port network:

$$B_i(\omega) = \sum_{j=1}^{N} S_{ij}(\omega)A_j(\omega), \ \forall i \in \{1, 2, \ldots, N\} \tag{3.6}$$

The set of complex coefficients, $S_{ij}(\omega)$, in (3.6) define the S-parameter matrix, or simply, the S-parameters, at that frequency. Equation (3.6), for the fixed set of complex S-parameters, determines the output phasors for any set of input phasors. The summation is over all port indices, so that incident waves at each port, j, contribute in general to the overall scattered wave at each output port, i. For now we consider all frequencies to be positive ($\omega > 0$). Note that contributions to a scattered wave at frequency ω come only from incident waves at the same frequency. This is not the case for the more general X-parameters, where a stimulus at one frequency can lead to scattered waves at different frequencies.

For definiteness and later reference, the S-parameter equations for a linear two-port are written explicitly in (3.7).

$$B_1 = S_{11}A_1 + S_{12}A_2$$

$$\tag{3.7}$$

$$B_2 = S_{21}A_1 + S_{22}A_2$$

The set of equations (3.6) represent a model of the network under study. For the purpose of creating a model for the network, all ports should be referenced to the same pin, as shown in Figure 3.2.

Such connectivity is the natural option for the measurement and modeling process of a 3-pin network (like a transistor) but it has to be extended in the general case of an arbitrary network and it is necessary for all networks, linear and/or nonlinear. This connectivity convention is considered by default (unless otherwise specified) for the remaining of this text.

Using the topological connection in Figure 3.2, the set of equations (3.6) represent a complete model of the network under test.

From (3.6) we note a stimulus (incident wave) at a particular port j will produce a response (scattered wave) at all ports, including the port at which the stimulus is applied.

3.4   Geometrical Interpretation of S-parameters

For linear two-port components described by (3.7), the scattering of A_2 at port 2 is given simply by $S_{22} A_2$, where S_{22} is just a fixed complex number. This means we can write for $\Delta B_2 = B_2 - S_{21} A_1$

$$\Delta B_2 = S_{22} A_2. \tag{3.8}$$

For fixed $|A_2|$, $S_{22} A_2 = S_{22} |A_2| e^{j\phi(A_2)}$, so by varying $\phi(A_2)$ from 0 to 2π radians, equation (3.8) traces out a circle with radius $|S_{22} A_2|$. This is shown in Figure 3.3.

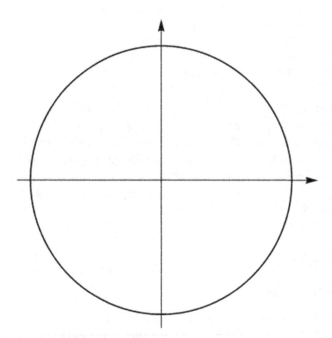

Figure 3.3 Locus of points produced from S-parameter equation (3.8) evaluated for all phases of A_2. The radius of the circle is $|S_{22} A_2|$.

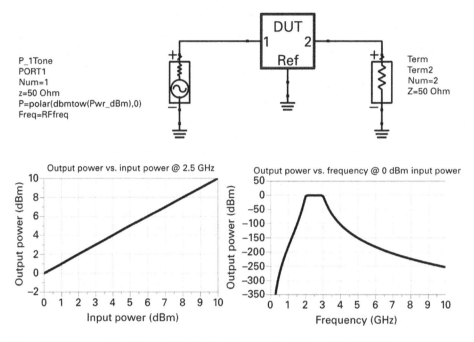

Figure 3.4 A linear network has a linear behavior versus input power level, but the dependence on frequency is usually not linear.

This fact is trivial in the case of S-parameters. We will see in the next chapter that the scattering of even a small-signal A_2 has a different (non-circular) geometry when a nonlinear component is driven with a large A_1 signal.

3.5   Nonlinear Dependence on Frequency

Equation (3.6) shows that the scattered waves are linear functions of the complex amplitudes (the phasors) of the incident waves, with the S-parameters being the coefficients of the linear relationship. These coefficients, however, have a frequency dependence that is usually not linear. That is, we generally have $S_{ij}(\lambda\omega) \neq \lambda S_{ij}(\omega)$ for λ a positive real number. For example, an ideal band-pass filter response is linear in the incident wave phasor, but the filter response is a nonlinear function of the frequency of the incident wave. This is shown in Figure 3.4.

3.6   S-parameter Measurement

3.6.1   Conventional Identification

By setting all incident waves to zero in (3.6), except for A_j, one can deduce the simple relationship between a given S-parameter (S-parameter matrix element) and a particular ratio of scattered to incident waves according to (3.9). Sometimes (3.9) is taken as the

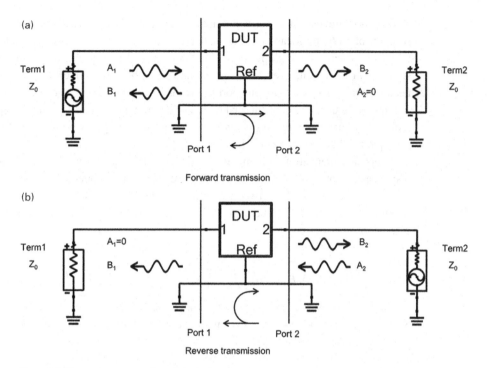

Figure 3.5 S-parameter experiment design.

definition of the S-parameters for a linear system, instead of our starting point (3.6). However, since we will show an alternative identification method for S-parameters in Section 3.6.2, we prefer to interpret (3.9) as a simple consequence of the more fundamental linearity principle, (3.6).

$$S_{ij}(\omega) = \frac{B_i(\omega)}{A_j(\omega)}\bigg|_{\substack{A_k = 0 \\ \forall k \neq j}} \tag{3.9}$$

Equation (3.9) corresponds to a simple graphical representation shown in Figure 3.5 for the simple case of a two-port component. In Figure 3.5 a, the stimulus is a wave incident at port 1. The fact that A_2 is not present ($A_2 = 0$) is interpreted to mean that the B_2 wave scattered and travelling away from port 2 is not reflected back into the device at port 2. Under this condition, the device is said to be *perfectly matched* at port 2. Two of the four complex S-parameters, specifically S_{11} and S_{21}, can be identified using (3.9) for this case of exciting the device with only A_1. Figure 3.5b shows the case where the device is stimulated with a signal, A_2, at port 2, and assumed to be perfectly matched at port 1 ($A_1 = 0$). The remaining S-parameters, S_{12} and S_{22}, can be identified from this ideal experiment.

3.6.2   Alternative Identification

The S-parameters may also be determined by a direct solution of (3.6) from multiple measurements of the B_i in response to excitations in which the A_j have multiple (relative) phases.

The S-parameters at a particular frequency may also be determined by a direct solution of (3.6) by measuring the scattered waves in response to known excitations by incident waves at multiple ports simultaneously. Exciting the device by multiple signals can be useful, even for linear components, because, for example, there is no need to switch a signal from one port to the others or to perfectly match the other ports as required in the ideal conventional case. For the nonlinear case of the next chapter, we will have to use *simultaneous excitations*. Let's use a simple two-port [$N = 2$ in (3.6)] example to illustrate the concept.

In this case there are four complex S-parameters to be determined from the measured responses at both ports, B_i, to the known excitations simultaneously at each port, A_j. Therefore, we need at least two sets of independent excitations, $A_i^{(n)}$, and the corresponding responses, $B_i^{(n)}$, where the superscript, n, labels the experiment number. In matrix notation we have

$$\begin{pmatrix} B_1^{(1)} & B_1^{(2)} \\ B_2^{(1)} & B_2^{(2)} \end{pmatrix} = \begin{pmatrix} S_{11} & S_{12} \\ S_{21} & S_{22} \end{pmatrix} \cdot \begin{pmatrix} A_1^{(1)} & A_1^{(2)} \\ A_2^{(1)} & A_2^{(2)} \end{pmatrix} \tag{3.10}$$

Equation (3.10) is a set of four complex equations for the four unknown S-parameters, S_{ij} given the four measured $B_i^{(n)}$ responses to the four stimuli $A_j^{(n)}$. The formal solution is given in matrix notation in (3.11).

$$S = BA^{-1} \tag{3.11}$$

More generally, for more robust results, there can be more data [columns for the first and third matrices in (3.10)] than unknowns and the formal solution is given in terms of pseudo-inverses according to (3.12). According to the discussion in Chapter 1, this provides the best S-matrix in the least squares sense:

$$S = BA^T \left(AA^T\right)^{-1} \tag{3.12}$$

3.6.3   S-parameter Measurement for Nonlinear Components

It is important to note that the ratio on the right-hand side of (3.9) can be computed from independent measurements of incident and scattered waves for actual components corresponding to any nonzero value for the incident wave, A_j. The value of this ratio, however, will generally vary with the magnitude of the incident wave. Therefore, the identification of this ratio with the S-parameters of the component is valid for any particular value of incident A_j only if the component behaves linearly, namely according to (3.6). In other words, the values of the incident waves, A_j, need to be in the linear region of operation for this identification to be valid. For nonlinear components, such as transistors biased at a fixed voltage, the scattered waves don't continue indefinitely to increase as the incident waves get larger in magnitude (this is compression). Therefore, different values of (3.9) result from different values of incident waves. A better definition of S-parameters for a *nonlinear component* is a modification of (3.9) given by (3.13). That is, for a general component, biased at a constant DC-stimulus, the S-parameters are related to ratios of output responses to input stimuli in the limit of small input signals. This emphasizes that S-parameters properly apply to nonlinear components only in the *small-signal* limit.

$$S_{ij}(\omega) \equiv \lim_{|A_j| \to 0} \frac{B_i(\omega)}{A_j(\omega)}\bigg|_{\substack{A_k = 0 \\ \forall k \neq j}} \qquad (3.13)$$

3.7   S-parameters as a Spectral Map

If there are multiple frequencies present in the input spectrum, one can represent the output spectrum in terms of a matrix giving the contributions to each output frequency from each input frequency.

An example in the case of three input frequencies is given by Eq. (3.14). Here we assume a single port, for simplicity, and therefore drop the port indices. It is clear from (3.14) that S-parameters are a diagonal map in frequency space. This means each output frequency contains contributions only from inputs at that same frequency. Or, in other words, each input frequency never contributes to outputs at any different frequency.

$$\begin{bmatrix} B(\omega_1) \\ B(\omega_2) \\ B(\omega_3) \end{bmatrix} = \begin{bmatrix} S(\omega_1) & 0 & 0 \\ 0 & S(\omega_2) & 0 \\ 0 & 0 & S(\omega_3) \end{bmatrix} \begin{bmatrix} A(\omega_1) \\ A(\omega_2) \\ A(\omega_3) \end{bmatrix} \qquad (3.14)$$

A graphical representation is given in Figure 3.6 for the case of forward transmission through a two-port network with matched terminations at both ports.

The interpretation of Figure 3.6, mathematically represented by (3.14), is that S-parameters define a particularly simple *linear spectral map* relating incident to scattered waves. S-parameters are diagonal in the frequency part of the map, namely they predict a response only at the particular frequencies of the corresponding input stimuli. It will be demonstrated in later chapters that the large-signal approaches provide for richer behavior.

For signals with a continuous spectrum, the diagonal nature of the S-parameter spectral map can be written as

$$B_i(\omega) = \sum_{j=1}^{N} \int S_{ij}(\omega)\delta(\omega - \omega')A_j(\omega')d\omega' \quad , \forall i \in \{1, 2, \ldots, N\}. \qquad (3.15)$$

Performing the integration over input frequencies in (3.15) results in Equation (3.6), the form usually given for S-parameters.

3.8   Superposition

Any linear theory, like S-parameters, enables the general response to an arbitrary input signal to be computed by *superposition* of the responses to unit stimuli (see the discussion of the impulse response in Chapter 1, Section 2). Superposition enables great simplifications in analysis and measurement. Superposition is the reason S-parameters

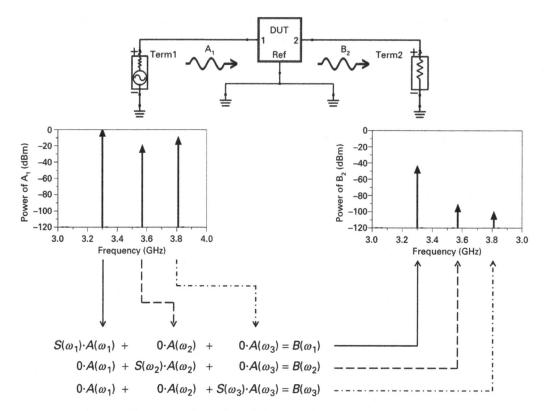

Figure 3.6 Linear spectral map through S-parameters matrix.

can be measured by independent experiments with one sinusoidal stimulus at a time, one stimulus per port per frequency using Eq. (3.9). The general response to any set of input signals can be obtained by superposition using Eq. (3.6).

An example of superposition is shown in Figure 3.7, with all signals represented both in time and frequency domains.

This example uses two signals, each containing two frequency components, as stimuli incident at each port, independently, with the other port perfectly matched. The example shows that the response to a linear combination of the stimuli is the same linear combination of the individual responses.

As always, the caveat is that the component actually behaves linearly over the range of signal levels used to stimulate the device. There is no *a priori* way to know whether a component will behave linearly without precise knowledge about its composition or physical or measured characteristics.

3.9   Time Invariance of Components Described by S-parameters

A DUT description in terms of S-parameters defined by (3.6) naturally embodies an important principle known as *time invariance*. Time invariance states that if $y(t)$ is the

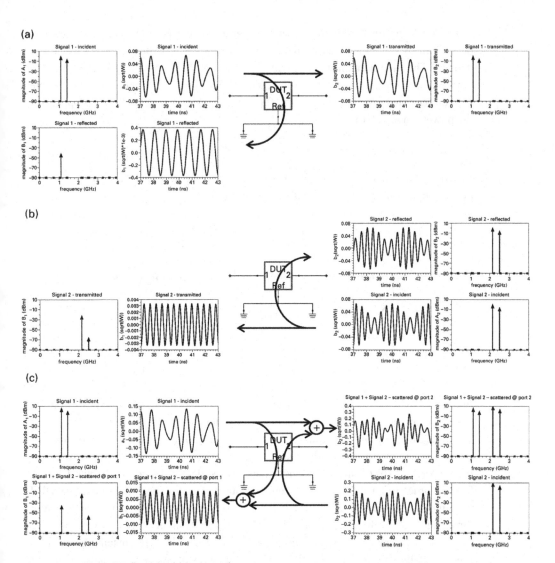

Figure 3.7 Superposition example.

DUT response to an excitation, $x(t)$, then the DUT response to the time-shifted excitation, $x(t - \tau)$, must be $y(t - \tau)$. This must be true for all time-shifts, τ. That is, if the input is shifted in time, the output is shifted by the corresponding amount but is otherwise identical with the DUT response to the nonshifted input. This is stated mathematically in (3.16), where O is the operator taking input to output:

$$\forall \tau \in \mathbb{R} \quad y(t) = O[x(t)] \quad \Rightarrow \quad y(t - \tau) = O[x(t - \tau)] \tag{3.16}$$

Time invariance is a property of common linear and nonlinear components, such as passive inductors, capacitors, resistors, and diodes, and active devices such as transistors. Examples of components not time-invariant (in the usual sense) are oscillators and other autonomous systems.

The proof follows from elementary properties of the Fourier Transform, where a phase shift by $e^{j\omega\tau}$ in the frequency domain corresponds to a time shift of τ in the time domain. The time-domain waves incident at the ports (the stimuli) are $a_j(t)$ and their Fourier transforms are $A_j^{(pk)}(\omega)$, as in equation (3.17). In (3.17), the symbol $\mathscr{F}$ denotes the Fourier transform. The superscript, (pk), in (3.17) refers to peak value (as opposed to the power-wave value).

$$\mathscr{F}\{a_j(t)\} = A_j^{(pk)}(\omega) \tag{3.17}$$

The time-domain waves scattered from the ports (the response) are $b_i(t)$ and their Fourier transforms are $B_i^{(pk)}(\omega)$, as in equation (3.18).

$$\mathscr{F}\{b_i(t)\} = B_i^{(pk)}(\omega) = \sum_{j=1}^{N} S_{ij}(\omega)A_j^{(pk)}(\omega) = \sum_{j=1}^{N} S_{ij}(\omega)\mathscr{F}\{a_j(t)\} \tag{3.18}$$

If all stimuli are delayed with the same time-delay, τ, the response becomes

$$\mathscr{F}^{-1}\left\{\sum_{j=1}^{N} S_{ij}(\omega)\mathscr{F}\{a_j(t-\tau)\}\right\} = \mathscr{F}^{-1}\left\{\sum_{j=1}^{N} S_{ij}(\omega)\mathscr{F}\{a_j(t)\}e^{i\omega\tau}\right\}$$
$$\tag{3.19}$$
$$= \mathscr{F}^{-1}\left\{B_i^{(pk)}(\omega)e^{i\omega\tau}\right\} = b_i(t-\tau)$$

Equation (3.19) proves that S-parameters are automatically consistent with the principle of time invariance. Therefore, any set of S-parameters describes a time-invariant system.

Unlike the case for S-parameters, a more general (e.g., nonlinear) relationship between incident A waves and scattered B waves is not automatically consistent with the property (3.16) of time invariance. This will be demonstrated in Chapter 4. Therefore, in order to have a consistent representation of a nonlinear time-invariant DUT, the time-invariance property must be manifestly incorporated into the mathematical formulation of the behavioral model relating input to output waves. A representation of a time-invariant DUT by equations not consistent with (3.16) means the model is fundamentally wrong and can yield very inaccurate results for some signals, even if the model "fitting" (or identification) appears good at time t.

3.10   Cascadability

Another key property of S-parameters is that the response of a linear circuit or system can be computed with perfect certainty knowing only the S-parameters of the constituent components and their interconnections. The overall S-parameters of the composite design can be calculated by using (3.6) in conjunction with Kirchhoff's Voltage Law (KVL) and Kirchhoff's Current Law (KCL) applied at the internal nodes created by connections between two or more components.

This is illustrated for a cascade of two-ports in Figure 3.8a. Each component is characterized by its own S-parameter matrix, $S^{(1)}$ and $S^{(2)}$, respectively. The output port

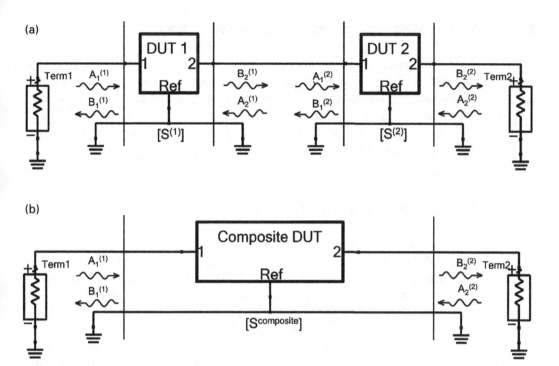

Figure 3.8 The cascade of 2 two-ports.

(subscript number 2) of the first component is connected to the input port (subscript number 1) of the second component, creating an internal node.

It is a straightforward exercise to write the two equations that follow from applying KVL and KCL at the internal node. These are given by equations (3.20) and (3.21)

$$B_2^{(1)} = A_1^{(2)} \tag{3.20}$$

$$A_2^{(1)} = B_1^{(2)} \tag{3.21}$$

Each two-port relates two input and two output variables. When cascaded, equations (3.20) and (3.21) can be used to compute the overall S-parameter matrix of the composite. The result is given in equation (3.22).

$$S^{composite} = \begin{pmatrix} S_{11}^{(1)} + S_{11}^{(2)} \dfrac{S_{12}^{(1)} S_{21}^{(1)}}{1 - S_{22}^{(1)} S_{11}^{(2)}} & \dfrac{S_{12}^{(1)} S_{12}^{(2)}}{1 - S_{22}^{(1)} S_{11}^{(2)}} \\[3ex] \dfrac{S_{21}^{(1)} S_{21}^{(2)}}{1 - S_{22}^{(1)} S_{11}^{(2)}} & S_{22}^{(2)} + S_{22}^{(1)} \dfrac{S_{12}^{(2)} S_{21}^{(2)}}{1 - S_{22}^{(1)} S_{11}^{(2)}} \end{pmatrix} \tag{3.22}$$

A network characterized by the $S^{composite}$ is equivalent to the cascade of the two DUTs, yielding the same response under the same stimuli. This is shown in Figure 3.8b.

3.11   DC Operating Point: Nonlinear Dependence on Bias

For devices such as diodes and transistors, the S-parameter values depend very strongly on the DC bias conditions defining the component operating conditions. For example, a FET transistor biased in the saturation region of operation will have S-parameters appropriate for an active amplifier (where $|S_{21}| \gg 1 \gg |S_{12}|$) whereas when biased "off" (say $V_{DS} = 0$) the S-parameters will represent the characteristics of a passive switch ($S_{21} = S_{12}$ with $|S_{ij}| < 1$).

3.12   S-parameters of a Nonlinear Device

It is possible to derive the S-parameters of a transistor starting from a simple nonlinear model of the intrinsic device. The process illustrates the concept of linearizing a nonlinear mapping about an operating point. This concept, suitably generalized, is a foundation for X-parameters and similar approaches in the next chapter.

The equivalent circuit of a simple, unilateral FET model is shown in Figure 3.9. More complete model topologies and equations are discussed in Chapter 5 and [7].

A current source, i_{DS}, represents the nonlinear bias-dependent channel current from drain to source, and a simple nonlinear two-terminal capacitor, q_{GS}, represents the nonlinear charge storage between gate and source terminals.

The large-signal model equations corresponding to this equivalent circuit are given by (3.23) and (3.24), where the stimulus applied to the device is the set of port voltages, v_{GS} and v_{DS}, and the device response is the set of port currents, i_{GS} and i_{DS}:

$$i_G(t) = \frac{dq_{GS}(v_{GS}(t))}{dt} \tag{3.23}$$

$$i_D(t) = i_{DS}(v_{GS}(t), v_{DS}(t)) \tag{3.24}$$

Equations (3.23) and (3.24) are evaluated for time-varying voltages assumed to have a fixed DC component and a small sinusoidal component at a single RF or microwave angular frequency, $\omega = 2\pi f$. Port 1 is associated with the gate and port 2 with the drain terminals, referenced to the source.

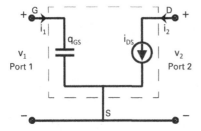

Figure 3.9 Simple nonlinear equivalent circuit model of a FET.

For a stimulus comprising a combination of DC and one sinusoidal signal per port, the voltages are formally expressed as shown in (3.25). The symbol δ is used to emphasize that such terms are considered sufficiently small that simple perturbation theory can be applied to simplify the ensuing analysis.

$$v_i(t) = v_i^{(DC)} + \delta V_i \cdot \cos\left(\omega t + \phi_i\right) \tag{3.25}$$

Expressions in (3.26) and (3.27) are obtained by substituting (3.25) for $i=1,2$, into (3.23) and (3.24), and evaluating the result to first order in the real quantities δV_i.

$$i_1(t) = i_1^{(DC)} + \delta i_1(t)$$

$$= \frac{d}{dt}\left(q_{GS}\left(v_1^{(DC)}\right) + \frac{dq_{GS}}{dv_1}\bigg|_{v_1^{(DC)}, v_2^{(DC)}} \delta V_1 \cdot \cos\left(\omega t + \phi_1\right)\right) \tag{3.26}$$

$$= c_{GS}\left(v_1^{(DC)}\right)\frac{d}{dt}\left(\delta V_1 \cdot \cos\left(\omega t + \phi_1\right)\right)$$

$$i_2(t) = i_2^{(DC)} + \delta i_2(t) = i_{DS}\left(v_1^{(DC)}, v_2^{(DC)}\right)$$

$$+ g_m\left(v_1^{(DC)}, v_2^{(DC)}\right) \cdot \delta V_1 \cdot \cos\left(\omega t + \phi_1\right) \tag{3.27}$$

$$+ g_{DS}\left(v_1^{(DC)}, v_2^{(DC)}\right) \cdot \delta V_2 \cdot \cos\left(\omega t + \phi_2\right)$$

Here the following definitions of the *linearized equivalent circuit elements* have been used:

$$c_{GS}\left(v_1^{(DC)}\right) = \frac{dq_{GS}}{dv_1}\bigg|_{v_1^{(DC)}}$$

$$g_m\left(v_1^{(DC)}, v_2^{(DC)}\right) = \frac{\partial i_{DS}}{\partial v_1}\bigg|_{v_1^{(DC)}, v_2^{(DC)}} \tag{3.28}$$

$$g_{DS}\left(v_1^{(DC)}, v_2^{(DC)}\right) = \frac{\partial i_{DS}}{\partial v_2}\bigg|_{v_1^{(DC)}, v_2^{(DC)}}$$

By equating the 0th order terms in δV_i, the operating point conditions are obtained:

$$i_1^{(DC)} = 0 \text{ A}$$

$$i_2^{(DC)} = i_{DS}\left(v_1^{(DC)}, v_2^{(DC)}\right) \tag{3.29}$$

Equating the first order terms in δV_i leads to

$$\delta i_1(t) = c_{GS}\left(v_1^{(DC)}\right)\frac{d}{dt}\left(\delta V_1 \cdot \cos\left(\omega t + \phi_1\right)\right)$$

$$\delta i_2(t) = g_m\left(v_1^{(DC)}, v_2^{(DC)}\right)\delta V_1 \cdot \cos\left(\omega t + \phi_1\right) + g_{DS}\left(v_1^{(DC)}, v_2^{(DC)}\right)\delta V_2 \cdot \cos\left(\omega t + \phi_2\right)$$

$$\tag{3.30}$$

Equation (3.30) is expressed in the frequency domain by defining complex phasors $\delta I(\omega)$ and $\delta V(\omega)$ according to (3.31):

$$\begin{aligned}
\delta i_i(t) &= \mathrm{Re}\left(\delta I_i(\omega)e^{j\omega t}\right) \\
\delta v_i(t) &= \mathrm{Re}\left(\delta V_i(\omega)e^{j\omega t}\right) = \mathrm{Re}\left(\delta V_i e^{j\phi_i}e^{j\omega t}\right)
\end{aligned} \tag{3.31}$$

Equations (3.30) can be rewritten for the complex phasors in matrix notation as

$$\begin{bmatrix} \delta I_1(\omega) \\ \delta I_2(\omega) \end{bmatrix} = Y\left(v_1^{(DC)}, v_2^{(DC)}, \omega\right) \cdot \begin{bmatrix} \delta V_1(\omega) \\ \delta V_2(\omega) \end{bmatrix} \tag{3.32}$$

where

$$Y\left(v_1^{(DC)}, v_2^{(DC)}, \omega\right) = \begin{bmatrix} j\omega c_{GS}\left(v_1^{(DC)}\right) & 0 \\ g_m\left(v_1^{(DC)}, v_2^{(DC)}\right) & g_{DS}\left(v_1^{(DC)}, v_2^{(DC)}\right) \end{bmatrix} \tag{3.33}$$

Equation (3.33) defines the (common source) admittance matrix of the model. The matrix elements are evidently functions of the DC operating point (bias conditions) of the transistor and also the (angular) frequency of the excitation.

Since the phasors representing the port voltage and currents in (3.32) can be reexpressed as linear combinations of incident and scattered waves using (3.5), it is possible to derive the expression for the S-parameters in terms of the Y-parameters (admittance matrix elements). This results in the well-known conversion formula (3.34) [3]:[2]

$$S = [I - Z_0 Y][I + Z_0 Y]^{-1} \tag{3.34}$$

Here I is the two-by-two unit matrix, Z_0 is the reference impedance used in the wave definitions in (3.4), Y is the two-port admittance matrix, and S is the corresponding S-parameter matrix. Substituting (3.33) into (3.34) results in an explicit expression, (3.35), for the S-parameters corresponding to this simple model in terms of the linear equivalent circuit element values given in (3.28):

$$S\left(v_1^{(DC)}, v_2^{(DC)}, \omega\right)$$

$$= \begin{pmatrix} \dfrac{1 - j\omega c_{GS}\left(v_1^{(DC)}\right)Z_0}{1 + j\omega c_{GS}\left(v_1^{(DC)}\right)Z_0} & 0 \\ \dfrac{-2g_m\left(v_1^{(DC)}, v_2^{(DC)}\right)Z_0}{\left(1 + g_{DS}\left(v_1^{(DC)}, v_2^{(DC)}\right)Z_0\right)\left(1 + j\omega c_{GS}\left(v_1^{(DC)}\right)Z_0\right)} & \dfrac{1 - g_{DS}\left(v_1^{(DC)}, v_2^{(DC)}\right)Z_0}{1 + g_{DS}\left(v_1^{(DC)}, v_2^{(DC)}\right)Z_0} \end{pmatrix}$$

$$\tag{3.35}$$

[2] See Exercise 3.3b for an alternative but equivalent formula.

In summary, the S-parameters of a nonlinear component can be derived or computed by linearizing the full nonlinear characteristics of the circuit equations around a static (DC) operating point defined by the voltage or current bias conditions. The S-parameters define a linear relationship between the incident and scattered waves at a fixed DC operating point of the device and fixed frequency for the incident waves. The S-parameters are an accurate description of how the device responds to signals, provided the signal amplitude is sufficiently small that the DC operating point is not significantly affected by the signal. This will almost always be the case for signals of sufficiently small amplitude.

3.13   Additional Benefits of S-parameters

3.13.1   Applicable to Distributed Components at High Frequencies

S-parameters can accommodate an arbitrary frequency-dependence in the linear spectral mapping. S-parameters therefore apply when describing distributed components for which lumped approximations are not very accurate or efficient. This is especially true for high-frequency microwave components when the typical wavelengths of the stimulus approach and become smaller than the physical size of the component. The simplest example is the case of linear transmission lines. Another common example is the case of an active device, for which measured S-parameters of a transistor can be much more accurate than those computable from the linearized lumped nonlinear model. This is especially true as the frequency approaches and exceeds the device cutoff frequency, f_T, beyond which a distributed representation is generally required.

3.13.2   S-parameters Are Easy to Measure at High Frequencies

S-parameters contain no more information than the familiar Y and Z-parameters of elementary linear circuit theory, yet they have great practical advantages. S-parameters are much easier to measure at high frequencies. Y and Z-parameters require short and open circuit boundary conditions, respectively, on the components for a direct measurement. Short and open circuit conditions are hard to achieve at microwave frequencies, and so are impractical. Moreover, such conditions presented to a power transistor can create oscillations that can destroy the component.

We will find in the next chapter that large-signal behavioral models combine the accuracy and the applicability to distributed components of a frequency-domain approach based on wave variables, with the ability to handle nonlinearities that go beyond the linear relationship assumed by equation (3.6).

3.13.3   Interpretation of Two-Port S-parameters

S-parameters relate to familiar quantities, such as gain, return loss, and output match. They provide insight into the component behavior. Table 3.1 lists the four complex

Table 3.1 S-parameters of a generic two-port and the corresponding figures of merit of a two-port amplifier.

S-parameter	Generic two-port (with input and output ports properly terminated)	Amplifier figure of merit
S_{11}	Input reflection coefficient	Return loss: $\mathrm{dB}|S_{11}|$
S_{12}	Reverse transmission coefficient	Isolation: $\mathrm{dB}|S_{12}|$
S_{21}	Forward transmission coefficient	Gain: $\mathrm{dB}|S_{21}|$
S_{22}	Output reflection coefficient	Output match: $\mathrm{dB}|S_{22}|$

S-parameters and their corresponding interpretation for generic linear two-ports [3]. The third column expresses common amplifier quantities in terms of the corresponding S-parameters.

Exercise 3.5 uses the S-parameters of a linear two-port component to derive an expression for the optimum impedance to match the device for maximum power transfer to the load. This approach will be generalized in the next chapter, using X-parameters, to compute a closed form expression for the optimal complex reflection coefficient to terminate a nonlinear device driven with a large input signal for optimal power transfer.

3.13.4   Hierarchical Behavioral Design with S-parameters

S-parameters can be measured at any level of the electronic technology hierarchy, from the transistor (in small-signal conditions) to the linear circuit, or all the way to the complete linear system. S-parameters provide a complete and accurate behavioral representation of the (linear) component at that level of the design hierarchy for efficient design at the next. In this context, complete means there are no possible responses of the linear component that are not accounted for by knowledge of the S-parameters, provided only that the S-parameters are sampled sufficiently densely and over a wide enough frequency range. Then, regarding accuracy, this means that if the S-parameters are measured accurately, the simulation of the linear component's response will also be accurate. S-parameters can dramatically reduce the complexity of a multicomponent linear design, by eliminating all internal nodes through the application of linear algebra as discussed in section 3.10. S-parameters protect the intellectual property (IP) of the design by providing only the behavioral relationship at the external terminals, with no information about the internal realization of the functionality in terms of circuit elements arranged in a particular topology (e.g., the schematic). S-parameters are therefore a *black-box* behavioral approach, requiring no *a priori* information about the component except that it is linear.

S-parameters are hierarchical. A system of two connected linear components can be represented by the overall S-parameters of the composite and used as a behavioral model at a higher level of the design hierarchy. The composite S-parameters can be obtained from direct measurement of the cascaded system, or by composition of the

S-parameters of the constituent components according to the procedure of section 3.10. Any subset of interacting linear components can be represented by an S-parameter block and inserted into a larger design. The process can be repeated iteratively from one level to the next level.

3.14   Limitations of S-parameters

Interestingly, S-parameters are still commonly used for nonlinear devices such as transistors and amplifiers. The problem, often forgotten or taken for granted, is that S-parameters only describe properly the behavior of a nonlinear component in response to *small signal* stimuli for which the device behavior can be approximated as linear around a fixed DC, or static, operating point. That is, only when the nonlinear device is assumed to depend linearly on all RF components of the incident signals is the S-parameter paradigm valid. S-parameters contain no information about how a nonlinear component generates distortion, such as manifested by energy generated at harmonics and intermodulation frequencies by the component in response to excitation by one or more tones (sinusoidal signals). S-parameters are inadequate to describe nonlinear devices excited by large signals despite many *ad hoc* attempts. Attempts to generalize S-parameters to the nonlinear realm have led to a wide variety of problems, including measurements that appear nonrepeatable, and the inability to make predictable design inferences from such measurements or simulations. Techniques such "hot S-parameter" measurements and modeling nonlinear components using "large-signal S-parameter analysis" are examples of incomplete, insufficient, and ultimately incorrect approaches. Under large-signal conditions, totally new phenomena, for which there are no analogues in S-parameter theory, appear and must be taken into account.

For example, when a nonlinear component such as a transistor or power amplifier is stimulated simultaneously by two sinusoidal signals at different frequencies, the output spectrum is not consistent with the superposition principle discussed in Section 3.8. Rather, the response of a nonlinear system to two or more excitations generally contains more nonzero frequency components than were present in the input signal (see Figure 3.10). These terms are generally referred to as *intermodulation distortion.*[3] It is clear that this phenomenon cannot be described by S-parameters. A more general framework is required to address this type of behavior, which is the subject of the next chapter.

3.15   Summary

S-parameters are linear time-invariant spectral maps defined in the frequency domain. They represent intrinsic properties of the DUT, enabling the hierarchical design of linear circuits and systems given only the S-parameters of the constituent functional blocks and their topological arrangement in the design. S-parameters can be defined, measured,

---

[3] In fact, there are even cases where "passive connectors" cause intermodulation distortion [6].

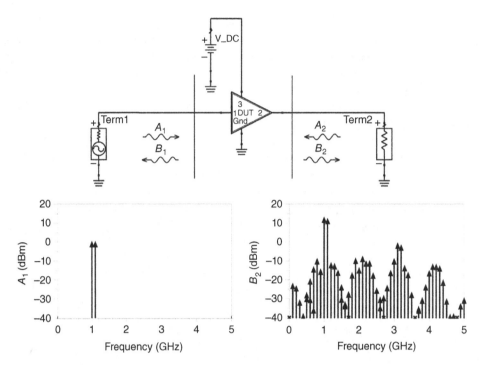

Figure 3.10 Intermodulation spectrum of typical transistor output in response to a simultaneous excitation by two sinusoidal signals at different frequencies. Superposition is not obeyed in general.

and calculated for nonlinear components, but they are valid descriptions only under small-signal conditions. Phenomena such as harmonic and intermodulation distortion, generated by nonlinear components, require a more comprehensive framework for their consistent description and application.

3.16   Exercises

Exercise 3.1   Prove (3.20) and (3.21) are equivalent to the circuit laws. Hint, start by expressing KVL and KCL in the time domain. Re-express these laws in wave variables using the time-domain versions of (3.1), and then transform them to the frequency domain for the corresponding phasors.

Exercise 3.2   Derive (3.22). Hint: eliminate the four internal variables appearing in (3.20) and (3.21). The matrix producing $B_1^{(1)}$ and $B_2^{(2)}$ from $A_1^{(1)}$ and $A_2^{(2)}$ is the required result.

Exercise 3.3   a.   Derive (3.34) from (3.1) and the definition of the Y-matrix.

b.   In general, the products of two matrices, A and B, to not commute. That is, for square matrices A and B of the same dimension, $AB \neq BA$ in general. Despite this fact, prove the expression (3.34) can also be written $S = [I - Z_0 Y][I + Z_0 Y]^{-1} = [I + Z_0 Y]^{-1}[I - Z_0 Y]$.

Exercise 3.4 Derive (3.35), the explicit form of the simple FET model S-parameter matrix, from the model of Figure 3.9.

Exercise 3.5 S-parameter expression for the complex reflection coefficient for delivered power to a load.

    a.   Show the expression for power delivered to a load from a two-port device described by a 2×2 S-parameter matrix is $P_{del} = |B_2|^2 - |A_2|^2$.

    b.   Substitute the S-parameter expression $S_{21}A_1 + S_{22}A_2$ for B_2 in (a) and compute the delivered power in terms of the S-parameters and the incident wave phasors and their complex conjugates.

    c.   The delivered power is maximized when the derivative of P_{del} with respect to A_2 is set equal to zero (A_1 is assumed fixed). Compute this derivative, solve the resulting equation, and show $A_2^{opt} = \frac{S_{22}^* S_{21} A_1}{1 - |S_{22}|^2}$. Assume the derivative of A_2^* with respect to A_2 is zero.

    d.   The complex valued reflection coefficient is defined by $\Gamma_2 = \frac{A_2}{B_2}$. Use the expression in (c) to evaluate $\Gamma_2^{opt} = \frac{A_2^{opt}}{B_2}$ and show the result reduces for A_1 fixed as in part (c) to $\Gamma_2^{opt} = S_{22}^*$. This procedure will be used in the next chapter to derive an expression for Γ_2^{opt} in the case when a nonlinear component is driven with a large A_1 and linear expression (3.7) does not apply.

    e.   Relax the assumption that A_1 is fixed and independent of A_2. Show that $\frac{\partial A_1}{\partial A_2} = \frac{\Gamma_S S_{12}}{1 - \Gamma_S S_{11}}$, where Γ_S is the source reflection coefficient. Use this result to derive the more general result, valid for nonunilateral devices: $\Gamma_2^{opt*} = S_{22} + \frac{S_{21} \Gamma_S S_{12}}{1 - \Gamma_S S_{11}}$

Exercise 3.6 Simpler derivation of $\Gamma_2^{opt} = S_{22}^*$ for the case A_1 fixed.

    a.   Start with $P_{del} = |B_2|^2 - |A_2|^2$ as in the previous exercise. Without substituting the S-parameter expression for B_2, differentiate with respect to A_2 and show that the delivered power is maximized when $\frac{dB_2}{dA_2} B_2^* = A_2^*$.

    b.   Divide the equation in (a) by B_2^* and show the result is

$$\frac{dB_2}{dA_2} = \Gamma_2^*.$$

    c.   Now substitute the S-parameter expression for B_2 in terms of A_1 and A_2, compute the derivative, and derive $\Gamma_2^{opt} = S_{22}^*$.

References

[1] D. Root, J. Verspecht, J. Horn, and M. Marcu, *X-Parameters: Characterization, Modeling, and Design of Nonlinear RF and Microwave Components*, Cambridge University Press, 2013, chapter 1.

[2] D. Rytting "Calibration and Error Correction Techniques for Network Analysis," IEEE Microwave Theory and Techniques Short Course, 2007. http://ieeexplore.ieee.org/servlet/opac?mdnumber=EW1062.

[3] G. Gonzalez, *Microwave Transistor Amplifiers*, 2nd ed., Prentice Hall, 1984.

[4] "S-parameter techniques for faster, more accurate network design," Hewlett-Packard application note AN 95–1, 1968.

[5] K. Kurokawa, "Power waves and the scattering matrix," *IEEE Trans. Microw. Theory Tech.*, Vol. MTT-13, No. 2, March 1965.

[6] J. J. Henrie, A. J. Christianson, and W. J. Chappell, "Linear–nonlinear interaction and passive intermodulation distortion," *IEEE Trans. Microw. Theory Techn.*, Vol. 58, No. 5, May 2010, pp 1230–1237.

[7] D. E. Root et al., "The intrinsic model," in *Nonlinear Transistor Model Parameter Extraction Techniques*, Rudolf, Fager, Root, eds., Cambridge University Press, 2011, chapter 5.

4 Nonlinear Frequency Domain Behavioral Models

4.1 Introduction and Overview

In this chapter, we consider the rigorous generalization of S-parameters to large-signal conditions. In the first part of the chapter we restrict the treatment to steady-state stimuli and responses or, equivalently, continuous wave (CW) conditions. Dynamics are introduced toward the end.

We saw in Chapter 1 that the behavior of real components generally violates linear superposition. Superposition fails even when the response frequency is always the same as the stimulus frequency. Additionally, real devices typically generate signal components at frequencies that are not present in the excitation spectra. Both of these cases are outside the simple and idealized realm of linear analysis and linear modeling, and require a more general treatment such as we consider here.

Like S-parameters, the general framework presented here is based on the notion of *time-invariant spectral maps*. (Recall time invariance was defined in Chapter 3 section 9). We saw, in Chapter 3, that linear scattering models automatically satisfy time invariance. We will see, in Section 4.3, that most nonlinear spectral maps do not satisfy the time-invariance condition. This means that to properly model nonlinear components (e.g. diodes and transistors) under large-signal stimulus conditions, we must impose the time-invariance property as a constraint on the mathematical formalism. Like S-parameters, composition rules for nonlinear system design follow from the behavioral models of constituent components based on the rules of nonlinear circuit theory. The models treated here can be combined with linear and nonlinear behavioral models, and also standard "compact" transistor models for large-signal frequency domain analysis such as harmonic balance, as described in detail in earlier chapters.

These large-signal behavioral models have a use model very similar to S-parameters but are much more powerful. They are rigorous supersets of S-parameters, to which they reduce in the small-signal limit. A conceptual diagram for this paradigm is shown in Figure 4.1.

Like S-parameters, the models considered here can be identified from measurements, derived from analytical models[1], or computed from numerical models in a circuit simulator. The types of required measurements are made on modern RF and microwave

---

[1] Although this may be difficult to do in closed form.

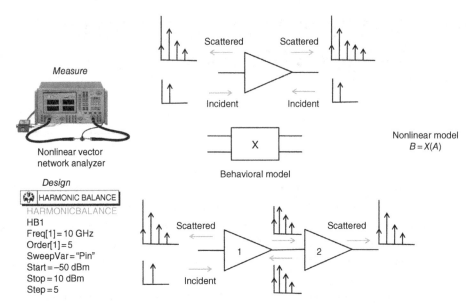

Figure 4.1 Large-signal frequency-domain behavioral modeling paradigm.

instruments called nonlinear vector network analyzers (NVNA) or large-signal network analyzers (LSNA) [1] [2] [3]. Based on good measurements, the resulting large-signal models of components, such as transistors, can be extremely accurate. Simulation-based nonlinear models can be created in the nonlinear simulator in analogy with how an S-parameter analysis produces a linear file-based behavioral model [4].

A major benefit of such high fidelity nonlinear frequency-domain behavioral models is that they can be shared between component vendors and system integrators freely, without the possibility that the component implementation can be reverse engineered. This feature protects vendor intellectual property (IP) and promotes its sharing and reuse. This is quite analogous to the case where S-parameters – measured or computed – are provided by manufacturers to represent the performance of a linear component, instead of a detailed equivalent circuit model that could betray a proprietary implementation or design approach.

The systematic set of approximations to the general theory, based on spectral linearization, is introduced in the chapter in a number of places, to enable practical engineering applications to important classes of problems. These simplifications can result in significant savings of measurement time and file-size, reduce hardware system complexity, and enable faster simulations at a minimal cost in accuracy. Multiple approaches to motivating, deriving, and interpreting the formalism are presented.

Finally, the important topic of dynamic memory is introduced, including its importance, some of its symptoms, and several major causes. A powerful extension beyond the static theory is presented in the guise of *dynamic X-parameters*. It is shown how this general approach can successfully model many important nonlinear dynamic systems showing a wide range of dynamic-to-the-envelope properties when stimulated with wideband modulated signals.

4.2   Signals and Spectral Maps on a Simple Harmonic Frequency Grid

An important application for nonlinear circuit simulation is the analysis of harmonic distortion generated by a nonlinear component, such as a transistor, in response to an excitation at a single fundamental frequency. These response signals, which generally have spectral energy at multiple harmonics of the fundamental frequency, propagate to other components in the circuit and can ultimately get reflected back into any or all of the ports of the device that created them, modifying the outputs in turn. An excitation class that covers most of these cases includes signals incident simultaneously at different ports with frequency components that fall on a grid of discrete non-negative integer harmonic multiples of the fundamental frequency. The DC component is included. A graphical representation is given in Figure 4.2.

We also assume that in response to signals with components at multiple harmonics of the fundamental frequency, the DUT produces scattered waves with nonzero components only on the same harmonic frequency grid. This is usually the case for microwave and RF applications. However, we know from Chapter 1 that it is possible for the DUT to be driven into a chaotic regime where this assumption fails. We also know that parasitic oscillations can be generated and appear in the output spectrum, falling outside the harmonic grid. A phenomenon that is not that uncommon in the PA context is the generation of oscillations at a frequency *half* that of the fundamental component of the driving frequency excitation, which also does not fall on the grid. We do not consider these cases further here – although the last phenomenon can be handled by minor adaptations of the methods presented here.

In Figure 4.2, the signal magnitudes are indicated by the lengths of the arrows. Each harmonic component of the signal also has a phase that can be well-defined relative to that of the fundamental frequency. This is a consequence of the fact that the frequency components are all commensurate. A set of K frequencies, $\{f_k\}$, $k = 1, \ldots, K$, are commensurate if there exist nonzero integers, n_k, such that (4.1) is satisfied. For more on commensurate signals see [5].

$$\sum_{k=1}^{K} n_k f_k = 0; \quad n_k \in \mathbb{Z}; \quad n_k \neq 0 \tag{4.1}$$

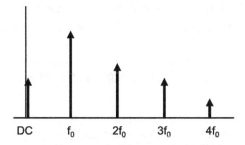

Figure 4.2 Harmonic signal grid on which nonlinear spectral maps are defined.

4.2.1   Notation for Wave Variables (Phasors) on Harmonic Grid

The complex-valued (for magnitude and phase) incident and scattered wave amplitudes (phasors) now require two indices. The first index refers to the port, and the second to the harmonic index relative to the fundamental. Incident and scattered waves will therefore be labeled according to (4.2):

$$A_{p,k}$$
$$B_{q,l} \tag{4.2}$$

For example, $A_{2,3}$ is an incident sinusoidal wave at port two at a frequency three times the fundamental, while $B_{1,2}$ is the scattered sinusoidal wave component at port 1 at the second harmonic of the fundamental.

By varying the complex amplitudes of the incident waves' phasor components at each harmonic of the fundamental frequency and measuring the complex amplitudes of the scattered waves at each port, p, and harmonic frequency index, k, we can build up the set of complex-valued nonlinear algebraic maps of the form (4.3). This set of maps[2] defines the behavior of the DUT under all steady-state conditions consistent with our previous assumptions.

$$B_{p,k} = F_{p,k}(A_{1,1}, A_{1,2}, \ldots, A_{2,1}, A_{2,2}, \ldots) \tag{4.3}$$

To be more complete we need to treat the DC components consistently with the RF components. In the end, it is the actual signals – waveforms – that matter, and the DC components are as important as the RF. Adding the bias stimulus response interactions to those of the RF, we obtain a system of nonlinear functions that define the steady-state behavior of the component. A simple example is given in (4.4). The superscripts in (4.4) simply label the type of value returned by the respective functions (e.g., complex-valued scattered wave, real-valued current, and real-valued voltage, respectively). This notation is standardized in [6].[3]

$$B_{p,k} = F_{p,k}^{(B)}(V_1, A_{1,1}, A_{1,2}, \ldots, I_2, A_{2,1}, A_{2,2}, \ldots)$$
$$I_1 = F_1^{(I)}(V_1, A_{1,1}, A_{1,2}, \ldots, I_2, A_{2,1}, A_{2,2}, \ldots) \tag{4.4}$$
$$V_2 = F_2^{(V)}(V_1, A_{1,1}, A_{1,2}, \ldots, I_2, A_{2,1}, A_{2,2}, \ldots)$$

In (4.4) there are many independent variables. These include the DC applied voltage at port one, V_1; the incident waves at port one at each positive harmonic frequency, $A_{1,1} - A_{1,N}$; the DC applied current at port 2, I_2; and the incident waves at port two, $A_{2,1} - A_{2,N}$.

The scattered RF waves at each port depend nonlinearly on the applied port biases, V_1 and I_2, and also on each complex-valued phasor of the incident RF signals at each

---

[2] Sometimes we will use the notation (4.3) to refer to a specific map labeled by the particular port, p, and particular harmonic index, k. In the present case, we are considering the entire set of maps for all values of the indices that describe the component. The meaning should be clear from the context.

[3] A slight simplification is made in the present work where only one letter appears in the superscript field.

port. Likewise, the DC responses of the DUT, current I_1 and voltage V_2, depend on all of the RF incident signal complex amplitudes, as well as applied biases V_1 and I_2, respectively. Note that this is different from the typical S-parameter situation where the DC conditions are taken to be independent of the applied RF signals since the RF signals are assumed to be arbitrarily small.

The system defined by equations (4.4) is really not much more than book-keeping at this point. For this two-port example there are $2N + 2$ nonlinear functions of $2N + 2$ variables. The model definition is not complete until all the functions $F^{(B)}$, $F_1^{(I)}$, and $F_2^{(V)}$ are specified mathematically or otherwise identified from data associated with the component. Even so, the structure of (4.4) makes clear that *cross-frequency stimulus-response interactions* are taken into account. For example, the third harmonic of the scattered wave at port two depends on the complex value of the second harmonic wave incident at port one. Thus harmonic time-domain load-pull under steady-state large-signal conditions is covered by the system of equations (4.4).

4.3   Time-Invariant Large-Signal Spectral Maps

A fundamental property of components such as transistors, power amplifiers, nonlinear resistors, etc.[4] is that of time invariance, discussed in earlier chapters. We saw in Chapter 3 that for linear components the S-parameter mathematical relationship between incident and scattered waves is automatically consistent with the time-invariance principle. This is not true in the nonlinear case, as we will see in section 4.3.1. Since a model inconsistent with the time-invariance property of the physical component *must* give wrong answers, it behooves us to properly ensure this property is correctly implemented.

Referring to the previous example of (4.4), we will show not all possible functions $F^{(B)}$, $F_1^{(I)}$, and $F_2^{(V)}$ are consistent with the mathematical representation of the time-invariance property. In fact, most are not. The restriction of such functions to those spectral maps obeying the time-invariance property results in X-parameters. This could be taken as the fundamental definition of X-parameters. We will derive these restrictions in Section 4.3.1

Other frequency-domain large-signal behavioral models also obey the time-invariance property. One of these, the Cardiff Model, will be described in section 4.9.2.

4.3.1   Derivation of Time-Invariant Spectral Maps

We don't need to restrict ourselves to periodic signals for the statement of time invariance, and we drop the port indices for simplicity. We assume the DUT produces an output signal, $b(t)$, when stimulated by an incident signal, $a(t)$. A time-invariant

---

[4] Except under pathological conditions that are not considered here.

DUT means that for an incident signal, $a(t - \tau)$ (the same signal as before but now delayed by an arbitrary time, τ), the resulting output signal *must* be $b(t - \tau)$ (the same response as before but delayed by precisely the same time, τ).

Returning to our case of periodic stimulus and response, we can write the time-invariance condition in terms of the Fourier coefficients of the incident and scattered periodic signals. We recall that a time-shift, τ, produces a phase-shift, $e^{jk\omega_0\tau}$, for the kth harmonic term of the Fourier series of a periodic signal. This results in the following frequency-domain expression of time invariance for one of the equations of (4.4).

$$B_{p,k} = F_{p,k}^{(B)}(V_1, A_{1,1}, A_{1,2}, \ldots, I_2, A_{2,1}, A_{2,2}, \ldots) \Rightarrow$$

$$B_{p,k}\, e^{-jk\omega_0\tau} = F_{p,k}^{(B)}\left(V_1, A_{1,1}\, e^{-j\omega_0\tau}, A_{1,2}\, e^{-2j\omega_0\tau}, \ldots, I_2, A_{2,1}\, e^{-j\omega_0\tau}, A_{2,2}\, e^{-2j\omega_0\tau}, \ldots\right)$$

$$(4.5)$$

Since (4.5) must hold for all real values of τ, we choose the particular value given in (4.6). The selection of τ according to (4.6) can be interpreted as selecting the phase reference to be one of the system's signals, and that signal is $A_{1,1}$. This has the effect of *phase-normalizing* each harmonic coefficient of the incident and scattered waves to the phase of $A_{1,1}$, and in particular results in (4.7).

$$\tau = \frac{\phi(A_{1,1})}{\omega_0} \tag{4.6}$$

$$A_{1,1}\, e^{-j\omega_0\tau} = |A_{1,1}|\, e^{j\phi(A_{1,1})} e^{-j\omega_0\tau} = |A_{1,1}| \tag{4.7}$$

Equation (4.5) can now be recast as (4.8), using the notation $P = e^{j\phi(A_{1,1})}$.

$$B_{p,k}\, P^{-k} = F_{p,k}^{(B)}\left(V_1, A_{1,1}\, P^{-1}, A_{1,2}\, P^{-2}, \ldots, I_2, A_{2,1}\, P^{-1}, A_{2,2}\, P^{-2}, \ldots\right)$$

$$\Rightarrow$$

$$B_{p,k} = F_{p,k}^{(B)}\left(V_1, A_{1,1}\, P^{-1}, A_{1,2}\, P^{-2}, \ldots, I_2, A_{2,1}\, P^{-1}, A_{2,2}\, P^{-2}, \ldots\right) P^k \tag{4.8}$$

$$\equiv X_{p,k}\left(V_1, |A_{1,1}|, A_{1,2}\, P^{-2}, \ldots, I_2, A_{2,1}\, P^{-1}, A_{2,2}\, P^{-2}, \ldots\right) P^k$$

Equation (4.8) shows that the principle of time invariance *restricts the functional form* of the spectral mappings to those that change only by a multiplicative phase factor when the arguments to the function are appropriately transformed (phase-shifted) according to their harmonic indices. Exercise 4.1 explores cases where (4.8) is not satisfied. Given (4.7), we can therefore define X-parameter functions as in the bottom of (4.8). Notice that the X-parameter functions, denoted for now by $X_{p,k}$, are defined on a set of independent variables of *dimension one fewer* than the original mappings $F_{p,k}^{(B)}$. This follows because only the real-valued quantity $|A_{1,1}|$, rather than the general complex quantity $A_{1,1}$, enters the domain of the $X_{p,k}$ functions. The final complex coefficient of the scattered wave is obtained by multiplying the value of the X-parameter function by the appropriate phase factor depending on the harmonic index of the scattered wave.

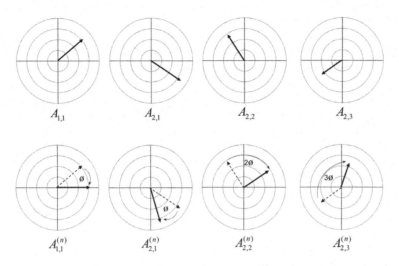

Figure 4.3 Phase normalization of incident waves to the reference stimulus.

4.3.2 Graphical Representation of Time-Invariant Spectral Maps

The evaluation of an X-parameter model makes explicit use of the time-invariant spectral mapping condition given by (4.8). For example, consider the case of a two-port DUT with signals incident at port one at the fundamental frequency, and simultaneously at port two with components at the fundamental frequency, second, and third harmonics. These four phasors are shown in the top row of Figure 4.3. The length of a vector indicates its magnitude, and the angle represents the phase.

Since $\phi(A_{1,1})$, the phase of $A_{1,1}$, is not zero, the X-parameter functions that appear in the bottom of (4.8) can't be evaluated. The prescription of (4.8) requires that we first phase-shift the kth spectral components of the incident waves by k times the phase of $A_{1,1}$ in the clockwise direction. This is accomplished by multiplying each variable $A_{p,k}$ by $P^{-k} = e^{-jk\phi(A_{1,1})}$. The second harmonic terms will therefore be phase-shifted twice as much as the fundamental component, and third harmonic terms three times as much, etc. DC terms (not shown in the figure) are not modified – consistent with DC being the "zeroth" harmonic, or $k = 0$. The resulting phasors, denoted by $A_{p,k}^{(n)} = A_{p,k}P^{-k}$, define the *reference stimulus*.[5] These are shown in the bottom of Figure 4.3, emphasizing the relative phase rotations of the various harmonic components. In the time domain, the reference stimulus is simply the original set of periodic incident waveforms *translated in time* with respect to the original waveforms by a time, τ, given by (4.6) [7].

The X-parameter functions of (4.8) can be evaluated at the reference stimulus because $A_{1,1}^{(n)}$ is a real number. We denote by *reference response* the values of the scattered waves returned by the X-parameter functions evaluated at the reference

---

[5] The superscript "n" denotes (phase-) normalization.

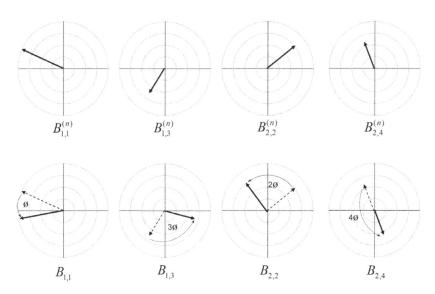

Figure 4.4 Phase denormalization from the reference response to the actual response.

stimulus. The corresponding phasors of the reference response are denoted $B_{p,k}^{(n)}$. Since the DUT is nonlinear, there will generally be scattered waves at each port with nonvanishing components at multiple harmonics beyond those of the incident signals. Values for four of these scattered phasors of the reference response are shown in the top part of Figure 4.4.

Finally, the desired $B_{p,k}$ values (corresponding to the actual $A_{p,k}$ incident waves) are obtained by a *phase denormalization* process applied to the reference response. This is accomplished by a phase-rotation in the counterclockwise sense (complex multiply by $P^k = e^{jk\phi(A_{1,1})}$) of the values returned by the X-parameter functions according to (4.8). In the time-domain, this corresponds to the appropriate delay of the periodic reference response by the precise amount, τ, by which the incident signals were delayed compared to the reference stimulus.

In summary, we have taken an arbitrary periodic stimulus, time-translated it to a reference condition where the phase of $A_{1,1}$ can be taken to be zero, evaluated the X-parameter functions, and time-translated the result back to obtain the final periodic response. Since this is a manifestly time-invariant process, we are guaranteed the DUT's time-invariant behavior is built in to the mathematical description through the X-parameter formalism.

4.4   Large-Signal Behavioral Modeling Framework in Wave-Space

The steady-state nonlinear behavior of RF and microwave components are defined in terms of time-invariant spectral maps defined in wave-space – with DC conditions specified in terms of real-valued currents and voltages that also depend on the RF

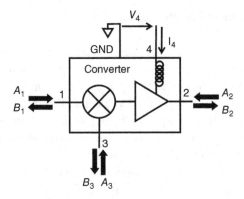

Figure 4.5 Description of a nonlinear microwave component in wave-space.

incident waves. Schematically, this is represented by Figure 4.5, for an example of a component that we label, simply, as a (frequency) converter.

The example of Figure 4.5 has three RF ports, labeled 1 through 3, as well as a DC port, labeled 4. The incident RF waves are indicated by A_p, where the port index p ranges from 1 to 3. The waves have multiple harmonic components, but these second indices are suppressed in the figure for simplicity. The incident waves and the DC applied voltage, V_4, are taken as the independent variables. The scattered waves, B_p, and the DC current, I_4, are given by X-parameter relations such as (4.8). For example, the RF outputs at port 2 are specified by the X-parameter expression (4.9), for $A_{p,k}^{(n)} = A_{p,k} P^{-k}$

$$B_{2,k} = X_{2,k}\left(V_4, A_{1,1}^{(n)}, A_{1,2}^{(n)}, \ldots, A_{2,1}^{(n)}, A_{2,2}^{(n)}, \ldots, A_{3,1}^{(n)}, A_{3,2}^{(n)}, \ldots\right) P^k \tag{4.9}$$

For port 4, the expression for the bias current response is given in (4.10). In the general case, there are both DC and RF input and outputs at each port.

$$I_4 = X_4^{(I)}\left(V_4, A_{1,1}^{(n)}, A_{1,2}^{(n)}, \ldots, A_{2,1}^{(n)}, A_{2,2}^{(n)}, \ldots, A_{3,1}^{(n)}, A_{3,2}^{(n)}, \ldots\right) \tag{4.10}$$

To model a particular device, the abstract constitutive relations (4.9) and (4.10) – as well as those for every X-parameter function corresponding to Figure 4.5 – must be identified from measured (or simulated) input–output characteristics of the component. In principle, this can be obtained by an exhaustive process of applying DC and RF sources at each port and varying the magnitudes and phases of all RF components and the value of the applied bias voltage, V_4, and acquiring the resulting responses. The outputs can be fitted with smooth functions of all the inputs or tabulated and inter-polated by the simulator during simulation. This process is limited by "the curse of dimensionality," however. That is, it becomes exponentially complex in the number of independent variables. To deal with this, we present systematic approximations that can be used to more conveniently apply the X-parameter principles to many important cases of practical interest, starting in Section 4.6.

4.5   Cascading Nonlinear Behavioral Model Blocks

How is it possible to combine, in analogy with linear S-parameters, multiple large-signal behavioral functional blocks to simulate a more complicated nonlinear circuit or system? We illustrate the concept by connecting port 2 of the converter block of Figure 4.5 to the input of a simple power amplifier (PA) block in Figure 4.6. The input port of the PA is labeled port 5, and the PA output is labeled port 6. The scattered waves at port 5 of the PA are defined by X-parameter constitutive relations that depend only on the incident waves (and biases, not shown) at ports 5 and 6.

To solve the circuit of Figure 4.6, Kirchhoff's Current Law (KCL) and Kirchhoff's Voltage Law (KVL), must be applied at the node connecting the two blocks. These conditions impose constraints on the sets of incident and scattered waves, given by the simple relationships (4.11). Specifically, at the solution of the circuit, the incident waves at port 2 must be equal to the scattered waves from port 5, and the scattered waves from port 2 must be equal to the incident waves at port 5. These equalities must hold separately for each harmonic index, k, $k = 1,...,N$. That is, (4.11) is a set of $2N$ equations, or if including DC terms, $2(N+1)$ equations. It is a simple matter to prove that (4.11) is equivalent to the circuit laws, KVL and KCL, when expressed in terms of wave variables using equations 3.1:

$$A_{2,k} = B_{5,k}$$
$$B_{2,k} = A_{5,k}$$

(4.11)

The solution for the entire circuit is given by eliminating the scattered wave variables and solving the set of self-consistent implicit nonlinear equations for the incident waves $A_{2,k}$ and $A_{5,k}$ for $k = 1,...,N$, given by (4.12). Contrary to the case of cascaded S-parameters considered in Chapter 3, because the present problem is nonlinear, no closed form solution can be found in general for the X-parameters of the cascade as a function of the X-parameters of each block. The solution to (4.12) is therefore obtained numerically by the nonlinear circuit simulator:

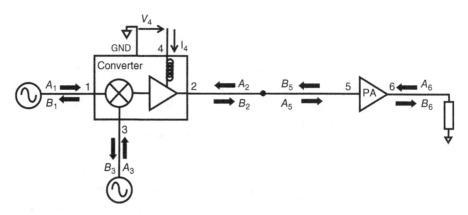

Figure 4.6 Cascading large-signal functional blocks in wave-space.

$$B_{2,k} = X_{2,k}^{(converter)}(A_{1,k}, A_{2,k}, A_{3,m}, V_4) = A_{5,k}$$

$$B_{5,k} = X_{5,k}^{(PA)}(A_{5,k}, A_{6,n}) = A_{2,k}$$

(4.12)

Of course most available circuit simulators first convert the wave variables in (4.12) to currents and voltages, using equations (3.2), and then solve the resulting implicit nonlinear equations in current-voltage space.

In fact, the simple relations (3.1) and (3.2) between wave variables and current-voltage variables enable functional blocks to be cascaded with nonlinear (and linear) functional blocks expressed in wave-space and/or conventional current-voltage space. That is, large-signal frequency-domain behavioral model functional blocks are completely compatible with distributed passive elements, lumped nonlinear elements defined in the time domain, and "compact" nonlinear transistor models. However, the present formulations of X-parameters and similar frequency-domain nonlinear behavioral models (e.g., the Cardiff Model – see Section 4.9.2) are presently limited to harmonic balance and circuit envelope analyses or their subsets (S-parameters, small-signal mixer analysis, etc.).

It should be stressed that the cascadability of these functional blocks depends only on the accuracy of the sampled constitutive relations of each block. The results for the cascade follow directly by applying only the circuit laws. No additional approximations are required. The cascadability becomes arbitrarily accurate by taking into account enough harmonic components of the incident and scattered waves at sufficiently many complex values of the independent incident wave variables.

4.6   Spectral Linearization

4.6.1   Simplification for Small Mismatch

The considerations used for the definition and application of X-parameters and related approaches described thus far apply to the steady-state behavior of time-invariant nonlinear components with incident and scattered waves on a harmonic frequency grid. The generality of the formalism comes at the cost of considerable complexity. Each spectral map is a nonlinear function of every applied DC bias condition and all the magnitudes and phases of each spectral component of every signal at every port of the component. Sampling such behavior in all variables for many ports and harmonics would be prohibitive in terms of data acquisition time, data file size, and model simulation time.

Fortunately, in most cases of practical interest, only a few large-signal components need to be considered with complete generality, while most spectral components can be considered small and can be dealt with effectively by methods of perturbation theory. In fact, as stated previously, S-parameters are the limiting case where *all* RF signals are considered small. Specifically, we can apply the following methodology, depicted in Figure 4.7.

- Identify the few large tones that drive the nonlinear behavior of the system.
- Identify the nonlinear spectral maps defined when only these few large tones are incident on the DUT. This nonlinear spectral map represents the specific steady-state of the system, designated the *large-signal operating point (LSOP)*.

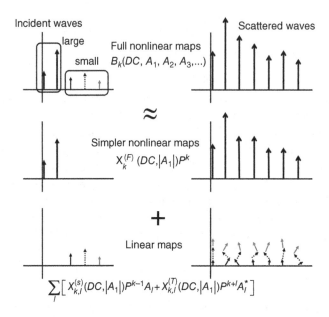

Figure 4.7 Illustration of spectral linearization approximation. The equations will be discussed in more detail in Section 4.6.2

- Linearize the spectral maps around this *LSOP* defined by the few important large tones.
- Treat the remaining small signals as small perturbations, using superposition appropriately, based on the above linearized maps around the *LSOP*.

This methodology approximates the complicated general multivariate nonlinear maps, B_k, with much simpler nonlinear maps, $X_k^{(F)}$, defined on the few preselected large tones only, and *simple linear maps*, $X_{k,k'}^{(S)}$ and $X_{k,k'}^{(T)}$, which account for the contributions of the many small amplitude tones. These approximations result in a dramatic reduction of complexity while providing an excellent description for many important practical applications. The details of the equations appearing in Figure 4.7 will be derived and discussed further in Section 4.6.2.

The simplest such approximation is to assume there are *no large RF signals* that need to be treated fully nonlinearly. The LSOP is therefore just the DC operating point. All the RF components can then be treated with perturbation theory, by linearizing around the DC operating point. This is just the S-parameter approximation of the previous chapter.

4.6.2   Simplest Nontrivial X-Parameter Case

Beyond S-parameters, the next simplest case of X-parameters involves one large RF signal that is treated generally while the remaining spectral components are treated as small. This is often the case for power amplifiers, at least those "nearly matched" to

50 Ω, where the main stimulus signal is a narrow-band modulation around a fixed carrier incident at the input port. The DUT response generally includes several harmonics. If the DUT is not perfectly matched at the output port, there will be small reflections at the fundamental and the harmonic frequencies back into the output port of the DUT. These small reflected signals will be treated as perturbations to simplify the description. Similarly, scattered waves at the harmonic frequencies at the input could be reflected back into the DUT if it is not perfectly matched at the source. These incident signals can also be treated as small in many applications. The validity of the approximation depends on whether these signals are small enough so that their contributions do not affect the LSOP. This is quite analogous to the condition on an RF signal to be small enough such that it does not change the DC operating point of a device for a valid S-parameter measurement, as discussed in Chapter 3. In Section 4.9 we will consider multiple large signals, in particular to deal with arbitrary load-dependence for large mismatch at the output port (load-pull) and two large incommensurate tones for mixer analysis.

4.6.2.1 Digression: Double-Sided Spectra and Spectral Linearization

So far we have simply assumed our phasor description corresponds to single-sided spectra. That is, the complex value, $A_{p,k}$ or $B_{p,k}$, is associated with $e^{jk\omega_0 t}$ for positive values of the harmonic index k, for ω_0 is a positive angular frequency. The spectral maps take complex amplitudes for positive frequencies into scattered wave amplitudes at positive frequencies (and DC). The simultaneous presence of signals incident at multiple frequencies on a nonlinear device, however, creates *self-mixing*, with signals scattered at frequencies corresponding to the sum and difference frequencies of the incident signals. For example, if we have signals simultaneously incident at positive frequencies f_1 and f_2, we will get scattered signal components at frequencies $f_1 \pm f_2$. This is fine for $f_1 > f_2$, and we can map incident phasors associated with positive frequencies into scattered phasors also associated with positive frequencies. But we must consider also the case where $f_1 < f_2$. If we restrict ourselves, as we have so far, to positive frequencies – single-sided spectra – we will have to be careful about how we consistently deal with some of the resulting mixing terms.

For this and other reasons, it is convenient to expand the domain of incident and scattered complex amplitudes to *double-sided* spectra. That is, we consider signals with positive and negative frequencies incident and scattered by the DUT.

We now have the situation depicted in Figure 4.8.

Formally, this means we have to consider the arguments to all spectral maps to include negative frequency components as well as positive. Thus we must generalize equations (4.3) to a form given by (4.13). Similar conditions apply to (4.4). We note (4.13) also depends on the DC bias but this is suppressed for simplicity.

$$B_{p,k} = F_{p,k}(A_{1,1}, A_{1,-1}, A_{1,2}, A_{1,-2}, \ldots, A_{2,1}, A_{2,-1}, A_{2,2}, A_{2,-2}, \ldots) \qquad (4.13)$$

Not only do the arguments of (4.13) include phasors corresponding to negative frequencies on a double-sided harmonic grid, but the index k for the scattered wave harmonic is now permitted to take negative integer values as well as positive integer

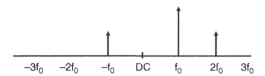

Figure 4.8 Double-sided spectra on a harmonic grid.

values. While this appears to increase the complexity, we will be able to use this approach as a simple starting point from which to derive the formalism of spectrally linearized X-parameters for small mismatch.

We now define spectral linearization of double-sided spectral maps. We assume all spectral components except $A_{1,1}$ and $A_{1,-1}$ are small so we can expand (4.13) in a complex Taylor series. We write, noting the summations omit terms with joint values of $(q = 1, l = \pm 1)$:

$$B_{p,k} = F_{p,k}(A_{1,1}, A_{1,-1}, A_{1,2}, A_{1,-2}, \ldots, A_{2,1}, A_{2,-1}, A_{2,2}, A_{2,-2}, \ldots)$$

$$\approx F_{p,k}(LSOPS) + \sum_{l=-N}^{N} \sum_{q=1}^{P} \left[\frac{\partial B_{p,k}}{\partial A_{q,l}} \bigg|_{LSOPS} \cdot A_{q,l} \right] \tag{4.14}$$

$$\equiv X_{p,k}^{(F)}(RefLSOPS)P^k + \sum_{l=-N}^{N} \sum_{q=1}^{P} \left[X_{p,k;q,l}^{(d)}(RefLSOPS) \cdot A_{q,l} P^{k-l} \right]$$

Here we have introduced the notation *large-signal operating point stimulus*, or *LSOPS*, to refer to the set of large-signal incident waves (and DC applied biases) in the absence of the small signals. Note the summation in (4.14) excludes the phasors $A_{1,\pm 1}$ that are arguments of the *LSOPS*. The *reference large-signal operating point stimulus*, *RefLSOPS*, is obtained from the *LSOPS* by phase-normalizing the arguments. This transformation brings out the phase factors P^k and $P^{k-k'}$ in the respective terms using the time-invariance principle discussed in Section 4.3. The derivation of the last equality of (4.14) is left as an exercise.

We have introduced the notation $X_{p,k}^{(F)}$ to denote the X-parameter function evaluated at the *RefLSOPS*. The symbols $X^{(d)}$ denote the partial derivatives of the double-sided spectral maps evaluated at the *RefLSOPS*.

Now we can get back to real signals by imposing conjugate symmetry. That is,

$$A_k = A_{-k}^* \text{ and } B_k = B_{-k}^* \text{ for k} < 0 \tag{4.15}$$

We now return to the single-sided spectral representation. We restrict ourselves to positive indices but invoke (4.15) to define the contributions from negative frequencies. We can therefore re-express (4.14) as (4.16) where $X^{(S)}$ and $X^{(T)}$ are defined for positive indices as the corresponding terms of $X^{(d)}$ for only positive indices. For notational simplicity, in this and what follows, we omit the double summation sign even though we are summing over two indices, q and l. In (4.16) the summation does not include the term for which both $q = 1$ and $l = 1$ together.

$$B_{p,k} = X_{p,k}^{(F)}(RefLSOPS)P^k$$

$$+ \sum_{\substack{q=1,...,P \\ l=+1 \\ not(1,1)}}^{N} X_{p,k;q,l}^{(S)}(RefLSOPS) \cdot A_{q,l} P^{k-l} + \sum_{\substack{q=1,...,P \\ l=+1 \\ not(1,1)}}^{N} X_{p,k;q,l}^{(T)}(RefLSOPS) \cdot A_{q,l}^* P^{k+l}$$

$$(4.16)$$

Equation (4.16) is the classical form of the spectrally linearized maps around a single large tone at the fundamental frequency at the input port of a nonlinear DUT [6]. This is defined in terms of single-sided spectra for positive frequencies only, since this is what is measured. Unlike S-parameters, there are independent contributions linear in A and also linear in A^*, with separate coefficients $X^{(S)}$ and $X^{(T)}$, respectively. The terms $X^{(F)}$, $X^{(S)}$, and $X^{(T)}$ are nonlinear functions of their arguments, namely the applied DC biases and (phase-normalized) complex amplitudes of the RF signals. We can now properly identify the terms of (4.16) with the corresponding labels in Figure 4.7.

Expressions similar in form to (4.16) hold when the *RefLSOPS* is more complicated than the case considered in this section, as we will see when considering two large tones for "load-dependent" X-parameters and for mixers in Section 4.9.

4.6.3 Identification of X-Parameter Terms from Measured Data

4.6.3.1 Ideal Experiment Design and Identification by Offset Phase Method

The objective is to identify (extract), for each specific *RefLSOPS* of the DUT, all of the X-parameter values, specifically $X_{p,k}^{(F)}$, $X_{p,k;q,l}^{(S)}$, and $X_{p,k;q,l}^{(T)}$, that enter equation (4.16). We will treat this algebraically in this section and provide a geometrical interpretation in Section 4.6.7.

It is assumed for now (this restriction will be removed in Section 4.6.3.2) that the source and the measurement instrument are ideal. Specifically, we assume an ideal incident tone at port one and that no signals generated by the DUT in response to the large excitation signal are reflected back into the DUT.

For each port number, p, and harmonic index k, the ideal experiment design for each *RefLSOPS* consists of a sequence of three distinct RF excitations, labeled by integer m and measurements of the corresponding scattered wave, $B_{p,k}^{(m)}$ for each excitation.

The first excitation is a single large tone of complex amplitude $A_{1,1}$, applied at port one, with measurement result denoted by $B_{p,k}^{(1)}$. The second excitation involves application of the same large-amplitude tone, $A_{1,1}$, and, simultaneously, the application at port q and harmonic index l of a small perturbation tone with complex amplitude $A_{q,l}^{(1)}$. The measurement result for this case is $B_{p,k}^{(2)}$. Finally, the third excitation is of the same type as the second experiment, but with a distinct phase for the small tone at port q and harmonic index l, which we label as $A_{q,l}^{(2)}$. The measurement result is $B_{p,k}^{(3)}$.

Using (4.16), we obtain three linear equations for the unknown X-parameter values $X_{p,k}^{(F)}$, $X_{p,k;q,l}^{(S)}$, and $X_{p,k;q,l}^{(T)}$, in terms of known excitations $A_{1,1}$, $A_{q,l}^{(1)}$, and $A_{q,l}^{(2)}$ and the measured values $B_{p,k}^{(1)}$, $B_{p,k}^{(2)}$, and $B_{p,k}^{(3)}$ shown in equation (4.17).

$$B_{p,k}^{(1)} = X_{p,k}^{(F)} P^k$$

$$B_{p,k}^{(2)} = X_{p,k}^{(F)} P^k + X_{p,k;q,l}^{(S)} P^{k-l} A_{q,l}^{(1)} + X_{p,k;q,l}^{(T)} P^{k+l} A_{q,l}^{(1)*} \qquad (4.17)$$

$$B_{p,k}^{(3)} = X_{p,k}^{(F)} P^k + X_{p,k;q,l}^{(S)} P^{k-l} A_{q,l}^{(2)} + X_{p,k;q,l}^{(T)} P^{k+l} A_{q,l}^{(2)*}$$

This procedure is then applied, separately, for small perturbation tones at other ports and harmonic indices (different values for the indices q and l), and the entire process repeated for each distinct *RefLSOPS* for which the X-parameter values are desired. The phases of the small tones, $A_{q,l}$, in (4.17) can be arbitrary,[6] but the system of equations is better conditioned when the phase difference between $A_{q,l}^{(1)}$ and $A_{q,l}^{(2)}$ is close to 90 degrees. More than two phases for the applied small tones, $A_{q,l}^{(m)}$, can be used as stimuli and the resulting augmented system (4.17) solved for the X-parameter values in a least squares sense in order to reduce the effect of measurement noise.

4.6.3.2   X-Parameter Identification in an Imperfect Environment

The method of Section 4.6.3.1 describes X-parameter identification in a perfect world. However, an actual microwave source, when used at high power conditions, may generate its own complex-valued harmonic distortion components, so the intended ideal excitations of the previous section may not be realized precisely. The output of the DUT in response to such a "dirty source" is therefore a combination of its response to a pure tone and that of the amplified spurious signals that contaminated the input. In addition, the measurement HW may not present a perfect 50 Ω environment to all signals generated by the DUT in response to the large-signal excitation. Both of these nonidealities, if not properly accounted for, create errors in the values of the X-parameters that we wish to attribute to the DUT.

An extension of the method described in Section 4.6.3.1 was developed in [8] for measuring and identifying X-parameters using nonideal HW. The method can be used to characterize the imperfect source and then "predistort" it so that it produces a pure, single-tone input, or, equivalently, correct the DUT response to a dirty source and imperfect match, in order to predict the DUT response to an ideal source or sources. The method is a practical way to calibrate out the imperfections of the source and measurement HW so as to properly return X-parameters that can be regarded as *intrinsic properties of the DUT on its own*, rather than the combined source-DUT-instrument system. A schematic representation of the idea is given in Figure 4.9.

Despite the imperfect environment, the complete set of $X^{(S)}$ and $X^{(T)}$ parameters can be identified from a global analysis of the measured data from the entire set of experiments of the type described in Section 4.6.3.1 for all values of small tone indices q and l. The desired stimulus (a large pure tone at the input) is subtracted from the actual measured stimulus that consists of one large tone and several complex harmonic components. This results in an "imperfection" signal with small components at each

---

[6] But not $\pm\pi$ radians.

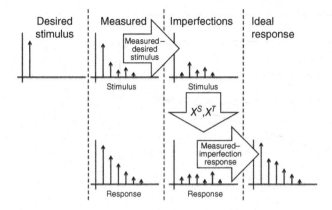

Figure 4.9 Method to identify X-parameters in an imperfect environment

harmonic of the fundamental. This imperfection signal is then used with the previously identified $X^{(S)}$ and $X^{(T)}$ parameters to calculate corrections to the measured response so as to return the actual DUT response to the ideal large stimulus.

This method is particularly effective in X-parameter applications where (far from 50 Ω) loads are controlled and presented to the DUT by the HW system (e.g. with a tuner or an active source at the output port), but where harmonic loads may be uncontrolled and vary as the fundamental load is changed. This will be described in the context of an example in section 4.9.1.5. The method, so applied, is used to calibrate out the harmonic terminations to help present X-parameters referenced to ideal excitations under known loads with perfectly matched harmonic terminations.

4.6.4   Example: *X*-Parameters of a Two-Port Amplifier with Small Mismatch

A simple and illustrative example of the basic X-parameter formulation (4.16) is a nearly matched two-port power amplifier driven by a large signal at port 1. Neglecting any harmonics, for simplicity, (4.16) reduces to (4.18).

$$B_{1,1}(A_{1,1}, A_{2,1}) = X_{1,1}^{(F)}(|A_{1,1}|)P + X_{1,1;2,1}^{(S)}(|A_{1,1}|)A_{2,1} + X_{1,1;2,1}^{(T)}(|A_{1,1}|)P^2 A_{2,1}^*$$
$$B_{2,1}(A_{1,1}, A_{2,1}) = X_{2,1}^{(F)}(|A_{1,1}|)P + X_{2,1;2,1}^{(S)}(|A_{1,1}|)A_{2,1} + X_{2,1;2,1}^{(T)}(|A_{1,1}|)P^2 A_{2,1}^*$$

(4.18)

Thus a two-port, in this simple "fundamental only" approximation, has six X-parameter functions, whereas a linear two-port is described by only four S-parameters. The six X-parameter functions in (4.18), and therefore the scattered waves, $B_{p,1}$, depend on $|A_{1,1}|$ (shown) and also depend on the DC bias conditions, but we suppress the DC arguments for simplicity.

In Figure 4.10, we plot the values of the three X-parameter functions, $X_{2,1}^{(F)}$, $X_{2,1;2,1}^{(S)}$, and $X_{2,1;2,1}^{(T)}$ in (4.18), that contribute to $B_{2,1}$, versus the input port drive, $|A_{1,1}|$. The function $X_{2,1}^{(F)}$ is divided by the magnitude of the incident wave to make the result unitless, like the other two functions shown.

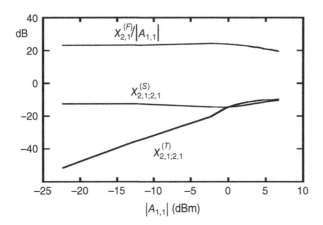

Figure 4.10 Some X-parameter functions of a power amplifier.

We observe two things about the various X-parameter functions. As the power of the incident signal diminishes, so does the value of the $X^{(T)}$ term. In fact, in the limiting case of vanishing power, the $X^{(T)}$ term goes all the way to zero, as we'll see in Section 4.8. This should be no surprise. After all, we know that in the small-signal limit, the S-parameter formalism of Chapter 3 is a complete description. Since the $X^{(T)}$ terms in (4.18) contribute to the scattered waves terms proportional to the conjugates of the incident phasors, and these conjugate terms don't appear in the S-parameter description, we logically conclude all $X^{(T)}$ terms must vanish in the small-signal limit.

On the other hand, at large incident power levels at port 1, the magnitudes of the $X^{(T)}$ terms can increase to the point where they can become comparable to, or even exceed, the magnitudes of the $X^{(S)}$ terms. The $X^{(T)}$ terms are therefore just as important as the $X^{(S)}$ terms at very high incident drive levels. This is true no matter how small the incident signals are at port two when the DUT is driven hard at port one.

4.6.4.1 $X^{(S)}$ and $X^{(T)}$ Functions in Terms of Double-Sided Sensitivities

If we want, we can relate $X^{(S)}$ and $X^{(T)}$ defined on positive frequencies, to the double-sided derivatives of the double-sided spectral maps as follows [9]:

Comparing (4.16) with (4.14) we have, for $k, l > 0$:

$$X_{p,k;q,l}^{(d)} = \frac{\partial\left(B_{p,k}P^{-k}\right)}{\partial\left(A_{q,l}P^{-l}\right)} = \frac{\partial B_{p,k}^{(n)}}{\partial A_{q,l}^{(n)}} = X_{p,k;q,l}^{(S)}$$

$$X_{p,k;q,-l}^{(d)} = \frac{\partial\left(B_{p,k}P^{-k}\right)}{\partial\left(A_{q,-l}P^{l}\right)} = \frac{\partial\left(B_{p,k}P^{-k}\right)}{\partial\left(A_{q,l}^{*}P^{l}\right)} = \frac{\partial B_{p,k}^{(n)}}{\partial A_{q,l}^{*(n)}} = X_{p,k;q,l}^{(T)}$$

(4.19)

In (4.19) we again use the superscript "(n)" to denote phase-normalized waves. Using (4.15), we also find

$$X^{(d)}_{p,-k;q,l} = \frac{\partial B^{(n)}_{p,-k}}{\partial A^{(n)}_{q,l}} = \frac{\partial B^{(n)*}_{p,k}}{\partial A^{(n)}_{q,l}} = \left(\frac{\partial B^{(n)}_{p,k}}{\partial A^{(n)*}_{q,l}}\right)^* = X^{(T)*}_{p,k;q,l}$$

$$\text{(4.20)}$$

$$X^{(d)}_{p,-k;q,-l} = \frac{\partial B^{(n)}_{p,-k}}{\partial A^{(n)}_{q,-l}} = \frac{\partial B^{(n)*}_{p,k}}{\partial A^{(n)}_{q,-l}} = \left(\frac{\partial B^{(n)}_{p,k}}{\partial A^{(n)}_{q,l}}\right)^* = X^{(S)*}_{p,k;q,l}$$

The double-sided expression (4.14) is an analytic function (in the sense of complex variable theory) of each complex phasor $A_{q,l}$. For single-sided expressions, however, the spectral map (4.16) is nonanalytic. That is, we must differentiate with respect to $A_{p,k}$ and $A^*_{p,k}$ independently in this expression. The coefficients, $X^{(S)}$ and $X^{(T)}$, correspond to *partial derivatives of the single-sided spectral map with respect to $A_{q,l}$ and $A^*_{q,l}$*, respectively.

4.6.5   Origins of the Conjugate Terms

The appearance in (4.16) of terms linear in both A and A^* is perhaps surprising. The preceding analysis shows that the origin of the A^* terms is the mixing of the small amplitude negative frequency components back into the positive domain. Of course there is no mixing when the DUT is perfectly linear, so these terms have no analogue in the simple S-parameter world. Simply making S-parameter measurements at high input power levels (e.g. so-called "hot S_{22}") cannot account for such terms and their measurable consequences [10].

Expression (4.16) allows us a more detailed interpretation of Figure 4.7. We suppress the port indices on the diagram for simplicity. The top line of Figure 4.7 shows a single-sided spectral map from four RF signal components (plus DC) into an output spectrum. The nonlinear function depends on the complex values of all independent variables. The approximation made now is to designate one of the incident RF signals as large and the remaining ones as small. Of the four incident signal components (apart from DC), the one at the fundamental frequency is supposed large and the others small. This is indicated by the groupings in the top left of Figure 4.7.

The approximate decomposition of the scattered waves, B_k, using (4.16), is indicated in Figure 4.7. The simple nonlinear map, designated by $X^{(F)}_k$, depends only on the DC and the single real number corresponding to the magnitude of the large tone. This is much simpler than the full nonlinear map that depends on all variables. The phase factor P^k multiplies the X-parameter function to ensure the map is properly time invariant. Note how close the output spectrum on the right column second row is to the full result (top row). The minor differences are now nearly completely accounted for by the contributions from the multiple linear maps, $X^{(S)}_{k;l}$ and $X^{(T)}_{k;l}$. Each small tone, A_l, contributes linearly to the response around all harmonics due to contributions weighted by $X^{(S)}$ terms and also linearly in A^*_l with contributions weighted by independent $X^{(T)}$ terms. Equivalently, there is a contribution at each harmonic, k, linear in each of the A_l and each of the A^*_l terms. The "wiggly" nature of the lines indicates the phase-dependent contributions of the complex numbers, added as vectors.

The length of the sum of all the vectors at each frequency depends on each of their phases and magnitudes; they add vectorially.

4.6.5.1 More on the Origin of Conjugate Terms: An Example with a Cubic Nonlinearity[7]

A concrete example is now provided as another illustration of the origins of the conjugate terms. We consider the simple case of a static algebraic nonlinearity (e.g., a polynomial) in the time domain and compute the mapping in the spectral domain. We start by considering a system described by a simple instantaneous nonlinearity containing both a linear and cubic term. We look at the following three cases, for which the analysis can be computed exactly. The first case is the linear response of this nonlinear system around a static (DC) operating point. This is the familiar condition for which linear S-parameters apply. The second case is the linearization of the system around a time-varying large-signal operating state, with the time variation and perturbation having the same fundamental period. The third case is a simple generalization of the second where the perturbation is applied at a distinct frequency compared to the fundamental frequency of the periodically driven nonlinear system.

The objective is to look at the linearized response of the system in the (single-sided) frequency domain and demonstrate that the relationship between the perturbation phasor and the response phasor is not an analytic function in cases 2 and 3, namely when the nonlinear system is driven. That is, these examples illustrate the simultaneous presence of the perturbation phasors and their complex conjugate phasors, with distinct coefficients, in the response of a nonlinear driven system to additional injected signals.

The nonlinearity is described by (4.21). The signal, $x(t)$, is written in (4.22) as the sum of a main signal and a perturbation term, assumed to be small. We will calculate, to first order in the perturbation, the response of the system (4.21) to the excitation (4.22). We will consider three cases.

$$f(x) = \alpha x + \gamma x^3 \tag{4.21}$$

$$x(t) = x_0(t) + \Delta x(t) \tag{4.22}$$

Case 1

$$x_0(t) = A$$
$$\Delta x(t) = \frac{\delta e^{j\omega t} + \delta^* e^{-j\omega t}}{2} \tag{4.23}$$

In this case, shown in (4.23), the main signal, $x_0(t)$, is just a constant value (a DC term) while the perturbation is a small tone at angular frequency ω. The constant A is real and δ is a complex number of small magnitude that allows for the phase of the perturbation to be arbitrary. The total signal is obviously real.

The first order response is computed in (4.24). The approximation becomes exact as $\Delta x(t) \rightarrow 0$

---

[7] This section is based on reference [11] and corrects minor errors in the original.

$$\Delta y(t) = f(x_0(t) + \Delta x(t)) - f(x_0(t)) \approx f'(x_0(t))\Delta x(t) \tag{4.24}$$

For this case, the conductance nonlinearity, $f'(x_0)$, is evaluated at the fixed value $x_0 = A$, where, using (4.21), we obtain (4.25), where $G(A)$ is defined as being,

$$f'(A) \equiv G(A) = \alpha + 3\gamma A^2 \tag{4.25}$$

Substituting (4.25) into (4.24) and using (4.23) we obtain (4.26):

$$\Delta y(t) = \left(\alpha + 3\gamma A^2\right) \cdot \left(\frac{\delta e^{j\omega t} + \delta^* e^{-j\omega t}}{2}\right) \tag{4.26}$$

If we look at the complex coefficient of the term proportional to $e^{j\omega t}$ we find it is given simply in (4.27):

$$\frac{\left(\alpha + 3\gamma A^2\right)}{2}\delta \tag{4.27}$$

Since (4.26) is a linear input–output relationship with constant coefficients, the complex Fourier component at the output frequency is linearly proportional to the complex Fourier coefficient of the input small-signal phasor. In fact we can read off the frequency-domain phasor relationship (4.28), or equivalently (4.29), where the gain, $G(A)$, is given by (4.25) and $X(\omega)$ is the complex input phasor corresponding to the positive frequency component of the small perturbation tone. We note in this case the gain, $G(A)$, depends nonlinearly on the DC operating point, through (4.25), but is constant in time.

$$Y(\omega) = G(A)X(\omega) \tag{4.28}$$

$$\frac{\Delta Y(\omega)}{\Delta X(\omega)} = G(A) \tag{4.29}$$

Case 2

$$x_0(t) = A\cos(\omega t)$$
$$\Delta x(t) = \frac{\delta e^{j\omega t} + \delta^* e^{-j\omega t}}{2} \tag{4.30}$$

This time we take $x_0(t)$ to be a pure sinusoid given in (4.30). There is no loss of generality by taking A to be real, which is equivalent to taking the phase of the main signal at time zero to be zero. We can interpret δ as the relative phase between the main signal and perturbation signal at the same frequency.

We go through the same procedure as in Case 1 to derive the conductance, but this time it is evaluated at the periodically time-varying operating condition (essentially the *LSOPS*) according to (4.31). The second form follows from the simple trigonometric identity $\cos^2(\omega t) = \frac{1}{2} + \frac{\cos(2\omega t)}{2}$.

$$f'(A\cos(\omega t)) = \alpha + 3\gamma A^2\cos^2(\omega t) = \left(\alpha + \frac{3\gamma A^2}{2}\right) + \frac{3\gamma A^2}{2}\cos(2\omega t) \tag{4.31}$$

Using (4.31) to evaluate (4.24) for this case, we obtain (4.32):

$$\Delta y(t) = \left[\left(\alpha + \frac{3\gamma A^2}{2} \right) + \left(\frac{3\gamma A^2}{2} \right) \left(\frac{e^{2j\omega t} + e^{-2j\omega t}}{2} \right) \right] \cdot \left[\frac{\delta e^{j\omega t} + \delta^* e^{-j\omega t}}{2} \right] \quad (4.32)$$

This time we get contributions proportional to $e^{j\omega t}$ and $e^{3j\omega t}$, and their complex conjugates, four terms in all. If we restrict our attention, as in case 1, to output terms proportional to $e^{j\omega t}$, we obtain the coefficient given in (4.33):[8]

$$\left(\frac{\alpha}{2} + \frac{3\gamma A^2}{4} \right) \delta + \left(\frac{3\gamma A^2}{8} \right) \delta^* \quad (4.33)$$

We observe that the output phasor at frequency ω is not proportional to the input phasor, δ, at frequency ω but instead has distinct contributions proportional to *both* δ and δ^*. That is, the linearization of the nonlinear system around the periodically time-varying operating point determined by the large tone, is not analytic in the sense of complex variable theory. If it were analytic, (4.33) would depend only on the complex variable, δ, and not on both δ and δ^*.

Taking the ratio of the complex output Fourier Coefficient to the input Fourier coefficient, we obtain the result (4.34), where $\phi(\delta)$ is the relative phase of the main and perturbation signal:

$$\frac{\Delta Y(\omega)}{\Delta X(\omega)} = \left(\alpha + \frac{3\gamma A^2}{2} \right) + \left(\frac{3\gamma A^2}{4} \right) e^{-2j\phi(\delta)} \quad (4.34)$$

Therefore, unlike linear S-parameters (and therefore unlike the case (4.29)), this small-signal ratio is not independent of the phase of the small perturbation tone. That is, the large tone creates a phase reference such that the response of the system to even a very small perturbation around the large-signal time-varying state depends explicitly on the relative phase of the perturbation tone and the large tone.

This should come as no surprise, if we think of the large tone as turning on and off, smoothly and periodically, an active device. Even the small-signal scattering of a second tone will clearly depend on where in the time-varying cycle (phase) the device happens to be when the scattering occurs.

Case 3

$$x_0(t) = A \cos(\omega t)$$

$$\Delta x(t) = \frac{\delta e^{j\omega_1 t} + \delta^* e^{-j\omega_1 t}}{2} \quad (4.35)$$

Here we allow the frequency, ω_1, of the perturbation tone to be distinct from the frequency, ω, of the main drive signal. We go through the analysis once again. The time-varying conductance is the same as before, with the only difference being the frequency of the small perturbation term in parentheses in the rightmost factor of (4.36):

---

[8] Equations (4.33), (4.34), and (4.38) correct errors in ref. [11].

$$\Delta y(t) = \left[\left(\alpha + \frac{3\gamma A^2}{2}\right) + \left(\frac{3\gamma A^2}{2}\right)\left(\frac{e^{2j\omega t} + e^{-2j\omega t}}{2}\right)\right] \cdot \left[\frac{\delta e^{j\omega_1 t} + \delta^* e^{-j\omega_1 t}}{2}\right] \tag{4.36}$$

Since ω and ω_1 are distinct, there are more frequency components generated then in the previous cases. We may write $\omega_1 = \omega + \varepsilon$ and look at the terms proportional to $e^{j(\omega+\varepsilon)t}$ and $e^{j(\omega-\varepsilon)t}$. We obtain, respectively, (4.37) and (4.38):

$$\left(\frac{\alpha}{2} + \frac{3\gamma A^2}{4}\right)\delta \tag{4.37}$$

$$\left(\frac{3\gamma A^2}{8}\right)\delta^* \tag{4.38}$$

Terms (4.37) and (4.38) represent the single-sided spectrum of the lower and upper sidebands of the intermodulation spectrum of the system (4.21) for excitation (4.22) defined by (4.35) around the fundamental frequency of the main signal.

We note that as the tone spacing, ε, goes to zero, both these contributions overlap (add) at the frequency ω of the drive signal, and we recover the results of case 2 (equation (4.33)).

The ability to separate the terms proportional to δ from those proportional to δ^* obtained by this "frequency offset" method remains true for the more general dynamic nonlinear system beyond our simple example. In the general case, the upper and lower sideband phasors depend explicitly on the frequency offset, ε (unlike the simple example here). Case 2 can be recovered using case 3 for each sideband for finite ε and then taking the limit $\varepsilon \to 0$. This indicates that it is possible to extract each upper and lower sideband term (per harmonic frequency component) from measurements of the system response to a small tone with a single, arbitrary phase rather than introduce two (or more) distinct phases to extract the two terms when they appear at the same frequency as in (4.17). We will formalize this concept in Section 4.6.6. On the other hand, there are now measurements to be made at distinct frequencies for the upper and lower sidebands, so this method does not lead to fewer measurements than that of case 2.

Examination of case 3 reveals that the complex conjugate term results from intermodulation or mixing, a consequence of nonlinearity, and disappears as the size of the drive signal decreases to zero. This is evident by evaluating (4.37) and (4.38) (or even (4.33) for case 2) as the drive $A \to 0$. The terms proportional to δ^* vanish and the terms proportional to δ reduce to the result we would get for a linear system with constant gain α.

In the limit $\omega \to 0$, case 3 reduces to case 1, equation (4.27), corresponding to the system linearized around a static operating point. This is most easily seen by taking the limit $\omega \to 0$ in (4.36). Thus, although the X-parameter model as discussed so far is representative of case 2 (perturbation signals at exact integer multiples of the fundamental drive signal), the origins of the different terms are made more obvious by examining the slightly more general case 3.

4.6.6   Spectrally LInearized X-Parameters in Terms of Sidebands – Offset Frequency Linear Responses

To provide a useful interpretation of the $X^{(S)}$ and $X^{(T)}$ terms, we consider the slightly more general example of a single large tone at frequency f_0 and a second small tone at $f_1 = 2f_0 + \varepsilon$ incident simultaneously on the nonlinear DUT [12]. Here ε is a small positive frequency offset. This is illustrated in Figure 4.11.

This second tone, injected at port q for definiteness, has small amplitude $\Delta A^+_{q,k'}$, at a frequency ε above the kth harmonic of the fundamental frequency, in the presence of the large tones. The "+" superscript indicates the frequency is slightly more positive than the harmonic. From standard mixer theory, we know there will be small sidebands, both upper and lower, appearing at each port around each of the harmonics of the original spectrum. The complex-valued spectral components are labeled $\Delta B^+_{p,k}$ and $\Delta B^-_{p,k}$ for the upper and lower sidebands, respectively, at port p and harmonic index k. This is shown in Figure 4.12.

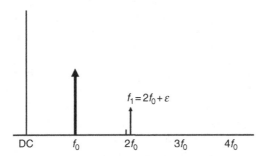

Figure 4.11 Large tone plus a small tone offset slightly from the second harmonic.

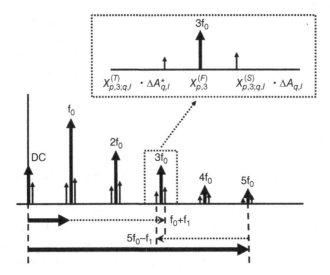

Figure 4.12 Sidebands created by mixing of time-varying nonlinear DUT with small tone stimulus offset from the second harmonic.

We observe that the response in the vicinity of any harmonic of the fundamental contains three separate contributions. We focus on the responses around the third harmonic for definiteness. The phase of $A_{1,1}$ is suppressed in the figure and the following discussion. The first of the three terms is that of the direct 3rd harmonic response of the DUT to the large tone in the absence of the perturbation. This is just the value of the single-tone response, $X_{p,3}^{(F)}$. The second contribution, just above the 3rd harmonic, is attributable to a second order mixing term, first order in the fundamental frequency of the large tone and first order in the positive frequency component of the perturbation tone. This is the value of $X_{p,3;q,2}^{(S)}$ times the value of $\Delta A_{q,2}$. The third contribution is a sixth order mixing term, fifth order in the DUT nonlinearity, and first order in the negative frequency component of the perturbation. This is the value of $X_{p,3;q,2}^{(T)}$ times the value of $\Delta A_{q,2}^{*}$ (neglecting the phase of $A_{1,1}$).

The mixing order can be inferred from the diagram as follows. The length of the solid horizontal arrows indicate the harmonic order of the fundamental frequency needed to be combined with either the positive or negative first power of the perturbation tone. The length of the right-pointing (positive frequency) and left-pointing (negative frequency) dashed horizontal arrows indicates the frequency multiple of the harmonic at which the perturbation tone is being applied.

There are only two combinations of solid and dashed horizontal lines such that, when added as vectors, fall within an ε neighborhood of the kth harmonic frequency. These terms are identified as the appropriate $X_{p,k;q,l}^{(S)}$ and $X_{p,k;q,l}^{(T)}$ parameters, the proportionality coefficients of the response at port p at the kth harmonic to the perturbation at port q at the lth harmonic in the presence of the large-signal(s) driving the nonlinear component.

As the frequency offset vanishes, these three terms in the vicinity of each harmonic coincide at the same frequency – or collide – and therefore must be added vectorially. Since the perturbations are small, we can consider their combined effects by a superposition of their effect at each frequency individually. The final result is just (4.16).

4.6.6.1 Interpretation of Sensitivity Terms as Sideband Stimulus Response Ratios

The above discussion provides an alternative definition of the $X_{p,k;q,l}^{(S)}$ and $X_{p,k;q,l}^{(T)}$ terms. In words, $X_{p,k;q,l}^{(S)}$ and $X_{p,k;q,l}^{(T)}$ are the limiting ratios of the upper and lower sideband responses, respectively, at angular frequency $k\omega_0$, to a small perturbation tone injected at angular frequency $l\omega_0 + \varepsilon$, as $\varepsilon \to 0^+$. This is formalized in (4.39). It should again be emphasized that the values of $X_{p,k;q,l}^{(S)}$ and $X_{p,k;q,l}^{(T)}$ depend on the *RefLSOPS*, and therefore on the incident DC bias conditions and value of the large RF signal.

$$\lim_{\substack{\left|\Delta A_{q,l}^{+}\right| \to 0 \\ \varepsilon \to 0}} \frac{\Delta B_{p,k}^{+}}{\Delta A_{q,l}^{+}} = \lim_{\substack{\left|\Delta A_q\right| \to 0 \\ \varepsilon \to 0}} \frac{\Delta B_p(k\omega_0 + \varepsilon)}{\Delta A_q(l\omega_0 + \varepsilon)} = X_{p,k;q,l}^{(S)}$$

$$\lim_{\substack{\left|\Delta A_{q,l}^{+}\right| \to 0 \\ \varepsilon \to 0}} \frac{\Delta B_{p,k}^{-}}{\Delta A_{q,l}^{+}} = \lim_{\substack{\left|\Delta A_q\right| \to 0 \\ \varepsilon \to 0}} \frac{\Delta B_p(k\omega_0 - \varepsilon)}{\Delta A_q(l\omega_0 + \varepsilon)} = X_{p,k;q,l}^{(T)}$$

(4.39)

We note from (4.39) that unlike some $X^{(S)}_{p,k;q,l}$ terms (those for which $k = l$), all $X^{(T)}_{p,k;q,l}$ terms are related to responses at frequencies distinct from the perturbation stimulus frequency. That means mixing must be occurring for there to be non-vanishing $X^{(T)}_{p,k;q,l}$ terms. Mixing can't happen in a linear device. In fact, we show in Section 4.8 that the value of the $X^{(T)}$ functions must go to zero when the amplitude of the driving signal goes to zero, or equivalently when the component is linear. This is another explanation for the fact that there are no S-parameter analogues to the X-parameter terms depending on the conjugate phasors of the incident signal components.

4.6.7   Geometrical Interpretation of $X^{(S)}$ and $X^{(T)}$ Terms

In the X-parameter case, for a nonlinear DUT, we have to deal with independent terms linear in both $A_{p,k}$ and $A^*_{p,k}$. We consider the case for $p = 2$ and $k = 1$ only for simplicity.

We define $\Delta B_{2,1}$ in (4.40) and (4.41) (compare equations 3.9 and 3.10 for the linear case) where we use the Δ symbol to emphasize these signals are small (whereas the drive signal, $A_{1,1}$ is large).

$$B_{2,1} = X^{(F)}_{2,1}P + \Delta B_{2,1} \tag{4.40}$$

$$\Delta B_{2,1} = X^{(S)}_{2,1;2,1}\Delta A_{2,1} + X^{(T)}_{2,1;2,1}\Delta A^*_{2,1}P^2 \tag{4.41}$$

We rewrite (4.41) as

$$\Delta B_{2,1} = \left|X^{(S)}_{2,1;2,1}\right|\left|\Delta A_{2,1}\right|e^{j\left(\phi\left(X^{(S)}\right)+\phi\left(\Delta A_{2,1}\right)\right)} + \left|X^{(T)}_{2,1;2,1}\right|\left|\Delta A_{2,1}\right|e^{j\left(\phi\left(X^{(T)}\right)-\phi\left(\Delta A_{2,1}\right)+2\phi\left(A_{1,1}\right)\right)}$$

$$= e^{\frac{j\left(\phi\left(X^{(S)}\right)+\phi\left(X^{(T)}\right)+2\phi\left(A_{1,1}\right)\right)}{2}}\left|\Delta A_{2,1}\right|\left[\left|X^{(S)}_{2,1;2,1}\right|e^{j\left(\frac{\phi\left(X^{(S)}\right)-\phi\left(X^{(T)}\right)}{2}+\phi\left(\Delta A_{2,1}\right)-\phi\left(A_{1,1}\right)\right)}\right.$$

$$\left.+\left|X^{(T)}_{2,1;2,1}\right|e^{-j\left(\frac{\phi\left(X^{(S)}\right)-\phi\left(X^{(T)}\right)}{2}+\phi\left(\Delta A_{2,1}\right)-\phi\left(A_{1,1}\right)\right)}\right]$$

$$= e^{\frac{j\left(\phi\left(X^{(S)}\right)+\phi\left(X^{(T)}\right)+2\phi\left(A_{1,1}\right)\right)}{2}}\left|\Delta A_{2,1}\right|\left[\left|X^{(S)}_{2,1;2,1}\right|e^{j\theta\left(\Delta A_{2,1}\right)} + \left|X^{(T)}_{2,1;2,1}\right|e^{-j\theta\left(\Delta A_{2,1}\right)}\right]$$

$$= e^{\frac{j\left(\phi\left(X^{(S)}\right)+\phi\left(X^{(T)}\right)+2\phi\left(A_{1,1}\right)\right)}{2}}\left|\Delta A_{2,1}\right|\left[\left(\left|X^{(S)}_{2,1;2,1}\right| + \left|X^{(T)}_{2,1;2,1}\right|\right)\text{Cos }\theta\left(\Delta A_{2,1}\right)\right.$$

$$\left.+j\left(\left|X^{(S)}_{2,1;2,1}\right| - \left|X^{(T)}_{2,1;2,1}\right|\right)\text{Sin }\theta\left(\Delta A_{2,1}\right)\right] \tag{4.42}$$

where $\theta(\Delta A_{2,1}) = \frac{\phi\left(X^{(S)}\right)-\phi\left(X^{(T)}\right)}{2} + \phi(\Delta A_{2,1}) - \phi(A_{1,1})$.

Defining $\quad X = \left|\Delta A_{2,1}\right|\left(\left|X^{(S)}_{2,1;2,1}\right| + \left|X^{(T)}_{2,1;2,1}\right|\right)\text{Cos }\theta(\Delta A_{2,1})\quad$ and $\quad Y = \left|\Delta A_{2,1}\right|$ $\left(\left|X^{(S)}_{2,1;2,1}\right| - \left|X^{(T)}_{2,1;2,1}\right|\right)\text{Sin }\theta(\Delta A_{2,1})$ we have[9]

---

[9] Except when $\left|X^{(S)}_{2,1;2,1}\right| = \left|X^{(T)}_{2,1;2,1}\right|$. This case is considered in Exercise 4.7.

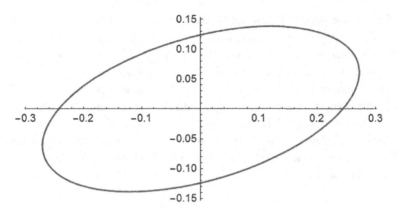

Figure 4.13 Ellipse obtained from X-parameter equation (4.41) by sweeping the phase of $\Delta A_{2,1}$.

$$\frac{X^2}{\left|\Delta A_{2,1}\right|^2\left(\left|X_{2,1;2,1}^{(S)}\right|+\left|X_{2,1;2,1}^{(T)}\right|\right)^2}+\frac{Y^2}{\left|\Delta A_{2,1}\right|^2\left(\left|X_{2,1;2,1}^{(S)}\right|-\left|X_{2,1;2,1}^{(T)}\right|\right)^2}=1 \qquad (4.43)$$

Which is just the standard expression for an ellipse with semi-major axis $\left|\Delta A_{2,1}\right|\left(\left|X_{2,1;2,1}^{(S)}\right|+\left|X_{2,1;2,1}^{(T)}\right|\right)$ and semi-minor axis $\left|\Delta A_{2,1}\right|\left(\left|X_{2,1;2,1}^{(S)}\right|-\left|X_{2,1;2,1}^{(T)}\right|\right)$. The phase factor multiplying the last line of (4.42) just rotates the ellipse around its center by an angle related to the phases of $A_{1,1}$ and the phases of the $X^{(S)}$ and $X^{(T)}$ terms. An example is shown in Figure 4.13.

The extraction from measurements of the X-parameter functions, using (4.17) and the discussion of 4.6.3, can therefore be interpreted geometrically as the identification of a set of ellipses like that given in Figure 4.13 for the particular case $p=2$ and $k=1$. For each port index p and harmonic index, k, there correspond ellipses with center displaced from the origin by the complex vector $X_{p,k}^{(F)}P^k$, with orientation and eccentricity determined by the values of $X_{p,k;q,l}^{(S)}$ and $X_{p,k;q,l}^{(T)}$ for different values of port stimulus q and harmonic stimulus index l.

4.7   Application: Optimum Load for Maximum Power Transfer

This example shows the power of X-parameters to predict, subject to specific assumptions enumerated below, the optimum load at which to terminate a nonlinear two-port for maximum power transfer. It uses the spectrally linearized approximation of Section 4.6 to enable a closed-form solution to the problem.

This problem was first solved in [13] but the results as presented appear complicated. Here we take an original approach and derive a compact result [14]. The methodology follows a similar approach we used in Chapter 3 to show that $\Gamma_2^{opt}=S_{22}^*$ for a linear component neglecting source mismatch, but the computation is more involved in the nonlinear case.

The significance of this application should not be underestimated. Up to now, the preferred experimental method to obtain Γ_2^{opt} was to actually do load-pull measurements and search the performance space for the optimum value. The present method provides an explicit solution for Γ_2^{opt} in terms of the measured X-parameters only.

We neglect the harmonics, and assume the DUT characteristics at port 2 can be described by the second equation of the simple X-parameter expression (4.18).

The objective is to obtain, for fixed available source power (equivalent to a fixed $A_{1,1}$), the complex value of the output reflection coefficient, $\Gamma_{2,1}$, that maximizes the power delivered by the DUT to the load.

The power delivered is described by (4.44) where we drop the harmonic index because we are restricting ourselves here to the fundamental frequency only. The wave indices refer to the port numbers.

$$P_{del}(A_1, A_2) = |B_2(A_1, A_2)|^2 - |A_2|^2 \tag{4.44}$$

We now substitute (4.18) for B_2 in (4.44) and multiply out the terms, using the fact the a squared magnitude of a complex number, Z, is simply $Z \cdot Z^*$. The result is given in simplified notation in (4.45).

$$P_{del}(A_2) = \left(FP + SA_2 + TP^2 A_2^*\right) \cdot \left(FP + SA_2 + TP^2 A_2^*\right)^* - A_2 A_2^* \tag{4.45}$$

In (4.45) we have used the simplifying notation:

$$F \equiv X_{2,1}^{(F)}(|A_1|)$$

$$S \equiv X_{2,1;2,1}^{(S)}(|A_1|) \tag{4.46}$$

$$T \equiv X_{2,1;2,1}^{(T)}(|A_1|)$$

Since A_1 is fixed, we suppress its argument, and we obtain the explicit expression for the power delivered as a function of A_2, A_2^*, and the DUT's X-parameter values at the fixed input power given in (4.46).

We first find the optimum value for A_2, and then compute the corresponding value for Γ^{opt}. The optimum occurs when (4.45) is stationary with respect to A_2. We therefore differentiate (4.45) with respect to A_2 and set the result equal to zero. We obtain the equation (4.47).

$$0 = FT^* P^{-1} + F^* SP^{-1} + 2ST^* A_2 P^{-2} + \left(|S|^2 + |T|^2 - 1\right) A_2^* \tag{4.47}$$

Equation (4.47) can be exactly solved (see Exercise 4.4 for details) to obtain

$$A_2^{opt} = \frac{N}{\Delta} P$$

with

$$N = FS^* \left(1 - |S|^2 + |T|^2\right) + F^* T \left(1 + |S|^2 - |T|^2\right) \tag{4.48}$$

$$\Delta = \left(1 - |S|^2 - |T|^2\right)^2 - 4|S|^2|T|^2$$

We now define $\Gamma^{opt} = \frac{A_2^{opt}}{B_2(A_2^{opt})}$, and using the bottom equation of (4.18) we evaluate (4.48). After some algebra, the final result is given in (4.49).

$$\Gamma^{opt} = \frac{X^{(S)*}\left(1 - \left|X^{(S)}\right|^2 + \left|X^{(T)}\right|^2\right) + e^{-j2\varphi\left(X^{(F)}\right)}X^{(T)}\left(1 + \left|X^{(S)}\right|^2 - \left|X^{(T)}\right|^2\right)}{\left(1 - \left|X^{(S)}\right|^2 - \left|X^{(T)}\right|^2\right) + 2e^{-2j\varphi\left(X^{(F)}\right)}X^{(S)}X^{(T)}} \qquad (4.49)$$

An alternate expression, including an insightful geometrical interpretation, is given in [14]. As we have seen in Section 4.6.4 and will revisit in Section 4.8, we know that as the magnitude of the driving signal becomes small, all $X^{(T)}$ terms vanish and the diagonal $X^{(S)}$ terms (in the harmonic indices) reduce to the linear S-parameters. In this case (4.49) reduces to $\lim_{A_1 \to 0} \Gamma^{opt} = \lim_{X^T \to 0} \Gamma^{opt} = X_{2,1;2,1}^{(S)*} \to s_{22}^*$ as expected. This is just the small-signal result from the previous chapter.

4.8 Small-Signal Limit of X-Parameters

We illustrate the small-signal limit for the simple case of a two-port amplifier being driven by a large single tone at the input port (port 1). We're starting from the spectrally linearized case, where $A_{2,1}$ has already been considered small. We are therefore interested in the limit of (4.50) as $|A_{1,1}| \to 0$.

$$B_{p,k} = X_{p,k}^{(F)}(|A_{1,1}|)P^k + \sum_{\substack{m \geq 1;\, q = 1,2 \\ \text{not } (m=1, q=1)}} X_{p,k;q,m}^{(S)}(|A_{1,1}|)P^{k-m}A_{q,m}$$

$$+ \sum_{\substack{m \geq 1;\, q = 1,2 \\ \text{not } (m=1, q=1)}} X_{p,k;q,m}^{(T)}(|A_{1,1}|)P^{k+m}A_{q,m}^* \qquad (4.50)$$

For simplicity, we neglect all harmonic terms and restrict our attention to the output port in (4.51).

$$B_{2,1} = X_{2,1}^{(F)}(|A_{1,1}|) + X_{2,1;2,1}^{(S)}(|A_{1,1}|)A_{2,1} + X_{2,1;2,1}^{(T)}(|A_{1,1}|)A_{2,1}^*P^2 \qquad (4.51)$$

We know (4.51) must reduce, for small $A_{1,1}$, to the S-parameter Equation 3.9. We motivate and then demonstrate this in the following.

It is shown in [15], using Volterra theory, that the leading asymptotic behavior of the various terms in (4.50) behave according to (4.52) for small $|A_{1,1}|$.

$$X_{p,k}^{(F)} \sim |A_{1,1}|^k$$

$$X_{p,k;q,l}^{(S)} \sim |A_{1,1}|^{|k-l|} \qquad (4.52)$$

$$X_{p,k;q,l}^{(T)} \sim |A_{1,1}|^{k+l}$$

Using (4.52) in (4.51) we obtain (4.53), where the term $|A_{1,1}|^2$ can be neglected compared to the others in the limit of small $|A_{1,1}|$.

$$B_{2,1} \approx \alpha|A_{1,1}|P + \beta|A_{1,1}|^0 A_{2,1} + \delta|A_{1,1}|^2 A_{2,1}^* \to \alpha A_{1,1} + \beta A_{2,1} \qquad (4.53)$$

Comparing (4.53) with the S-parameter equation for a two-port, we obtain the expected result, (4.54), consistent with 3.9.

$$\alpha = S_{21}$$
$$\beta = S_{22} \tag{4.54}$$

For weakly nonlinear systems, the considerations of section 4.6.6 show that for an $X^{(T)}$ term to exist, the DUT must have at least a third-order nonlinearity. Higher order nonlinear systems produce more non-vanishing $X_{p,k;q,l}^{(T)}$ terms. In general, $X_{p,k;q,l}^{(T)}$ become more significant when the driving signal becomes large.

4.9   Two Large Signals

4.9.1   Two Large Signals at the Same Frequency: Load-Pull

In this section we relax the approximation that of all the RF signal components of the incident waves, only $A_{1,1}$ is considered large.

A diagram of the situation is given in Figure 4.14. The DC bias conditions are not shown in the Figure.

In this case, we consider both $A_{1,1}$ and $A_{2,1}$ to be large, with all additional incident signals at the harmonic frequencies assumed to be small. We choose to take the full nonlinear dependence of $A_{2,1}$ into account, treating it on the same footing as $A_{1,1}$.[10] This covers the case of active and passive fundamental load-pull, where the incident wave at the fundamental frequency at port two could be generated by a very large reflection or by an independent source at the output port, for example.

4.9.1.1   LSOP for Two Large Signals at Same Frequency

The large-signal operating point stimulus (*LSOPS*) for this case is given by (4.55), and the corresponding *RefLSOPS* by (4.56). The second expression of 4.65 is defined explicitly in terms of the relative phase between the two signals. Here DCS_p is the set of DC stimulus conditions on the various ports, p, either voltage or current biases.

$$LSOPS = \left(DCS_p, A_{1,1}, A_{2,1}\right) \tag{4.55}$$

$$ReFLSOPS = \left(DCS_p, |A_{1,1}|, A_{2,1}P^{-1}\right) = \left(DCS_p, |A_{1,1}|, |A_{2,1}|, \theta_{2,1}\right) \tag{4.56}$$

4.9.1.2   Harmonic Superposition Approximation

The spectral linearization process is performed similarly to the process described in Section 4.6 to deal with small mismatch at the output port at the fundamental frequency. In the present case, however, we linearize around a *ReFLSOPS* corresponding to large tones incident simultaneously at port one and port two. This enables us to model the complete nonlinear effect of large mismatch at the fundamental frequency at port two. The result, in single-sided spectral terms, is given in (4.57) with the argument of the functions given by (4.56).

---

[10] Another approach is to take polynomial orders of $A_{2,1}$ greater than unity in the spectral mapping as in [16].

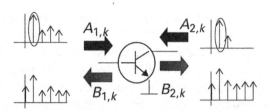

Figure 4.14 Two large RF tones (circled) at the same fundamental frequency and small tones at harmonic frequencies incident at ports 1 and 2 of a nonlinear DUT.

$$B_{p,k} = X_{p,k}^{(F)}(RefLSOPS)P^k$$

$$+ \sum_{\substack{q=1,...,P \\ l>1}}^{N} X_{p,k;q,l}^{(S)}(RefLSOPS) \cdot A_{q,l} P^{k-l} + \sum_{\substack{q=1,...,P \\ l>1}}^{N} X_{p,k;q,l}^{(T)}(RefLSOPS) \cdot A_{q,l}^{*} P^{k+l}$$

$$(4.57)$$

We note (4.57) is very similar in form to (4.16), with only two differences. The first is that the *RefLSOPS* stimuli are different as just described above. The second difference, related to the first, is that in (4.57) the sums begin with the second harmonics on both ports while in (4.16) the sums include the terms linear in $A_{2,1}$ and $A_{2,1}^{*}$.

The same experiment design and verification methods, introduced in Section 4.6.3 for equation (4.16) in the one-tone case can be used to identify the X-parameter values appearing in (4.57).

The effects on the scattered waves of signals incident at the harmonics are taken into account by the approximation (4.57) as a summation of contributions linear in $A_{p,l}$ and $A_{p,l}^{*}$ for $l \geq 2$. This explains why this approximation is called the *harmonic superposition approximation* [10]. The advantage of this approximation is that there are only three real parameters – apart from the DC conditions – in the arguments of the X-parameter functions in (4.57) that determine the *LSOP* of the DUT. This low-dimensional space of independent RF variables is much easier to control when characterizing the DUT then if all RF signal components, including harmonics, are specified independently, as in the more general formulation of (4.8).

The harmonics generated by the DUT are determined by the DC bias conditions, the magnitudes of the large RF signals $|A_{1,1}|$ and $|A_{2,1}|$, and the relative phase, θ_{21}, between the two RF signals. The $X_{p,k}^{(F)}P^k$ terms of (4.57) specify these scattered waves when no additional energy is injected back into the DUT at harmonic frequencies. That is, the $X_{p,k}^{(F)}P^k$ terms describe the DUT under "harmonic matching" at all ports. The effects of additional harmonic energy, as might be generated by harmonic reflections from actual non 50 Ω termination impedances, are accounted for by the terms of (4.57) proportional to the $X^{(S)}$ and $X^{(T)}$ functions. Since these terms are linear in the harmonic signal variables, $A_{q,l}$ and $A_{q,l}^{*}$, we call the $X^{(S)}$ and $X^{(T)}$ terms *harmonic sensitivities*. The value of these sensitivity terms, however, depend nonlinearly on the device *LSOPS* variables.

The harmonic superposition approximation breaks down only when an incident signal at the second or higher integer harmonic of the fundamental frequency is large enough to change significantly the DUT's LSOP. The approximation breaks down gradually and is DUT-dependent. This is strictly analogous to the familiar condition that in order to obtain valid linear S-parameters of a two-port (such as a transistor), the RF signals A_1 and A_2 must be sufficiently small that the DC operating point is not changed. The condition is also strictly analogous to the case of a "nearly matched" nonlinear DUT in Section 4.6 where $|A_{2,1}|$ must be sufficiently small that it does not change the LSOP and can be treated as a first order perturbation. If it is found that the harmonic superposition approximation is not sufficiently accurate for a particular application, one can include the harmonic signal component as a general argument to the full X-parameter function (see Section 4.9.1.7), or take polynomial orders of the signal component higher than linear in the approximation of the scattered waves [16].

4.9.1.3   Load-Dependent X-Parameters

Equations (4.55)–(4.57) describe the scattered waves as functions of the incident waves. In many practical experimental situations, a passive load tuner is used to reflect the signal generated by the DUT at the output port back into the device as an incident wave. In this case it is often more convenient to take the *complex output reflection coefficient*, $\Gamma_{2,1}$, as one of the independent variables, instead of the incident wave at port two. The relationship is given by (4.58).

$$A_{2,1} = \Gamma_{2,1}B_{2,1} \tag{4.58}$$

The *LSOPS* for this case is now specified by (4.59) and the *ReFLSOPS* by (4.60). It is left as an exercise to show (4.60) follows from (4.59).

$$LSOPS = \left(DCS_p, A_{1,1}, \Gamma_{2,1}\right) \tag{4.59}$$

$$ReFLSOPS = \left(DCS_p, |A_{1,1}|, \Gamma_{2,1}\right) \tag{4.60}$$

The X-parameter framework is able to re-parameterize relations (4.57) in terms of X-parameter functions defined in terms of $\Gamma_{2,1}$. The results are given in (4.61) and discussed and derived in more detail in [17].

$$
\begin{aligned}
B_{p,k} = {} & \tilde{X}_{p,k}^{(F)}(|A_{1,1}|, \Gamma_{2,1})P^k \\[4pt]
& + \tilde{X}_{p,k;2,1}^{(S)}(|A_{1,1}|, \Gamma_{2,1}) \cdot \left[\Gamma_{2,1} \cdot \left(B_{2,1} - \tilde{X}_{2,1}^{(F)}(|A_{1,1}|, \Gamma_{2,1})\right)\right] \cdot P^{k-1} \\[4pt]
& + \tilde{X}_{p,k;2,1}^{(T)}(|A_{1,1}|, \Gamma_{2,1}) \cdot \left[\Gamma_{2,1} \cdot \left(B_{2,1} - \tilde{X}_{2,1}^{(F)}(|A_{1,1}|, \Gamma_{2,1})\right)\right]^* \cdot P^{k+1} \\[4pt]
& + \sum_{\substack{q=1,\dots,P \\ l=2\dots,N}} \left[\tilde{X}_{p,k;q,l}^{(S)}(|A_{1,1}|, \Gamma_{2,1}) \cdot A_{q,l} \cdot P^{k-l} + \tilde{X}_{p,k;q,l}^{(T)}(|A_{1,1}|, \Gamma_{2,1}) \cdot A_{q,l}^* \cdot P^{k+l}\right]
\end{aligned}
\tag{4.61}
$$

The functions with the *tilde* superscripts have the same meaning as before, but the new symbol indicates they have a different functional form when expressed in terms of the new variable, $\Gamma_{2,1}$, rather than $A_{2,1}$. This formulation of the harmonic superposition approximation is sometimes called *arbitrary load-dependent X-parameters.*[11]

We note another difference between (4.61) and (4.57). In (4.61) we see in the second and third rows the appearance of X-parameter sensitivity functions with respect to $A_{2,1}$ and $A_{2,1}^*$, whereas these terms do not appear in (4.57).

For passive load-pull conditions, even having a refection coefficient, $|\Gamma_{2,1}| \approx 1$ may not produce an incident wave $A_{2,1}$, through (4.58), large enough to invalidate the harmonic superposition approximation. The approximation fails only when both the reflection coefficient is large and the DUT has a strong (nonlinear) dependence on the resulting reflected wave.

4.9.1.4 GaN Packaged Transistor X-Parameter Model: Validating Harmonic Tuning Predictions Using Harmonic Load-Pull

The degree to which the controlled harmonic superposition approximation is accurate and useful has been investigated many times. Here we present but one result [18] [19].

A load-dependent X-parameter model was developed for a 10W GaN packaged transistor. The part was a Cree CGH40010 GaN HEMT. A fixed number of 9 load states was used in a passive fundamental load-pull configuration for X-parameter characterization. The harmonic terminations were uncontrolled during the data acquisition for the X-parameter model; the tuner introduces various values of harmonic impedances as it moves from one fundamental complex load state to the next.

Validation measurements were made using a full multiharmonic load-pull system. Three output tuners were used for the validation measurements, one tuner each to control, independently, the first, second, and third harmonic impedances. Nine tuner states for each harmonic were used to present the set of independently specified fundamental and harmonic impedances to the device output, for a total of $9^3 = 729$ load states. An ideal (fundamental) load-dependent X-parameter measurement would require only 45 measurements at a given input power varying the fundamental load impedance over 9 states. This is a significant reduction in measurements for nearly the same accuracy.

In the validation process, the set of harmonic impedances that are controlled in the independent harmonic load-pull experiments, are taken into the nonlinear simulator and presented to the X-parameter model. The model was constructed by controlling only the fundamental load, and the harmonic sensitivity functions were measured at nominally 50 Ω. The model uses the sensitivity functions in (4.61) to predict the effects of tuning the harmonic loads to any value, even outside the Smith chart. Comparison is made of the measured and simulated time-domain voltage and current waveforms, the dynamic load-lines, and the power added efficiency (PAE) from small to large input power at many values of the specified harmonic load impedances. A typical comparison is shown in Figure 4.15. Figure 4.16 depicts the case when the second harmonic impedances are swept around the edge of the Smith chart, meaning the reflection coefficient has magnitude nearly unity in these cases. More results are presented in [19]. The results

---

[11] Even though a reflection coefficient, rather than a load, is being controlled.

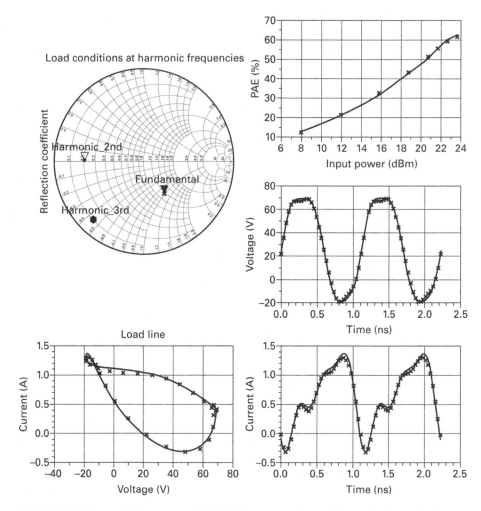

Figure 4.15 Validation of X-parameter GaN HEMT model using multiharmonic load-pull. Results from simulations using X-parameter model extracted into controlled fundamental impedance, only, (line) versus measurements (symbols).

demonstrate that the X-parameter model, extracted by controlling only the fundamental impedance value and without controlling harmonic impedances, nevertheless predicts accurately the DUT response to the *independently tuned* (controlled) *fundamental and harmonic impedances* over the entire Smith Chart.

4.9.1.5 Doherty PA Design from Active Source Injection Load-Dependent X-Parameters

Passive tuners are of little utility when characterizing a transistor when it is biased in an "off" state – one where there is no signal being generated at port two in response to an input at port 1. Of course, linear S-parameters are perfectly well-defined when a transistor is off, and the port 2 characteristics are directly obtained by injecting small waves from an RF source into port 2 and measuring the reflections. For the same

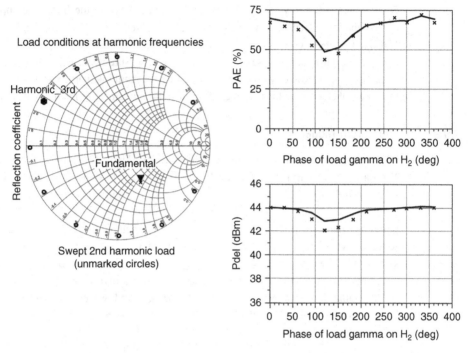

Figure 4.16 PAE and Pdel versus phase of swept second harmonic load impedance at the edge of the Smith Chart for fixed fundamental and third harmonic load impedances. Results from simulations using X-parameter model extracted into controlled fundamental impedance, only, (line) versus measurements (symbols).

reasons, active source-injection measurements are very valuable for large-signal characterization and nonlinear behavioral modeling of transistors over their entire region of operation. This is actually indispensable for such applications as Doherty amplifier design using frequency-domain behavioral models, where the Class C peaking amplifier is biased in the off state; it only turns on with sufficient large-amplitude input drive. Nevertheless, the behavior of the peaking amplifier output port is critical to how the two PAs "mutually load-pull" one another to achieve the high efficiency output.

A complete Doherty amplifier was designed and validated from X-parameter models obtained from bare GaN transistors using an active source-injection characterization system in [20]. Detailed experimental comparisons were presented for the harmonic performance of the peaking amplifier that essentially sees "active loads" with reflection coefficients of magnitudes greater than unity until it turns fully on. The harmonic tuning prediction capabilities of X-parameters is essential to the design. The models were based on the harmonic superposition approximation and obtained without controlling the harmonic impedances during the characterization. The final performance of the manufactured Doherty amplifier agreed extremely well with the simulations.

4.9.1.6   Harmonic Load-Dependent X-Parameters

It is reiterated here that, should it be desired or necessary, the X-parameter formalism is capable of handling the full nonlinear dependence of incident harmonic signal components (or, equivalently, the harmonic loads) beyond the harmonic superposition approximation. That is, the LSOPS can be taken to be, for example, $LSOPS = (DCS_p, A_{1,1}, A_{2,1}, A_{2,2}, A_{2,3})$, where three harmonics at the output port are considered, as opposed to the simpler case of (4.55). Of course measurement systems to control multiple large signal input waves (or complex impedances at multiple harmonics) are considerably more complicated than those that control only the fundamental frequency signals.

4.9.2   The Cardiff Model

We derive one version of the Cardiff model ([16], [21], [22]) from a general X-parameter description where the LSOP is assumed to depend on the dc biases (not shown), and large-signal incident waves at port 1 (incident power) and also at port 2, both signals at the fundamental frequency. Assuming for now that the device is perfectly matched at the harmonics of the fundamental, from (4.8) we can write the scattered waves at port p at each of the harmonics according to (4.62). Here we add a subscript to the notation $P_1 = e^{j\phi(A_{1,1})}$ for reasons to become evident shortly. All wave variables are phase-normalized to $A_{1,1}$ to ensure time invariance.

Neglecting for now the dependence of the scattered waves on incident waves with harmonic components[12], we start from a simple version of the X-parameter formalism, (4.62).

$$B_{p,k} = X_{p,k}^{(F)}\left(|A_{1,1}|, A_{2,1}P_1^{-1}\right)P_1^k \tag{4.62}$$

We now define $P_2 = e^{j\phi(A_{2,1})}$ and the relative angle, ϕ_{21}, between $A_{2,1}$ and $A_{1,1}$ according to (4.63).

$$e^{j\left(\phi(A_{2,1})-\phi(A_{1,1})\right)} = P_2 P_1^{-1} = \frac{P_2}{P_1} \equiv e^{j\phi_{21}} \tag{4.63}$$

Writing $A_{2,1}P_1^{-1} = |A_{2,1}|P_2P_1^{-1} = |A_{2,1}|e^{j\phi_{21}}$, we obtain (4.64).

$$B_{p,k} = X_{p,k}^{(F)}\left(|A_{1,1}|, |A_{2,1}|, e^{j\phi_{21}}\right)P_1^k \tag{4.64}$$

Clearly $X_{p,k}^{(F)}$ in (4.64) is periodic in the angle $\phi_{2,1}$ for fixed values of the incident wave magnitudes. Therefore, we can expand (4.64) in a Fourier series, (4.65), where we take only a finite number ($2M + 1$) of terms into account.

$$X_{p,k}^{(F)} = \sum_{m=-M}^{M} C_{p,k,m}(|A_{1,1}|, |A_{2,1}|)e^{jm\phi_{2,1}} \tag{4.65}$$

Re-writing (4.65) using (4.63) we obtain a form of the Cardiff Model (4.66).

$$X_{p,k}^{(F)} = \sum_{m=-M}^{M} C_{p,k,m}(|A_{1,1}|, |A_{2,1}|)\cdot\left(\frac{P_2}{P_1}\right)^m \tag{4.66}$$

---

[12] This means the harmonics in the scattered waves are entirely generated by the DUT when stimulated at the fundament frequency, and there is no reflection of these harmonics back into the DUT.

The Cardiff Model coefficients, $C_{p,k,m}$, themselves nonlinear functions of the RF signal magnitudes and DC bias conditions, can be extracted from active load-pull measurements at fixed input power and fixed injected power into port two (like that mentioned in Section 4.9.1.6 and discussed in this context in [16]) by varying the relative phase of the injected signals over 360 degrees and computing the simple integral according to (4.67). In (4.67), only the angular dependence on $X^{(F)}$ is indicated since all other independent variables are kept constant.

$$C_{p,k,m} = \frac{1}{2\pi} \int_{-\pi}^{\pi} X_{p,k}^{(F)}(\phi) e^{-mj\phi} d\phi \tag{4.67}$$

The Cardiff model can therefore be considered a representation of the $X^{(F)}$ terms of the X-parameter model in the Fourier space complementary to the dependence on the relative angle between the large signals. The descriptions are otherwise equivalent.

There are $P \cdot N$ X-parameter functions of three real variables in (4.64) for P ports and N harmonics. There are $P \cdot N \cdot (2M + 1)$ functions of two real variables, ($|A_{1,1}|$ and $|A_{2,1}|$), in (4.66). If the expression (4.66) can fit the measured data sufficiently accurately over all conditions for a small value of the integer M (that is only a small number of C-coefficient functions are needed), this formulation can be efficient.

4.9.2.1 Extension to Harmonic Load Dependence

Just as X-parameters apply quite generally to conditions of arbitrarily large input signal values at harmonic frequencies (see Section 4.9.1.7), the Cardiff Model can be similarly generalized beyond (4.66) and applied to cases where harmonic loads are designed to be very large [22]. In conjunction with automated harmonic "time-domain" load-pull measurement systems, the Cardiff Model has been used with great success in designs based on waveform engineering concepts [23].

4.9.2.2 Other Frequency Domain Nonlinear Behavioral Models

Classical Volterra and related frequency domain nonlinear behavioral models have been used for many years and are described in [24] and references therein. References [25], [10], [26], [11], [12], [27], [21], [16], and [22] are generally more suitable for large-signal microwave problems, however. These references also share the common attribute of proper implementation of time-invariance and can all be interpreted as being based on mixing phenomena, and therefore also share commonality with classical "conversion matrix" approaches [28] and [29]. They all appropriately take into account the dependence of the DUT output characteristics on the relative phase of multiple commensurate signals, unlike the older and incomplete "Hot S-parameter" techniques [30].

4.9.3 X-Parameters for Incommensurate Tones

In this section, our treatment of X-parameters is extended to include two large incommensurate incident signals. The signals can be incident on the same port, or on different ports of the DUT. This extension is necessary to cover the important cases of

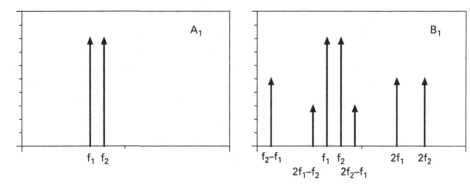

Figure 4.17 Incident and simplified scattered wave spectrum for nonlinear two-port excited by two large incommensurate tones.

intermodulation distortion created by a power amplifier when stimulated by two incident signals at the input port, and also to mixers where a large LO and large RF signal are incident at different ports, creating a frequency-translated signal (the IF) at the third port.

Two sinusoidal signals, with (non-zero) frequencies f_1 and f_2, are incommensurate if the ratio of their frequencies is irrational. More formally, re-expressing (4.1), f_1 and f_2 are incommensurate if there are no two non-zero integers, n and m, such that $nf_1 + mf_2 = 0$. This is summarized in (4.68). The generalization to N incommensurate tones is straightforward but the notation is somewhat tedious so we won't consider it further, here.

$$n, m \in \mathbb{Z} \text{ such that } nf_1 + mf_2 = 0 \Rightarrow n = m = 0 \tag{4.68}$$

An example of two incident incommensurate sinusoidal waves and the resulting spectrum of scattered waves from a nonlinear DUT is given in Figure 4.17. Only a finite number of spectral tones in the output spectrum are shown.

Because the output spectrum generated by a nonlinear DUT can be reflected back into the device due to mismatch, the spectral maps introduced for this case must be well-defined for all intermodulation frequencies of the large incident signals. In fact, in this case all nonzero signals can be consistently defined precisely on the discrete infinite set of all intermodulation frequencies, f, given by (4.69), where n and m are integers. This set takes the place of the harmonic grid used for periodic signals up to this point. The case where f evaluates to a negative number is treated in Section 4.9.7.

$$f = nf_1 + mf_2 \tag{4.69}$$

Each wave variable is specified by a complex amplitude, or phasor, as before. The first index of the phasor is again the port index, p, a positive integer indicating the port at which the stimulus is applied. The second index is composed of the two integers of (4.69) specifying the signal frequency. This is shown in (4.70) for incident and scattered waves. Note that there can be more ports than large tones. For example, a mixer may have 3 ports but there may be only two incident large tones, one at the LO and the other at the RF.

$$A_{p,[n,m]}$$
$$B_{p',[n',m']} \tag{4.70}$$

As an example, $A_{3,[2,-1]}$ is the incident wave at port 3 corresponding to a frequency of $2f_1 - f_2$. Other examples are given in [31].

Two tones with randomly chosen real frequencies will be incommensurate with probability unity. This is related to the greater cardinality of irrational numbers compared to rational numbers. In this sense commensurate signals are "special cases" and therefore less representative.

4.9.4   Time Invariant Maps on a Two-Tone Intermodulation Spectrum

Importantly, it is only for the special case of commensurate tones that there is a well-defined time-invariant concept of cross-frequency phase. Cross-frequency phase for commensurate tones was the basis of the particular phase-normalization process for the spectral maps on the harmonic grid described in Section 4.3. For incommensurate signals, things are fundamentally different, requiring an alternative approach.

As time elapses, to each successive zero-crossing of the phase of the first tone at frequency f_1, there corresponds a unique value of the phase, ϕ_2, of the second tone at frequency f_2.[13] This infinite sequence of distinct ϕ_2 values is dense on the interval $[0, 2\pi)$, with no finite gaps. This means that all possible real values of phase are approached arbitrarily closely. A corollary is that it is possible to take the phase of the first fundamental tone equal to zero, and find the phase of the second fundamental tone arbitrarily close to zero just by waiting a long enough time.

An important consequence of this fact, related to taking appropriate limits in the space of such signals, is that an incommensurate two-tone steady-state signal, can be uniquely identified in the frequency domain by specifying only the magnitudes of the individual signals. This is one less degree of freedom compared to the commensurate case considered in the first part of this chapter, where the two magnitudes plus a relative phase angle were required.

Consider now the two fundamental signals as well as signals at any other intermodulation frequency. Each intermodulation frequency is manifestly commensurate with respect to the two fundamental tones. This follows trivially from (4.71) and (4.72).

$$nf_1 + mf_2 - lf_3 = 0 \tag{4.71}$$

$$f_3 = freq\{A_{p,[n,m]}\} = nf_1 + mf_2 \tag{4.72}$$

The time-invariant mapping is achieved by the following phase-normalization procedure. Apply independent phase-shifts to each of the phasors associated with the fundamental tones so they each have zero phase. This is shown in (4.73), where $P_{[1,0]}$ and $P_{[0,1]}$ are the phases associated with the two large tones, respectively.

---

[13] This is not the case for commensurate signals, where there are only a finite number of possible values for the phase of the second signal, ϕ_2[32].

This translates the two large signal incident waves to the reference excitation state:

$$A_{p,[1,0]} \rightarrow A_{p,[1,0]}P_{[1,0]}^{-1} = \left| A_{p,[1,0]} \right| \text{ and } A_{p',[0,1]} \rightarrow A_{p',[0,1]}P_{[0,1]}^{-1} = \left| A_{p',[0,1]} \right| \qquad (4.73)$$

Any other signal component at an intermodulation frequency labeled *[n,m]* is commensurate with respect to the fundamental tones, and therefore has a well-defined cross-frequency phase with respect to both of the fundamental tones. The same reasoning invoked earlier in this chapter leads to the conclusion that the phase of any signal at a frequency defined by the intermodulation indices *[n,m]* must be coherently phase-shifted by the sum of the integral powers of the individual phases of the large tones according to the respective indices. This is shown in (4.74) for an incident signal phasor.

$$A_{q,[n,m]} \rightarrow A_{q,[n,m]}P_{[1,0]}^{-n}P_{[0,1]}^{-m} \qquad (4.74)$$

The multi-tone signal defined by phasors (4.74) corresponding to zero phase at each fundamental tone through (4.73), will define our reference excitation.

Since the scattered waves (the response of the nonlinear system) also have components only on the same intermodulation spectrum, we must phase-shift the scattered waves by the same terms to achieve the reference response. This is given in (4.75).

$$B_{q,[n,m]} \rightarrow B_{q,[n,m]}P_{[1,0]}^{-n}P_{[0,1]}^{-m} \qquad (4.75)$$

The time-invariance property is achieved by requiring that the spectral mapping functions be invariant with respect to the phase normalization transformations (4.74) and (4.75). This is expressed in (4.76). The phase factors in (4.75) have been taken to the right hand side of (4.76). Only those functions $F_{q,[n,m]}$ with the specific properties expressed by the first two lines of (4.76) are admissible spectral mappings for a time-invariant DUT being excited by signals at all intermodulation frequencies associated with two fundamental tones at incommensurate frequencies.

$$
\begin{aligned}
B_{q,[n,m]} &= F_{q,[n,m]}\left(DCS, A_{p_1,[1,0]}, A_{p_2,[0,1]}, \ldots, A_{p',[n,m]} \right) \\
&= F_{q,[n,m]}\left(DCS, A_{p_1,[1,0]}P_{[1,0]}^{-1}, A_{p_2,[0,1]}P_{[0,1]}^{-1}, \ldots, A_{p',[n,m]}P_{[1,0]}^{-n}P_{[0,1]}^{-m}, \ldots \right) \cdot P_{[1,0]}^{n}P_{[0,1]}^{m} \\
&\equiv X_{q,[n,m]}\left(DCS, \left| A_{p_1,[1,0]} \right|, \left| A_{p_2,[0,1]} \right|, \ldots, A_{p',[n,m]}P_{[1,0]}^{-n}P_{[0,1]}^{-m}, \ldots \right) \cdot P_{[1,0]}^{n}P_{[0,1]}^{m}
\end{aligned}
$$

$$(4.76)$$

Noting that the first two RF arguments of the $F_{q,[n,m]}$ functions at the reference excitation are real by (4.73), the introduction of the X-parameter function is defined according to the second and third lines of (4.76). This means the DUT X-parameter functions are defined on a manifold with two fewer real dimensions than arbitrary multitone spectral mappings, a big simplification, without any approximation.

4.9.5   Discussion: Phase-Shifts and Time-Shifts

Finite and independent phase-shifts are used to coherently shift (phase-normalize) signals to the *RefLSOPS* and then to shift back (phase denormalize) the scattered waves appropriately. This is the procedure used to implement time invariance in the incommensurate case. It is noteworthy to consider that the actual time-shift necessary to reach the *RefLSOPS* (where the two fundamental signals have zero phase) may be infinite. This fact can be established, rigorously, considering sequences of time-shifts, and the convergence of continuous maps in function spaces. This level of detail is beyond the scope of this introduction.

4.9.6 Spectral Linearization

The nonlinear maps (4.76) depend on many variables and are therefore very complex to deal with in general. Just as in earlier sections, useful approximations can be developed to simplify things, dramatically, for practical applications.

Here it is considered that out of all the possible signal components, represented by $A_{p,[n,m]}$, incident on the DUT, only the fundamental tones, $A_{p_1,[1,0]}$ and $A_{p_2,[0,1]}$ are large, while all other components at intermodulation frequencies are considered small. This is the case, for example, in mixers with nearly matched ports. It is also the case for power amplifiers with ports nearly matched to 50 Ω. In these cases, the rich intermodulation spectra will be reflected, but the terms will be small enough that the spectral linearization process will still be a good approximation for taking their effects into account.

Equation (4.76) is therefore spectrally linearized around the reference LSOP stimulus, *RefLSOPS*, given by (4.77).

$$RefLSOPS = \left(DCS_p, \left|A_{p_1,[1,0]}\right|, \left|A_{p_2,[0,1]}\right|\right) \tag{4.77}$$

Spectral linearization of (4.76) with respect to all other signals is applied around the *RefLSOPS* of (4.77). The now familiar linearization procedure presented earlier is applied and the resulting X-parameter model equations take the form (4.78)–(4.80). The sum is over all ports, q, and all integers n' and m'. These equations are the analogues of (4.16) for the harmonic grid case.

$$
\begin{aligned}
B_{p,[n,m]} = {}& X^{(F)}_{p,[n,m]}(refLSOPS)P^n_{[1,0]}P^m_{[0,1]} \\
& + \sum_{q,n',m'} X^{(S)}_{p,[n,m];q,[n',m']}(refLSOPS)P^{n-n'}_{[1,0]}P^{m-m'}_{[0,1]}A_{q,[n',m']} \\
& + \sum_{q,n',m'} X^{(T)}_{p,[n,m];q,[n',m']}(refLSOPS)P^{n+n'}_{[1,0]}P^{m+m'}_{[0,1]}A^*_{q,[n',m']}
\end{aligned} \tag{4.78}
$$

$$
I_p = X^{(I)}_p(refLSOPS) + \sum_{q,n',m'} \mathrm{Re}\left(X^{(Y)}_{p;q,[n',m']}(refLSOPS) \cdot P^{-n'}_{[1,0]}P^{-m'}_{[0,1]}A_{q,[n',m']}\right) \tag{4.79}
$$

$$
V_p = X^{(V)}_p(refLSOPS) + \sum_{q,n',m'} \mathrm{Re}\left(X^{(Z)}_{p;q,[n',m']}(refLSOPS) \cdot P^{-n'}_{[1,0]}P^{-m'}_{[0,1]}A_{q,[n',m']}\right) \tag{4.80}
$$

The X-parameter sensitivity functions, $X^{(S)}$ and $X^{(T)}$, can be identified, as before, as the partial derivatives of the (nonanalytic single-sided) spectral maps (4.76) with respect to the phase-normalized incident wave phasors. Similar interpretations follow for the $X^{(Y)}_{p;q,[n',m']}$ and $X^{(Z)}_{p;q,[n',m']}$ terms.

Examples

A two-tone multi-port model (with no explicit DCS dependence), that describes a mixer or amplifier can be defined by (4.81).

$$
\begin{aligned}
B_{p,[n\ m]} = {}& X^{(F)}{}_{p,[n\ m]}\left(\left|A_{p_1,[1,0]}\right|, \left|A_{p_2,[0,1]}\right|\right) P_{[1,0]}^{n} P_{[0,1]}^{m} \\
&+ \sum_{q;j,k} X^{(S)}{}_{p,[n,m];q,[j,k]}\left(\left|A_{p_1,[1,0]}\right|, \left|A_{p_2,[0,1]}\right|\right) P_{[1,0]}^{n-j} P_{[0,1]}^{m-k} A_{q,[j,k]} \\
&+ \sum_{q;j,k} X^{(T)}{}_{p,[n,m];q,[j,k]}\left(\left|A_{p_1,[1,0]}\right|, \left|A_{p_2,[0,1]}\right|\right) P_{[1,0]}^{n+j} P_{[0,1]}^{m+k} A^{*}{}_{q,[j,k]}
\end{aligned}
\tag{4.81}
$$

For $p_1 = 1$ and $p_2 = 2$, this example describes a mixer with the RF port 1 and LO at port 2. The IF corresponds to port 3. The *RefLSOPS* is specified only by the magnitude of the RF and LO signals (assumed incommensurate).

For both $p_1 = 1$ and $p_2 = 1$, (4.81) describes a PA with a two-tone stimulus at the input port (port 1). In this case, the port index, p, of the scattered B-waves, goes from 1 to 2. The *RefLSOPS* is specified by the magnitudes of the complex amplitudes of the two input tones.

4.9.7   When Intermodulation Frequencies Are Negative

X-parameters operate in the frequency domain, mapping the spectrum of the input stimulus to the spectrum of the DUT response. The input and output signals, considered in the time domain, are real. That means the spectra in the frequency domain are double-sided with conjugate symmetry, hence contain redundant information by a factor of two. For efficiency, only half the spectra needs to be considered in the equations and only those X-parameters mapping the chosen spectral components need to be stored in the files. In many applications, it is typical that only positive frequency and DC components are considered. For X-parameters, the frequencies of the signals are determined by the indices, n and m, and by the values of the fundamental frequencies, f_i, through (4.69). If the frequencies, f_i, are swept, the frequency of the intermodulation term corresponding to a particular set of n and m can become negative. Specifically, for the choice $f_1 = 3\ GHz$ and for f_2 swept from 2.9 GHz to 3.1 GHz in two steps, the intermodulation term [1, −1] will correspond first to 100MHz and then to −100MHz. It is much more convenient to have a consistent indexing scheme independent of the numerical values of the f_i. Therefore, the convention is adopted such that the spectral components considered correspond to those sets of n and m where the first non-zero value in the set of these integers is positive. For any two index vectors, [n, m] and [−n,−m], each of which corresponds to the same $|f|$, one will be consistent with this convention and the other will be excluded. If the frequency corresponding to the convention on indices is negative, the corresponding positive frequency is considered for the signal and the complex conjugate of the corresponding phasor is taken.

4.9.7.1   Examples

Wave variables and $X^{(F)}$ functions with intermodulation indices [2, 3] are selected for processing and storage whereas those indexed by [−3, 5] are not, because in the latter case, the first non-zero integer is negative. So in this case, the corresponding inter-modulation vector [3, −5] is considered instead. If $3f_1 - 5f_2$ is positive, the complex value of the phasor at that frequency is considered. If $3f_1 - 5f_2$ is negative, the complex conjugate of the corresponding phasor is considered and its contribution attributed to the spectral component with the corresponding positive frequency.

X-parameter functions labeled by two sets of port and intermodulation vectors have rules to remove redundancy that can be deduced in a similar way. The details are not described here, but the following is given as an example.

The parameter $X^{(S)}_{p,[2,3];q,[0,2]}$ is an allowed (and saved) quantity, but the parameter $X^{(S)}_{p,[-3,2];q,[0,2]}$ is not allowed.

4.9.8   X-Parameter Models of Mixers

Consider the following example of an incommensurate two-tone X-parameter model for a commercial double-balanced mixer [Mini-Circuits LAVI-22VH+ (TB-433)]. A schematic of the component is given in Figure 4.18.

The RF signal is applied at port 1, the LO at port 2. The conversion gain of a mixer with IF matched to a perfect 50 Ω termination is given by the simple X-parameter expression (4.82). Here we assume the IF signal of interest is at the difference of the RF and LO frequency (where LO frequency is greater than the RF frequency).

$$Conv.\ Gain = \frac{\left|X^{(F)}_{3,[1,-1]}\right|}{\left|A_{1,[1,0]}\right|} \tag{4.82}$$

The X-parameters of the Mini-Circuits mixer were measured and the conversion gain computed from (4.82) and compared with values from the manufacturer's datasheet for

Electrical schematic

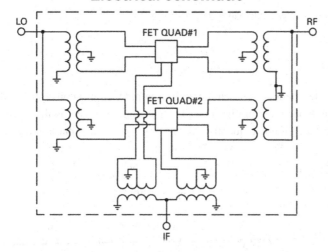

Figure 4.18 Mini-Circuits double-balanced mixer for 3-port X-parameter model.

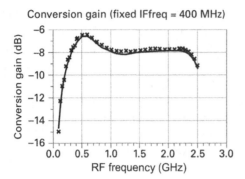

Figure 4.19 Comparison of measurement-based X-parameter mixer model and manufacturer's datasheet for conversion gain (in dB) versus frequency.

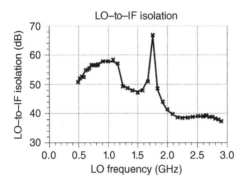

Figure 4.20 Comparison of measurement-based X-parameter mixer model and manufacturer's datasheet for LO-IF isolation (in dB) at the IF port versus frequency.

the same part number in Figure 4.19. The LO-IF isolation term, at the IF port, can be computed from a similar expression given in (4.83).

$$LO - IF\ iso = \frac{|A_{2,[0,1]}|}{|X_{3,[0,1]}^{(F)}|} \tag{4.83}$$

The X-parameter model has much more information than just conventional data sheet information as demonstrated in Figure 4.20. An example is given by the phase of the IF signal, which is a well-defined quantity when normalized to the phase of the RF and LO signals using (4.78). For the Mini-Circuits mixer, this information is shown in Figure 4.21. The expression to compute it is given simply by (4.84).

$$IF\ Phase = Phase\left[X_{3,[1,-1]}^{(F)}\right] \tag{4.84}$$

Other mixer performance measures, including more exotic quantities such as the second harmonic of the LO at the IF port, $X_{3,[0,2]}^{(F)}$, are listed in Table 4.1 and discussed in [31].

Table 4.1 Mixer behavior in terms of X-parameters. Large RF at port 1, large LO at port 2, IF output at port 3.

Mixer terminology	X-parameter expressions or terms
Conversion gain (difference frequency)	$\dfrac{\left\|X^{(F)}_{3,[1,-1]}\right\|}{\left\|A_{1,[1,0]}\right\|}$
LO leakage at IF port for perfectly matched IF	$X^{(F)}_{3,[0,1]}$
Mismatch terms at IF frequency and IF port	$X^{(S)}_{3,[1,-1];3,[1,-1]} \qquad X^{(T)}_{3,[1,-1];3,[1,-1]}$
IF-to-LO isolation terms	$X^{(S)}_{2,[1,-1];3,[1,-1]} \qquad X^{(T)}_{2,[1,-1];3,[1,-1]}$
RF to LO isolation terms for the second harmonic of LO	$X^{(S)}_{3,[3,2];1,[0,2]} \qquad X^{(T)}_{3,[3,2];1,[0,2]}$

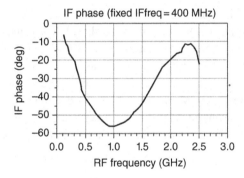

Figure 4.21 Phase of IF signal versus frequency as contained in the X-parameter mixer model.

4.10   Memory

4.10.1   Introduction to Memory

X-parameters were introduced in this chapter as frequency domain nonlinear mappings. The X-parameter functions are defined on the complex amplitudes of the incident CW waves, returning the complex amplitudes of the scattered CW waves. For fixed power, bias, and load conditions these input and output amplitudes are constant (in time) complex numbers. As such, X-parameters were defined only for steady-state conditions.[14]

Real applications deal with stimulus signals that vary in time, and the corresponding time-varying responses of DUTs to such stimuli. In particular, signals used in wireless communications can have independent time-varying values for the magnitudes and the phases associated with each complex amplitude of the multiple carriers involved. These

---

[14] We are not including here the option of acquiring the X-parameter within an RF pulse but nevertheless interpreting them with respect to a steady-state model of the nonlinear DUT.

time-dependent modulations contain the information associated with the signal. Much of communication systems design is devoted to trying to preserve the information content of these signals while efficiently amplifying, transmitting and then demodulating them at the receiver to recover the information.

Section 4.10.2 introduces the notion of dynamics into the signals and into the mappings between input and output stimuli that define the DUT as a nonlinear dynamic system. The class of signals considered is introduced in the envelope domain. Quasi-static applications of static X-parameters are introduced as methods for estimating the DUT's response to modulated signals and limitations are discussed. Examples are provided of phenomena that require the abandonment of the static nonlinear system description. This introduces the notion of memory, where the response of the system depends not just on the instantaneous value of the input signal, but also on the history of the input signal. The concept of modulation-induced baseband memory is articulated.

Section 4.11.5 is devoted to introducing dynamic X-parameters. This is a fundamental extension of X-parameters to modulation-induced baseband memory effects. An outline of the theory is presented starting from a few basic assumptions that can be independently validated. Applications to actual DUT complex stimulus-response characterization and modeling for practical wideband modulated signals are presented to validate the approach. More information on dynamic X-parameters can be found in [33–35]. Simplifications are described in [36–38], and another somewhat similar approach is described in [39]. Another approach is presented in [40].

4.10.2   Dynamics

4.10.2.1   Modulated Signals in the Envelope Domain

Modern communication signals can be represented efficiently by a sum of high frequency carriers with complex amplitudes that vary in time. We've seen this in Chapter 2. These signals can be described mathematically by (4.85). Here $A_n(t)$ is the time-varying complex amplitude associated with the nth carrier having frequency ω_n. The frequencies may or may not be commensurate. The signal $A(t)$ can have components at each port, but port indices are suppressed for simplicity.

The scattered waves from a nonlinear DUT stimulated with signal $A(t)$ can also be assumed to take a form equivalent to that of (4.85). This is shown in (4.86). Here $\{\omega_m\}$ constitutes the set of all intermodulation frequencies produced by the DUT in response to the signal. There can be more spectral components generated by the nonlinear DUT, and therefore more time-varying envelopes, $B_m(t)$, than were present in the incident signals.

Complex envelopes are represented by their magnitudes and phases, each of which varies in time. This is detailed in (4.87), for the incident envelopes, with a similar expression, (4.88), holding for the envelope of each scattered wave.

The dynamic modeling problem is then how to define the mappings from the $\{A_n(t)\}$ of the incident waves to the $\{B_m(t)\}$ of the scattered waves.

$$A(t) = \text{Re}\left\{ \sum_n A_n(t) e^{j\omega_n(t)} \right\} \tag{4.85}$$

$$B(t) = \text{Re}\left\{ \sum_m B_m(t) e^{j\omega_m(t)} \right\} \tag{4.86}$$

$$A_n(t) = |A_n(t)| e^{j\phi(A_n(t))} \tag{4.87}$$

$$B_m(t) = |B_m(t)| e^{j\phi(B_m(t))} \tag{4.88}$$

4.10.2.2 Quasi-Static X-Parameter Evaluation in the Envelope Domain

In this section, the case is considered where a static X-parameter model can be evaluated in the envelope domain to calculate the DUT's response to a modulated input signal, under the *quasi-static approximation*. This approximation asserts that at any time, t, the actual output time-dependent envelopes can be approximated by applying the static X-parameter mappings to the values of the input amplitudes at that same instant of time.

To see how this works, we start with an X-parameter model that was identified from measurements presumed to have been taken under steady-state conditions. For example, consider a simple one-tone X-parameter steady-state mapping defined by (4.89). All mismatch terms are neglected, for simplicity, and therefore only the $X_{p,k}^{(F)}$ terms are considered. The time variation of the input envelope is specified either by an explicit time-dependent I-Q signal or with one of several modulated source components available in the simulator. The settings for the envelope simulation are chosen to allow adequate time-sampling of the envelope corresponding to the input signal and to allow a suitably long period for the simulation.

In the quasi-static approximation, the DUT response, at time t, is computed to be the value of the static X-parameter mapping, as described in (4.89), applied to the value of the input envelope at the same time instant, t. This results in (4.90). This procedure produces time-varying envelopes for each of the harmonics, indexed by integer k, produced by the DUT in response to the modulated amplitude around the carrier at the fundamental frequency.

$$B_{p,k} = X_{p,k}^{(F)}(|A_{1,1}|) e^{j\phi(A_{1,1})} \tag{4.89}$$

$$B_{p,k}(t) = X_{p,k}^{(F)}(|A_{1,1}(t)|) e^{j\phi(A_{1,1}(t))} \tag{4.90}$$

The quasi-static approximation is an extrapolation from steady-state conditions to a dynamic (time-varying) condition. It is certainly valid for sufficiently slowly varying $A_n(t)$, since it reduces to the static mapping as the instantaneous envelope amplitudes become constant. It is an accurate approximation to the actual time-dependent response provided that any underlying system dynamics are sufficiently fast that at any instant of time the DUT is nearly in its steady-state condition determined by the value of the input amplitude at that same time. That is, the system adiabatically tracks the input from one

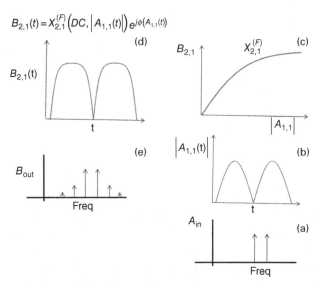

$$B_{2,1}(t) = X_{2,1}^{(F)}\left(DC, \left|A_{1,1}(t)\right|\right) e^{j\phi(A_{1,1}(t))}$$

Figure 4.22 Evaluation of a quasi-static X-parameter model in the envelope domain. (a) The two-tone input in the frequency domain. (b) Equivalent time-varying amplitude around a single CW tone. (c) Evaluation of the static X-parameter map at each time-point over the variation of the envelope. (d) The distorted time-varying output amplitude. (e) Spectrum of the output envelope in the frequency domain.

steady-state to the next, parameterized by the time. The quasi-static approximation breaks down as the signal modulation rate increases and becomes comparable to or faster than timescales for which other dynamical effects become observable. In particular, electro-thermal effects, bias-line interactions, and other phenomena come into play for which a more elaborate dynamical description of the DUT is required.

The way in which the quasi-static approximation is used by the simulator to compute the response to modulated signals is depicted in Figure 4.22 for the simple example of a sinusoidally amplitude modulated signal. The two-tone input (a) is represented in the envelope domain as a time-dependent amplitude (b) around a single carrier at a frequency halfway between the two CW tones. At each time sample, the corresponding value of $A_{1,1}$ is used to evaluate the static X-parameter map shown in (c) using (4.90). This evaluation produces a distorted output envelope waveform as shown in (d). Transforming the output time-varying envelope back into the frequency domain (e) yields the intermodulation spectrum produced by the quasi-static model corresponding to the original modulated input signal of (a).

Experimental validation of the described quasi-static approach is provided by comparing amplitude values of the intermodulation products as calculated by Fourier decomposing the time-varying output envelope computed using (4.90) with actual measured values, and this for several different input power levels. Figure 4.23 shows the computed output power, per tone, as a function of the input power, per tone (solid line), together with actual measured data acquired from the example 2-tone experiments (dots).

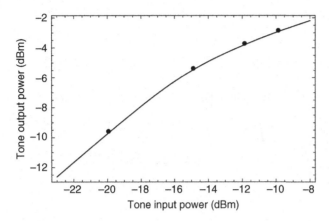

Figure 4.23 Measured and calculated tone output power using quasi-static X-parameter model.

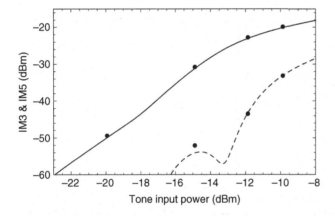

Figure 4.24 Measured and calculated IM3 and IM5 using quasi-static X-parameter model.

Figure 4.24 shows the calculated values of IM3 and IM5 respectively, together with the measured data points. The measured and calculated values correspond very well, thereby proving the validity of the quasi-static approach for this case.

There is an important limitation of using the quasi-static X-parameter map (4.90) for intermodulation distortion analysis. Since (4.90) has no explicit dependence on frequency, it follows that the intermodulation spectrum simulated this way does not depend on the spacing of the two tones [41]. The quasi-static approximation is valid in the limit of slow, or narrow-band modulation. But for more widely separated tone spacing, its validity decreases.

4.10.3   ACPR Estimations by Quasi-Static Approach

In a similar manner to the case of the sinusoidal amplitude modulation considered above, it is possible to estimate other nonlinear figures of merit (FOM), such as ACPR

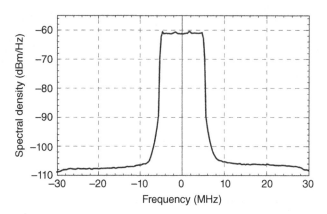

Figure 4.25 Input signal spectrum.

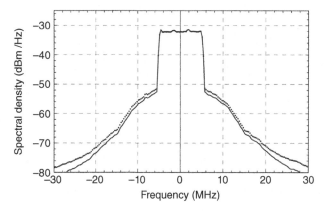

Figure 4.26 Output spectra: static X-parameter model (. . .) & measurements (____)

and others common to digital communications circuits, from a simple static X-parameter model in an envelope analysis. For digitally modulated signals, it is easy to use built-in modulated sources available in the simulator. The simulator simply evaluates the static X-parameter model at each sampled time according to (4.90). This is illustrated using an actual device, namely a 25W GaN MMIC power amplifier (CREE CMPA2560025F). The device is first characterized by a static X-parameter model. This is equivalent to measuring the AM-AM AM-PM characteristic of the device. Next an LTE input signal with a 2.5 GHz carrier frequency is applied to the device. The spectrum of the input signal is shown in Figure 4.25.

The output signal is predicted by applying the static X-parameter model to the measured input envelope. Next, the spectra of the modeled output signal is compared to the actual measured output spectrum. The result is shown in Figure 4.26. One notes significant spectral regrowth in both the measured and the modeled output spectrum.

The relevant FOM is computed from the simulated output spectrum according to the specific protocol appropriate to the modulation format. Examples of quasi-statically

Table 4.2 ACPR estimated from static X-parameters vs. independent measurements

	Output power	Lower sideband ACPR	Upper sideband ACPR
X-parameter model	37.9 dBm	18.6 dB	18.7 dB
Measurements	38.0 dBm	19.0 dB	19.1 dB

estimated ACPR from X-parameters for a real amplifier and independent experimental validation are provide in Table 4.2.

We note that there is a good correspondence between the measurements and the values derived from the static X-parameter model.

4.10.4 Limitations of Quasi-Static Approach [41]

It can be shown that the quasi-static evaluation of a static X-parameter model will produce an intermodulation spectrum with identical levels for upper and lower sidebands from a two-tone input signal. Moreover, the simulated levels are independent of the modulation rate – the frequency separation, Δf, between the two incident tones. That is, there is no "bandwidth dependence" to the distortion spectrum. In the envelope domain, it can be shown that the time-varying output envelopes are symmetric with respect to their peaks.

In the limit of slowly varying input envelopes, (4.90) reduces to evaluating a set of independent steady-state mappings at different power levels. That is, (4.90) reduces to (4.89), provided the latter was characterized over the full range of amplitudes covered by the time-varying input signal. Simulations for modulated signals become exact as the modulation rate, or signal bandwidth (BW), approaches zero (the narrowband limit).

For the two-tone case, it is possible to measure true steady-state X-parameters as functions of an LSOP that depends on both large input tones, using the approach discussed in 4.9.3. Two-tone X-parameters provide exact intermodulation spectra, in magnitude, phase, and their dependence on the frequency separation of the tones.

It should not be surprising that full information about the DUT response to two steady-state tones cannot be obtained from measurements taken with an LSOP set only by a single incident CW wave. It is clear there is more information in a set of two-tone measurements than a single tone measurement.

At high modulation bandwidths, the actual time-dependent scattered waves are no longer accurately computable from (4.90). That is, the quasi-static approximation of going from (4.90) to (4.89) becomes less valid. Any FOMs, such as adjacent channel power ratio (ACPR), derived from the scattered waves simulated under the quasi-static approximation, will therefore not be in complete quantitative agreement with the actual performance characteristics of the DUT.

4.10.5 Advantages of Quasi-Static X-Parameters for Digital Modulation

Key nonlinear FOMs are usually scalars (numbers). For example, IP3, ACPR, and EVM are numbers, typically computed from measurements made with a spectrum analyzer.

The FOMs of particular components, such as individual stages of a multistage amplifier, cannot generally be used to infer the overall FOM for the composite system. That is, just knowing the FOMs of individual parts is not enough to design a nonlinear RF system for lowest ACPR.

X-parameters, on the other hand, enable at least a quantitative estimate of any nonlinear FOM for the full system just from knowledge of the X-parameters of the constitutive components. The X-parameters of the composite system follow from the nonlinear algebraic composition of the component X-parameters according to the general treatment presented earlier in this chapter. The quasi-static approximation can be used easily in the envelope domain to evaluate the overall system response to modulated signals of various kinds.

Improved estimates of DUT behavior in response to wideband modulated signals requires a more careful treatment of dynamical behavior that begins in section 4.11.5.

4.10.6   Manifestations of Memory

Before defining memory and dealing with it precisely, the topic is introduced with a set of frequently observed manifestations, or *signatures*, of memory. These can be considered as evidence of underlying dynamical interactions that cause the DUT to behave differently from quasi-static inferences. None of these manifestations of memory is sufficient, on their own, to "characterize" the dynamics in a complete way, but rather they are evidence of the existence of dynamics for which a general treatment is required.

Real amplifiers, for the case of rapidly varying input envelopes, exhibit behaviors more complicated than those predicted by quasi-static techniques. For example, the intermodulation spectrum of a power amplifier in response to a large-amplitude two-tone excitation will show asymmetric upper and lower levels, with levels depending strongly on bandwidth (frequency separation of the tones). These effects are induced by nonlinear behavior and are especially visible under high levels of compression. Some simulated results are shown in Figure 4.27.

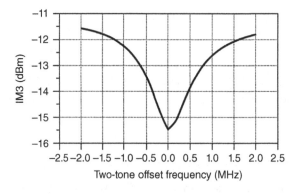

Figure 4.27 Frequency-dependent and asymmetric intermodulation spectrum.

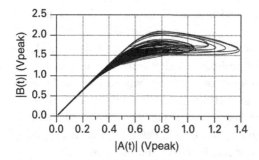

Figure 4.28 Multivalued instantaneous AM-AM amplifier characteristics

For this simulation, a constant input amplitude of 6 dBm per tone is chosen, whereby the two-tone frequency spacing is swept from nearly 0 Hz up to 4 MHz. The carrier frequency for the experiment is fixed at 2.6 GHz. The device is an RFIC amplifier whereby the nonlinear memory effects are caused by the presence of a 6 nH inductor in series with the DC power supply. Note that the X-axis scale indicates the "offset frequency" of each of the two tones, thereby corresponding to half the tone spacing. The power at the output, per tone, is about 9 dBm. The magnitude of intermodulation distortion depends significantly on the separation frequency of the two excitation tones. Also evident is that the lower frequency sideband, at the maximum separation frequency, is about 1dB larger than the upper sideband.

Another manifestation of memory is that the instantaneous AM-AM characteristic exhibits *multivaluedness*. These effects are illustrated by performing a simulation whereby one uses bandwidth limited noise as the input signal. The instantaneous output envelope amplitude $|B(t)|$ versus the instantaneous input envelope amplitude $|A(t)|$ is depicted in Figure 4.28.

It is evident that the output power is not uniquely determined by the input power at the same time instant. Rather, the output depends on the history of the input signal.

4.11 Causes of Memory

Primary causes of memory effects include dynamic self-heating, bias modulation effects, charge storage (junction capacitances) and transit time effects in nonlinear devices (transistors and diodes), and trapping effects in some semiconductor devices (e.g., drain-lag in GaN FETs).

4.11.1 Self-Heating

Self-heating involves the self-consistent coupling of electrical power dissipation in the device (transistor or amplifier) to a temperature change that in turn modulates the gain and therefore changes the output waveforms. Typically, self-heating in a transistor is modeled by two coupled equivalent circuits, one electrical and one thermal. The electrical circuit has current sources and capacitors, the values of which depend on voltages in the electric circuit and also the dynamic junction temperature, T_j of the thermal circuit. The thermal circuit calculates the junction temperature, T_j, as a response to electrical power dissipation computed in terms of the voltages and currents of the electrical circuit, and thermal resistance and thermal capacitance parameters in the thermal circuit [40]. More on self-heating models will be presented in Chapter 6.

A hallmark of dynamic-thermal effects is that they vary at timescales typically many orders of magnitudes slower than those of the RF signals. As the modulation rate becomes comparable to thermal relaxation timescales, significant dynamical effects from the mutual electrical and thermal coupling become observable.

4.11.2   Bias Modulation

Bias modulation occurs when a nonlinear device produces energy at baseband frequencies by intermodulation of higher frequency RF components. The low-frequency signal components thus produced can pass through the bias lines, modulating the voltages and currents, causing them to vary in time, which in turn modulates the RF scattering properties of the DUT. Bias modulation is explained in detail in Section 6.5.2 of [41], along with applications of a quasi-static approach, with feedback, based on static X-parameters measured at many fixed DC operating voltages.

4.11.3   Importance of Memory

Memory effects make it much more difficult to quantify nonlinearity, since the standard figures of merit such as IM3 and ACPR, depend sensitively on the modulation bandwidth and format. In particular, these FOMs cannot be *a priori* calculated just from knowledge of static X-parameters (e.g. AM-AM and AM-PM characteristics). It is therefore more difficult to design circuits and systems trading off efficiency for linearity with nonlinear components demonstrating significant memory effects.

Efficiency is of such paramount importance that the power amplifier market is ready to sacrifice open-loop linearity, preferring to externally linearize these nonlinear devices, primarily with digital predistortion (DPD) and other techniques. Predistorting components exhibiting strong memory effects is one of the major problems facing the PA industry today. Control, elimination, or (ultimately) exploitation of memory effects would enable a dramatic improvement in performance from existing technologies. In all these cases, a good model of memory is a key component of a solution.

4.11.4   Modulation-Induced Baseband Memory and Carrier Memory

Two distinct kinds of memory effects are significant for power amplifiers. They will be referred to as *carrier memory* effects and *modulation-induced baseband memory* effects.

Carrier memory is caused by dynamical effects that physically occur at the carrier frequency, typically at RF and microwave frequencies. Group velocity effects, and gain variation with frequency fall into this category. Simply sweeping the frequency of a single RF tone across the band of interest for an amplifier and noting the frequency response indicates RF carrier memory. Such effects have to be taken into account when one wants to construct a model that remains accurate across a bandwidth that is significant when compared to the RF carrier frequency. Note that this kind of memory does not require any nonlinear effects. It shows up for a linear filter as well as for an amplifier under small signal operating conditions if there is a significant variation of the scattering parameters across the modulation bandwidth. For this reason, this kind of memory is often also called *linear* memory. It is sometimes called *short-term* memory, referring to the fact that the timescale of this kind of memory effect is typically in the nanosecond range, comparable to the period of the RF carrier frequency. When a high-Q resonating filter structure is involved, however, carrier memory can introduce transients that can last a long time. Under such conditions the terminology *short-term memory* is misleading and, therefore, is not used in this treatment. We will not discuss carrier memory any further here.

Modulation-induced baseband memory is of a different nature. It is caused by dynamical effects that physically happen in the baseband, at timescales many orders of magnitude slower than the RF stimuli, like microseconds, milliseconds or even seconds. Examples include time-variation of the bias settings, dynamic self-heating, and trapping effects. Although the physical effects happen relatively slowly, their influence shows up at the high-frequency carrier through the process of modulation. Dynamic temperature variations and dynamic bias settings may slowly modulate the compression and AM-PM characteristics of the amplifier. Such modulation-induced baseband memory effects can become significant even at relatively small modulation bandwidths, like for example a few kHz or MHz. A general way to treat such memory effects in the framework of X-parameters is described in the Section 4.11.5. Modulation-induced baseband memory is sometimes called *long-term memory*, complementary to the term *short-term memory*. As explained above, this terminology is misleading as some modulation-induced baseband memory effects may have shorter settling times than carrier memory effects.

4.11.5 Dynamic X-Parameters

4.11.5.1 Overview

The dynamic X-parameter method was introduced in [33] to extend X-parameters to include modulation-induced baseband memory effects. Starting from a few basic assumptions that can be validated by independent experiments, a full theory, rigorously derived, was developed. The approach results in a new term subsuming all long-term memory effects that simply adds to the existing static X-parameter expressions as developed previously. The core of the new term is a "memory kernel" function that is uniquely identified from a new type of complex envelope transient measurements that can be measured on an NVNA or vector signal analyzer (VSA). The approach is extremely powerful in that within the original assumptions the resulting model is valid for any type of modulated signal, independent of format, power level, or details of the probability density function (PDF) defining the statistics of the signal. That is, the resulting behavioral model is highly transportable [42]. The dynamic X-parameter method generically abstracts the intrinsic dynamics and nonlinearities of the DUT. It can therefore be directly used to accurately predict the DUT response to a wide variety of modulated signals, including those with high peak-to-average ratios (PARs) and wide modulation bandwidths, without the need to re-extract the model for different formats or average power conditions. Independent experimental validation of the model, applied to a real power amplifier subjected to very different types of modulated signals confirms the theory. The model and its identification has a compelling and intuitive conceptual interpretation as well.

4.11.5.2   Assumptions

There are only a very few basic assumptions behind the dynamic X-parameter approach. The first is that there are some unknown, or hidden, dynamical variables, in addition to the known incident waves, on which the scattered waves (and output bias conditions) also depend. The values of these hidden variables, if known, would define, together with the incident signals, a unique map from this augmented set of inputs to the responses of the system. The second basic assumption is that these unknown variables depart from their quasi-statically mapped steady-state value only slightly when the system is modulated. That is, the system is assumed to be always close to its "instant-aneous" steady-state value. The small departure of the hidden variables from their quasi-statically mapped steady-state values is used to linearize the mapping around the quasi-static response. A third assumption is added about how the nonlinear dependence of the device on the incident waves, at all prior times, excites the dynamics of the hidden variables at the present time. The technical details presented in [33] and [41] are easy to follow, but we only report the results here.

Considering only the fundamental envelope at the output port, denoted here simply by $B(t)$, the dynamic X-parameter model equation is given by the compact and simple form (4.91). Here the DC stimulus and response terms are neglected and there is considered to be only one large time-dependent incident wave envelope, $A(t)$, modulating the DUT.

$$B(t) = \left(X^{(F)}(|A(t)|) + \int_0^\infty G(|A(t)|, |A(t-\tau)|, \tau)d\tau \right) e^{j\phi(A(t))} \qquad (4.91)$$

The term in the integral is a complex-valued function of three real variables, designated the (3D) *memory kernel*.

4.11.5.3   Discussion

The memory kernel, $G(x,y,t)$, depends nonlinearly on the present instantaneous amplitude of the input signal envelope, the past value of the input signal envelope, and explicitly on the difference between the present and past input envelope times. This can be seen by examining the integrand in (4.91).

Remarkably, the memory kernel can be identified uniquely from a specific set of nonlinear time-dependent envelope-domain measurements. Once the memory kernel is identified, it can be substituted into (4.91) and the contributions to the scattered waves from all hidden variables can be accounted for, no matter how many variables are assumed to be actually present. There is no need to actually identify any of them explicitly.

4.11.5.4 Identification of the Memory Kernel

The idea is that stepped RF envelope measurements can be made, from each steady state condition to all other steady state conditions to cause each hidden variable to depart from its initial steady-state value and relax, over time, to its final steady state value. This transient envelope behavior is quite analogous to the simulator methods for modulated signals considered in Chapter 2.

In general, the response to such steps between steady-state envelope values depends on both the initial state and the final state of the transition. Schematic examples of three such envelope transitions are given in Figure 4.29, together with the time-dependent response of the scattered wave, $B(t)$.

In fact, the memory kernel is related to the time-derivative of these step responses, parameterized by the initial and final steady-state values of the incident waves. That is, the kernel is a type of nonlinear impulse response.

Specifically, we have the identification of the memory kernel, $G(A_2, A_1, t)$, in terms of $B^{step}_{A_1 \to A_2}(t)$, the time-dependent response to a step in input amplitude from A_1 to A_2 at $t = 0$, given by (4.92). The memory kernel is plotted for one of the stepped envelope transitions and transient envelope responses in Figure 4.30.

$$G(A_2, A_1, t) = -\frac{d}{dt} B^{step}_{A_1 \to A_2}(t) \cdot e^{-j\phi(A_2)} \qquad (4.92)$$

4.11.5.5 Validation of Dynamic Memory Model

The above methodology was applied to a Mini-Circuits ZFL11AD+ amplifier. The carrier frequency used was 1.75 GHz. Fifteen different values of the input amplitudes

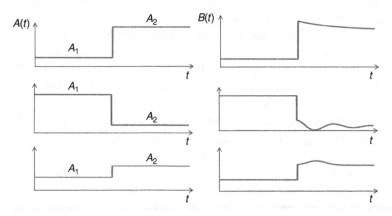

Figure 4.29 Envelope step inputs and responses. The step excitations occur at $t = 0$.

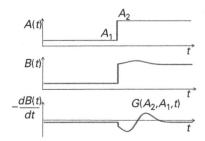

Figure 4.30 Identification of memory kernel from step responses.

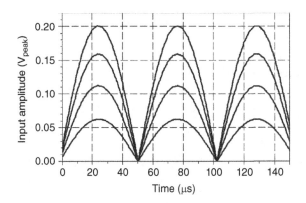

Figure 4.31 Dynamic X-parameter model validation: Measured input envelope amplitudes for a two-tone frequency difference of 19.2 kHz.

from $0.01V_{peak}$ to $0.2V_{peak}$ were selected for the set of step envelope measurements. The data used to derive the model is shown in [41].

The model is validated by comparing the actual measured DUT response to simulations using the model stimulated with the same signals in the circuit simulator. That is, the actual measured signals are brought into the Keysight ADS simulator to excite the model of the DUT for a more accurate comparison. Any difference between measured response and simulated response is therefore due entirely to the model, and not due to slight differences between the ideal and actual modulated stimuli. The modulated signals are applied using a Keysight ESG. The response measurements for validation are made on the Keysight NVNA. The measured response data is taken back into ADS to make the comparison with the simulated performance.

Figure 4.31 shows the amplitude of the input envelopes for two-tone experiments at a frequency spacing of 19.2 kHz and for 4 different power levels. In Figure 4.32 the measured and simulated output waveforms are shown. The output envelopes corresponding to higher power levels are clearly not symmetric about their peak values, despite the fact that the input amplitudes are symmetric. The simulated and measured output amplitudes agree very well. The memory model is therefore able to go far beyond the quasi-static application of the static X-parameter model discussed earlier,

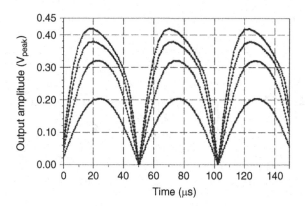

Figure 4.32 Measured and modeled output envelope amplitudes (measurement: solid, model: dots).

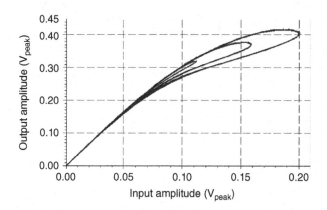

Figure 4.33 Measured dynamic compression characteristics at four different power levels.

where in particular, it was stated in 4.10.4 that the time-varying output waveforms predicted by a quasi-static model were symmetric about their peak. The improvement in accuracy shown in Figure 4.32 is purely attributable to the second term in (4.91) involving the memory kernel.

Figure 4.33 shows the measured amplitudes of the output envelopes versus the instantaneous input amplitude – the so-called dynamic compression characteristic. For the higher input power levels, one can clearly see the looping (hysteresis) of the characteristics. One concludes that the dynamic X-parameter model accurately models one of the key symptoms of memory, namely multiple-valued output versus input power characteristics, sometimes called *dynamic AM-AM distortion*.

The memory model can also predict the detailed dependence of DUT gain and intermodulation distortion characteristics on the modulation band-width. In Figure 4.34, the measured and simulated gains of the lower and upper signals are compared as a function of tone spacing over a wide range of frequencies. The sharp resonances and asymmetric values are well predicted.

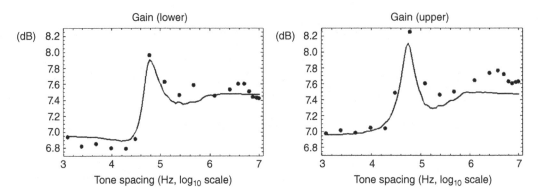

Figure 4.34 Gain versus tone-spacing: measured and simulated with dynamic X-parameters.

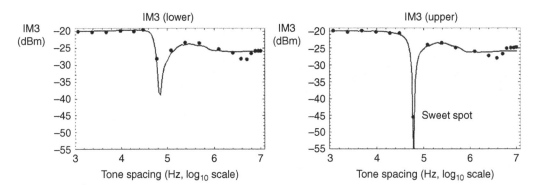

Figure 4.35 Magnitude of IM3 versus tone-spacing: measured and simulated with dynamic X-parameters

In Figures 4.35 and 4.36, a comparison is presented between the measured and simulated frequency dependence of the two third-order intermodulation sidebands. The model is able to very accurately predict both the magnitudes and the phases of the intermodulation products, including the sharp resonance. This "IMD sweet spot" – an operating condition where the device behaves more linearly than other nearby conditions, is clearly identified by the memory model.

Another demonstration of the sweet spot is evident in Figure 4.37, which shows measured (a) and simulated results (b) at two different offset frequencies of a two-tone stimulus-response measurement. The 60 kHz tone-spacing (dashed lines in both plots in the figure) correspond to a peak in the gain and minimum in the power of the third-order distortion product. The instantaneous gain, while still showing considerable hysteresis, is evidently more linear than the characteristics at 120 kHz tone spacing (the solid lines in both plots) that compresses more with incident power.

Another stringent test of the memory model is how well it can predict the measured DUT characteristics in response to a wideband digitally modulated signal. The same dynamic X-parameter model obtained for the Mini-Circuit amplifier was used to predict

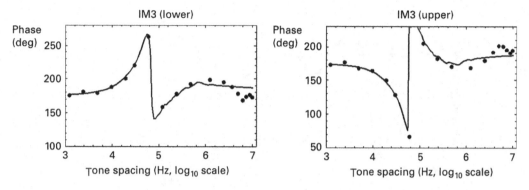

Figure 4.36 Phase of IM3 versus tone-spacing measured and simulated with dynamic X-parameters.

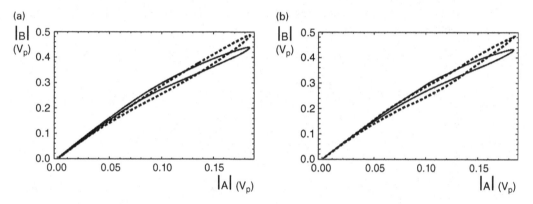

Figure 4.37 Dynamic compression at 60 kHz tone spacing (dashed trajectories) and 120 kHz tone spacing (solid trajectories). (a) Measured and (b) simulated from dynamic X-parameter model.

the modulated output envelope and spectral re-growth in response to a WCDMA signal. Comparisons of the dynamic X-parameter model simulation with measurements are shown in Figures 4.38 and 4.39. Also compared are the predictions of the static (CW) X-parameter model. The CDMA signal was generated and uploaded to a Keysight ESG, and the waveforms were measured on the NVNA in envelope mode.

4.11.5.6 Discussion

It should be noted that both the two-tone signal (sinusoidally modulated carrier) and the WCDMA signals are completely different types of signals compared to the complex step excitations used to identify the memory kernel. Nevertheless, the dynamic X-parameter model is able to very accurately reproduce the actual characteristics of the measured DUT responses to each signal. In the case of the two-tone signal, the model is able to resolve the narrow resonance in gain and distortion with frequency separation of the tones. For the digitally modulated signal, the model does an excellent job describing the DUT output time-dependent waveforms in the envelope domain and

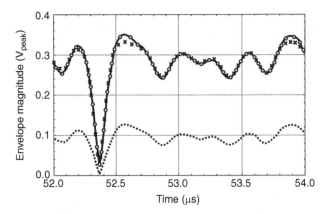

Figure 4.38 Validation of dynamic X-parameter and quasi-static X-parameter model simulations for CDMA input signal: waveforms (solid = measured, o = dynamic X-parameters, x = static X-parameters, ... = input signal)

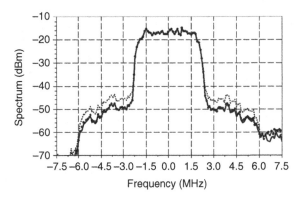

Figure 4.39 Validation of dynamic X-parameter and quasi-static X-parameter model simulations for CDMA input signal: Spectral regrowth (solid = measured, o = dynamic X-parameters, ... = static X-parameters)

spectral regrowth in the frequency domain. In both cases, the simulations with the memory model are in much better agreement with the detailed measurements than the predictions of the CW X-parameter model evaluated in the quasi-static approximation. This is a powerful validation of the claim of transportability of the memory model.

Dynamic X-parameter theory has also been used to *predistort* or otherwise *linearize* nonlinear components that also exhibit substantial memory effects. The reader is referred to [43] and [44] for a treatment.

4.12   Summary

This chapter presented a framework for large-signal steady-state behavioral modeling in the frequency domain based on time-invariant nonlinear spectral mappings. The foundations were presented and derived from fundamental principles. Several approaches, concrete examples, and distinct interpretations were provided to help convey an intuitive understanding of the various terms that arise in the formalism. Application examples for amplifiers and mixers were provided illustrating the power of the techniques. An explicit closed-form solution was derived for the problem of predicting the maximum delivered power to a load (assuming a perfectly matched source) based on a simplified form of the X-parameter model for a nearly matched power amplifier driven by a large input signal. The limiting case where the large-signal formalism reduces properly to the linear S-parameters of the previous chapter was presented. X-parameter applications in the quasi-static approximation were described and their applicability and limitations reviewed. Envelope dynamics were introduced and the phenomena and importance of memory effects discussed. Finally, an important extension beyond quasi-static theory was presented by introducing the "dynamic X-parameter" formalism for modulation-induced baseband memory effects. It was demonstrated that this generic approach can deal with a very wide range of nonlinear dynamic memory effects in real devices for important applications involving diverse types of wideband modulated signals.

4.13   Exercises

Exercise 4.1   A nonlinear spectral map defined by (4.93) is proposed to model a time-invariant DUT. Prove (4.93) is not time invariant.

$$B_{2,1} = G \cdot A_{1,1} - \gamma A_{1,1}^3 \tag{4.93}$$

Prove an alternative model, given by (4.94), defines a time-invariant map.

$$B_{2,1} = G \cdot A_{1,1} - \gamma |A_{1,1}|^2 A_{1,1} \tag{4.94}$$

Reduce (4.94) to the form given by (4.95) and find an explicit expression for $X_{2,1}(.)$.

$$B_{2,1} = X_{2,1}(|A_{1,1}|)P^1 \tag{4.95}$$

Exercise 4.2   Prove (4.11) are equivalent to the circuit laws. Hint, start by expressing KVL and KCL in the time domain, transform them to the frequency domain, and then to the wave domain.

Exercise 4.3   Derive the second equality of (4.14) using the time-invariance property.

Exercise 4.4   Show the solution of (4.47) is in fact (4.48). Hint, take the complex conjugate of equation (4.47). Argue that these two equations, considered as a 2×2 system for both A_2 and A_2^* at the optimum, are linearly independent. Using the explicit inverse of a 2×2 matrix, compute the formal solution for A_2^{opt}. Simplify the expression to obtain (4.48).

Exercise 4.5   Section 4.6.6 defined the $X^{(S)}$ and $X^{(T)}$ parameters in terms of upper and lower sideband responses to small perturbations tones injected at frequencies just greater than positive harmonics of the fundamental frequency. What happens when the perturbation term is applied at frequencies just smaller than the harmonics?

Exercise 4.6   Derive (4.78–4.80).

References

[1] W. Van Moer and L. Gomme, "NVNA versus LSNA: enemies or friends?" *IEEE Microw. Mag.*, vol. 11, no. 1, 2010, pp. 97–103.

[2] P. Blockley; D. Gunyan, and J. B. Scott, "Mixer-based, vector-corrected, vector signal/ network analyzer offering 300kHz-20GHz bandwidth and traceable phase response," *IEEE MTT-S International Microwave Symposium Digest, 2005*, June 2005, p. 4.

[3] www.keysight.com/find/NVNA

[4] www.keysight.com/find/X-parameters

[5] D. E. Root, J. Verspecht, J. Horn, and M. Marcu, *X-Parameters: Characterization, Modeling, and Design of Nonlinear RF and Microwave Components*, Cambridge University Press, 2013, (referred to below simply as *X-Parameters*), section 2.8.

[6] *X-Parameters*, chapter 3.

[7] *X-Parameters*, chapter 2.

[8] J. Horn, D.E. Root, D. Gunyan, and J. Xu, "Method and apparatus for determining a response of a DUT to a desired large signal, and for determining input tones required to produce a desired output," US Patent 7924026, April 12, 2011.

[9] R. Biernacki, M. Marcu, D. E. Root, "Circuit optimization with X-parameter models," *International Microwave Symposium Digest*, Honolulu, Hawaii, June 2017.

[10] J. Verspecht, M. V. Bossche, and F. Verbeyst, "Characterizing components under large signal excitation: defining sensible 'large signal S-parameters'?!" *49th ARFTG Conference Digest*, 1997, pp. 109–117.

[11] J. Verspecht and D. E. Root, "Poly-harmonic distortion modeling," *IEEE Microw. Mag.*, June 2006.

[12] *X-Parameters*, section 4.4.2.1.

[13] A. M. Pelaez-Perez, S. Woodington, M. Fernández-Barciela, P. J. Tasker, and J. I. Alonso, "Application of an NVNA-based system and load-independent-parameters in analytical circuit design assisted by an experimental search algorithm," *IEEE Trans. Microw. Theory Techn.*, vol. 61, no. 1, January 2013.

[14] D. E. Root, J. Verspecht, and J. Xu, "Closed-form solutions to large-signal PA problems: Wirtinger calculus applied to X-parameter expressions," *2017 IEEE European Microwave Integrated Circuits Conference*, Nuremberg, Germany, October 2017.

[15] *X-Parameters*, Appendix B.

[16] S. Woodington, R. Saini, D. Williams, J. Lees, J. Benedikt, P. J. Tasker, "Behavioral model analysis of active harmonic load-pull measurements," *IEEE MTT-S International Microwave Symposium Digest* , June 2010, pp. 1688–1691.

[17] *X-Parameters*, sections 5.2 and 5.3.

[18] J. Horn, D. E. Root, and G. Simpson, "GaN device modeling with X-parameters," *IEEE Compound Semiconductor Integrated Circuit Symposium* (CSICS), October 2010.

[19] *X-Parameters*, section 5.5.1.

[20] T. S. Nielsen, M. Dieudonné, C. Gillease, and D. E. Root, "Doherty power amplifier design in gallium nitride technology using a nonlinear vector network analyzer and X-parameters," *IEEE CSICS Digest*, October 2012.

[21] H. Qi, J. Benedikt, and P. J. Tasker, "Nonlinear data utilization from direct data lookup to behavioral modeling," *IEEE Trans. Microw. Theory Techn.*, vol. 57, no. 6. pp. 1425–1432, June 2009.

[22] P. J. Tasker and J. Benedikt, "Waveform inspired models and the harmonic balance Emulator," *IEEE Microw. Mag.*, April 2011, pp. 38–54.

[23] P. J. Tasker, "Practical waveform engineering," *IEEE Microw. Mag.*, vol. 10, issue: 7, 2009, pp. 65–76.

[24] M. Schetzen, *The Volterra and Wiener Theories of Nonlinear Systems*, New York: Wiley, 1989.

[25] F. Verbeyst and M. Vanden Bossche, "VIOMAP, the S-parameter equivalent for weakly nonlinear RF and microwave devices," *IEEE MTT-S International Microwave Symposium Digest*, May 1994, vol. 3, pp. 1369–1372.

[26] D. E. Root, J. Verspecht, D. Sharrit, J. Wood, and A. Cognata, "Broad-band, poly-harmonic distortion (phd) behavioral models from fast automated simulations and large-signal vectorial network measurements," *IEEE Trans. Microw. Theory Techn.*, vol. 53, no. 11, November 2005, pp. 3656–3664.

[27] D. E. Root, J. Horn, L. Betts, C. Gillease, J. Verspecht, "X-parameters, the new paradigm for measurement, modeling, and design of nonlinear rf and microwave components," *Microwave Engineering Europe*, pp. 16–21, December 2008.

[28] S. Maas, *Nonlinear Microwave and RF Circuits*, 2nd ed., London: Artech House, 2003.

[29] G. D. Vendelin, A. M. Pavio, U. L. Rohde, *Microwave Circuit Design using Linear and Nonlinear Techniques*, 2nd ed., Wiley, 2005.

[30] S. R. Mazumder, P. D. van der Puije, "Two-signal method of measuring the large-signal S-parameters of transistors," *IEEE Trans. Microw. Theory Techn.*, vol. 26, no. 6, pp. 417–420, June 1978.

[31] *X-Parameters*, section 5.7.

[32] S. Strogatz, *"Nonlinear Dynamics and Chaos: With Applications to Physics Biology, Chemistry, and Engineering,"* Perseus Press, 1994.

[33] J. Verspecht, J. Horn, L. Betts, D. Gunyan, R. Pollard, C. Gillease, and D. E. Root, "Extension of X-parameters to include long-term dynamic memory effects," *IEEE MTT-S International Microwave Symposium Digest, 2009.*, June 2009, pp. 741–744.

[34] J. Verspecht, J. Horn, and D. E. Root, "A simplified extension of X-parameters to describe memory effects for wideband modulated signals," *IEEE Microwave Measurements Conference (ARFTG)*, May 2010.

[35] *X-Parameters*, chapter 6.

[36] *X-Parameters*, Appendix D.

[37] R. Biernacki, C. Gillease, and J. Verspecht, "Memory effect enhancements for X-parameter models in ADS," *IEEE International Microwave Symposium, MicroApps*, 2014.

[38] A. Soury and E. Ngoya, "Handling long-term memory effects in X-parameter model," *IEEE International Microwave Symposium*, 2012.

[39] E. Ngoya and A. Soury, "Envelope Domain Methods for Behavioral Modeling," in chapter 3 of *Fundamentals of Nonlinear Behavioral Modeling for RF and Microwave Design*, Norwood, MA: Artech House, 2005.

[40] D. E. Root, D. Sharrit, and J. Verspecht, "Nonlinear behavioral models with memory" in *2006 IEEE International Microwave Symposium Workshop (WSL) on Memory Effects in Power Amplifiers*, June 2006

[41] *X-Parameters*, chapter 6.

[42] D. E. Root, J. Wood, and N. Tufillaro, "New techniques for nonlinear behavioral modeling of microwave/RF ICs from simulation and nonlinear microwave measurements," in *40th ACM/IEEE Design Automation Conference Proceedings*, Anaheim, CA, USA, June 2003, pp. 85–90.

[43] J. Verspecht, D. Root, T. Nielsen, "Digital predistortion method based on dynamic X-parameters," *82nd ARFTG Microwave Measurement Conference*, November 2013, pp.1, 6, 18–21.

[44] J. Verspecht, T. Nielsen, and D. Root, "Digital predistortion method based on dynamic X-parameters," *IEEE International Microwave Symposium Workshop (WMI-6)*, Honolulu, Hawaii, June 2017.

5 Linear Device Modeling

5.1 Introduction: Linear Equivalent Circuit Models of Transistors

This chapter introduces the major principles and methods of linear device modeling. Specifically, we present an introduction to models of transistors based on linear equivalent circuits. The equivalent circuit description attempts to associate the electrical model description with the physical structure of the device.

The models are linear in the sense we have described in detail in Chapters 1 and 3, specifically that the response of such a model to a superposition of stimuli is equal to the superposition of responses to the stimuli applied independently. As we have learned in Chapter 1, linearity implies that an RF sinusoidal signal applied at a particular frequency will produce a sinusoidal response at (only) the same frequency. We can then completely describe the linear model in terms of any one of several equivalent conventional small-signal frequency-domain descriptions, such as admittance, impedance, or scattering parameters.

Although the response of linear device models is proportional to the complex amplitudes (phasors) of the input stimuli, as with linear behavioral models, the explicit dependence on frequency of the response is *not* linear. That is, the response at frequency $2f$ is not twice the response at frequency f. Each linear circuit element contributes to the network description an admittance (or impedance, or S-parameter), a complex valued frequency-dependent number that defines the proportionality of complex-valued response to stimulus. Resistors contribute real-valued impedances independent of frequency, while ideal capacitors and inductors contribute admittances and impedances, respectively that are purely imaginary and proportional to frequency. Note that the capacitance, or the inductance, values themselves are independent of frequency. For a model composed exclusively of lumped elements, the model response is a rational function (i.e., it is given by the ratio of two polynomials) of frequency. The quantitative response also depends, of course, upon the numerical values of the elements (e.g., the resistances and capacitances).

This chapter also presents an introduction to parameter extraction. The parameters of the linear equivalent circuit model are the values (e.g., the resistance or capacitance) of each of the corresponding elements of the equivalent circuit. Parameter extraction is the process of determining the numerical values for these parameters, from measured or simulated data. The typical data used for linear RF and microwave device model parameter extraction are S-parameters. This is natural because the measurements are (ideally) linear, like the models. We will see, however, that even for purely linear microwave and RF device models, several distinct DC bias conditions must be applied. This is necessary to put the device in the appropriate operating condition for a useful linear model and to provide distinct states of device operation to help in the extraction of parasitic element values necessary to correctly determine all the device model parameters from data at the externally accessible measurement ports.

The modeling and parameter extraction processes are highly related. The full transistor model is constructed from linear electrical element building blocks (e.g., resistors and capacitors) by putting them together in a particular network – the equivalent circuit topology – in order to represent the overall device behavior. Parameter extraction is, in a rough sense, the reverse process, namely going from information at the device external terminals and using knowledge of the presumed topology to determine the numerical values of the individual electrical circuit parameters (ECPs), (e.g., the capacitance and resistance values of the primitive elements).

For this chapter we stay exclusively in the frequency domain. We will meet linear models again in Chapter 6 where we show how they arise, generally, as first-order approximations to nonlinear models in the small-signal limit. Recall we already met a particularly simple linear FET model derived from time-domain considerations in Chapter 3 (Section 3.12).

5.2   Linear Equivalent Circuit of a FET

A simple picture of a Field Effect Transistor (FET) is shown in Figure 5.1.

Metal electrodes, labelled Source, Gate, and Drain, are shown on the top of the figure, typical of a GaAs pHEMT device (but not drawn to scale). The source and drain electrodes make ohmic contacts with a highly doped layer of semiconductor. The gate metal makes a Schottky contact with a different semiconductor layer. The active semiconductor channel, through which most of the current flows, is represented by the lighter region below. The substrate, taken as semi-insulating, is shown as the bottom layer.

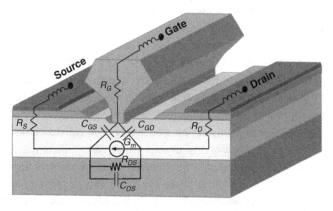

Figure 5.1 Simple FET structure with equivalent circuit elements superimposed.

The type of circuit elements and their arrangement in the equivalent circuit are consequences of our physical understanding of the device physics and its structure. The Schottky contact presents a barrier to (DC) current flow, so there is no resistor in the electrical path from the gate terminal to the channel in this simple idealization. The ohmic contacts permit current flow through the semiconductor, so there are resistances in the path from the source and drain terminals. The metal electrodes themselves are represented by inductors that affect the high-frequency RF and microwave signals. The Schottky contact sets up a built-in electric field at the gate metal-semiconductor interface that depletes charge carriers in the semiconductor depending also on the applied bias voltages at the terminals. This is modeled by parallel plate capacitors, labelled C_{GS} and C_{GD}. The current source element represents the main current flow in the channel and is responsible for the transistor action. Although not clear from the figure, it is electrically controlled by the local potential differences that result from applied bias conditions at the external terminals taking into account the voltage dropped across the resistors. Coupling between drain and source terminals is represented by the C_{DS} element.

There also exists capacitive coupling among some of the physical features of the device structure that should be modeled by additional circuit elements to complete the idealized electrical model. A planar schematic that illustrates the model in somewhat more detail is shown in Figure 5.2.

The model of Figure 5.2 is further divided into the intrinsic model, within the dashed box, with the remaining elements belonging to the extrinsic or parasitic model. The extrinsic part is represented by shells of parasitic shunt capacitance elements, within which is a shell of series-connected resistances and inductances. We will make use of this structure in the parameter extraction process. Various topologies are used [1], some more elaborate, but for simplicity, we will restrict ourselves to this.

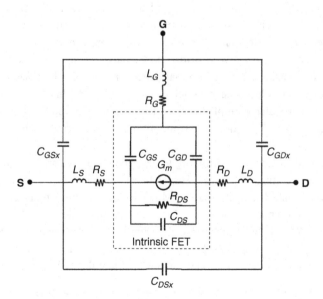

Figure 5.2 Simplified equivalent circuit of a FET with parasitic and intrinsic ECPs.

The shunt capacitances represent metal-to-metal interelectrode parasitic effects. The resistors R_S and R_D, considered parasitic elements in this treatment, represent the "access resistance" – that portion of the semiconductor that is not controlled by the gate potential. The gate resistor, R_G, is largely attributable to the resistance of the gate metallization. Inductances are also associated with the electrode metal and become important at high frequencies.

The intrinsic transistor is the heart of the semiconductor device. The intrinsic transistor model accounts for the bias-dependent transconductance responsible for the transistor action, channel resistance, and the bias-dependent junction capacitances. Although not visible in the structure of Figure 5.1, the physical mechanisms of self-heating, and in some cases, charge trapping, can be considered part of the intrinsic device. These phenomena are modeled by additional coupled equivalent circuits that will be introduced in Chapter 6.

5.2.1   Bias Considerations

The isolated structure of the transistor shown in Figure 5.1 provides no information about the DC operating condition at which the linear model is defined. Different device operating regimes, specified by different bias conditions, substantially affect the numerical values of the ECPs of the intrinsic model. Implicitly, this means a linear device model is defined for a specific bias condition.

Linearity requires that the ECPs do not vary with the applied RF signal. And, as these ECPs are dependent on the DC biases, these DC biases do not change with the RF signal. The DC conditions merely parameterize the linear dependence of the RF input-output relations. Since physical transistors are nonlinear components, as we learned in Chapter 1, a linear transistor model is a valid approximation to the actual device behavior provided the device is stimulated by a sufficiently small RF signal. This is why linear models of transistors are often referred to as *small-signal* models.

There is no precise way to know, *a priori*, how small such signals need to be for a linear model to be considered valid. Formal procedures for the linearization of the nonlinear equations to produce linear models in the small-signal limit are described in Chapter 6. Recall a particularly simple linear FET model was derived from time-domain considerations in Chapter 3 (Section 3.12), and ultimately converted to an S-parameter representation.

Linear device models are useful primarily to describe the device small-signal behavior in an active bias operating condition.[1] At such bias conditions, we can usually neglect forward and reverse gate leakage currents, which would require a more complicated equivalent circuit representation for Figure 5.2 with additional elements. However, we will see in Section 5.5.2.1 that deliberately choosing bias conditions outside the range of normal small-signal operation can be useful for parasitic element value extraction, necessary to ultimately get the best model for simulation at the standard conditions.

---

[1] Although passive "switch" bias conditions are also useful.

5.2.2   Temperature Considerations

Temperature, analogously to bias, affects the linear equivalent circuit element values. But just as with bias, for a linear model we can usually take the device temperature to be fixed and independent of the RF signal. As long as the device is operating in the small-signal regime, this assumption is valid for most RF applications. However, as we will see in Chapter 6, there can still be a frequency-dependence to the device small-signal electrical response caused by a *temperature modulation* produced by a small-signal stimulus *if the applied signal is varied slowly enough*. Since timescales for this phenomena are typically of the order of milliseconds or longer, and the RF timescales are typically in the nanosecond range, we neglect this phenomenon in this chapter. We deal with electro-thermal considerations in Chapter 6.

5.3   Measurements for Linear Device Modeling

5.3.1   Terminal Mappings to RF Measurement Ports

The device structure and equivalent circuit shown in Figures 5.1 and 5.2 has three external terminals, one each for the gate, drain, and source, respectively. Microwave measurements are made on actual layouts – physical realizations of the transistor embedded or "wired" into a particular two-port structure. The resulting measurement data will usually be in the form of a 2×2 matrix, such as a two-port S-parameter or Y-parameter representation. There are several ways a two-port matrix description, and therefore the actual physical structure of the test FET for measurement purposes, can be mapped to and from the three-terminal schematic of Figure 5.2. This depends on the choice of a common terminal and the choice of the numerical ordering of the ports.

Examples of the two most common physical realizations of actual test FET layout structures, and their wiring diagrams mapping the three terminals to distinct two-port configurations, are presented in Figures 5.3–5.6 [2]. These are *common source* and *common gate* layouts, respectively, and typically have the ports ordered as shown. The relationships between the ECP values of the FET equivalent circuit and the data are completely different when the two-port parameters are provided in common source or common gate configuration.

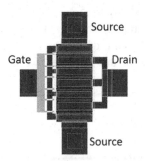

Figure 5.3 Common-source layout of a multifinger FET.

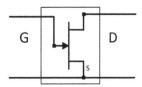

Figure 5.4 Wiring diagram of a three-terminal device consistent with the common-source layout of Figure 5.3. Port one is the gate and port two is the drain.

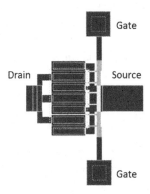

Figure 5.5 Common-gate layout.

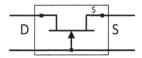

Figure 5.6 Wiring diagram of a three-terminal model corresponding to the common-gate layout of Figure 5.5. Port one is the drain and port two is the source.

To make this point more concrete, imagine, for the moment, that we can neglect the inductances and resistances in the extrinsic equivalent circuit of Figure 5.2. We could then identify each intrinsic ECP from the two-port parameters in either common source or common gate configuration. For example, the formula for the feedback capacitance, C_{GD}, in terms of common source Y-parameters is given in (5.1) and the corresponding formula in terms of common-gate Y-parameters is shown in (5.2), where the superscripts are used to label the respective two-port configuration. The formulas (5.1) and (5.2) correspond to the port orderings specified in the diagrams of Figures 5.3–5.6. Other port orderings lead to different but related formulas. In the Appendix we show explicit formulas to relate these descriptions under ideal circumstances.

$$C_{GD} = \frac{-\mathrm{Im}\, Y_{12}^{(CS)}(\omega)}{\omega} \tag{5.1}$$

$$C_{GD} = \frac{\mathrm{Im}\, Y_{11}^{(CG)}(\omega) + \mathrm{Im}\, Y_{12}^{(CG)}(\omega)}{\omega} \tag{5.2}$$

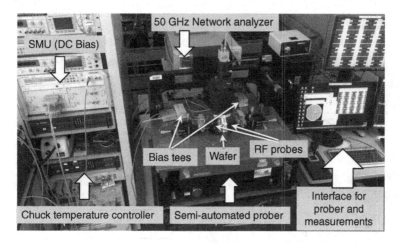

Figure 5.7 Test bench for S-parameter measurements for modeling.

It is therefore essential for parameter extraction to know the particular two-port realization of the physical device, including the ordering of the ports.

5.4   On-Wafer Measurements and Calibration

Transistor S-parameters for modeling purposes are best made on-wafer using vector network analyzers (VNA). A typical test bench is shown in Figure 5.7. A semi-automated wafer probe station is shown with a wafer, micro-manipulators for probing, microwave ground-signal-ground (GSG) probes and a microscope for visualization. A 50 GHz VNA with built-in sources is visible behind the probe station. Bias voltages are controlled and monitored with multiple source monitor units (SMUs) as shown on the left. A temperature controller is also shown that is used to set the temperature of the chuck on which the wafer is placed, providing controllable thermal boundary conditions for the transistor.

5.4.1   Linear Device Modeling Flow

The linear modeling flow is the process of making well-calibrated microwave measurements on one or more FET test patterns and using the information to obtain the ECPs for the corresponding model, sometimes including the geometrical scaling rules for the ECPs in the process.

5.4.1.1   Test FET Layout

A typical common source layout of a 6 finger by 50 μm (6 × 50) test FET structure for modeling and extraction is shown in Figure 5.8. The test structure contains more than just the FET. It contains ground-signal-ground pads (at the extreme left and right) for on-wafer probing and carefully designed nearly 50 Ω transmission lines (the narrow

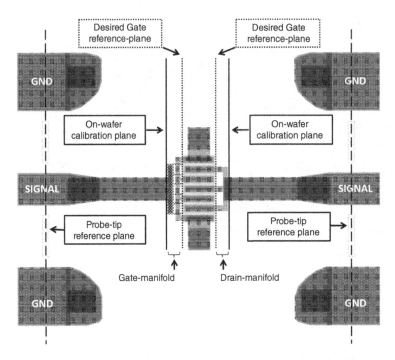

Figure 5.8 Test FET pattern for model extraction of a common-source FET.

structures attached to the signal pads) to achieve single-mode EM wave propagation for the microwave signals to the device.

The S-parameter (and DC) measurements at the probe tips represent the characteristics of the entire structure. In an actual circuit, say a MMIC, the pads and transmission lines are not present. It is therefore essential to characterize and model just the device at the well-defined reference planes where the component is defined, so that the model representing the device can be placed into a circuit for design purposes.

5.4.1.2   Calibration

Calibration, for RF and microwave network analysis measurements, is a very important topic, but will only be summarized briefly here. More details can be found in [3–5]. Poorly calibrated measurements lead to poor models. On-wafer calibration standards are usually best [4], and we will restrict ourselves to this approach.

We need well-calibrated S-parameters at the on-wafer calibration planes of Figure 5.8. The best way to do this is to define on-wafer calibration standards that can be used to move the reference planes from the probe tips to these planes. Four calibration patterns, defining known short, open, load, and thru (SOLT) standards, for RF and microwave calibrations, are shown in Figure 5.9.

The pads and transmission lines are identical to those of the FET test pattern of Figure 5.8. The first three standards are placed at the calibration planes the same distance apart as in the test FET structure. Only the "zero-delay" thru pattern is different

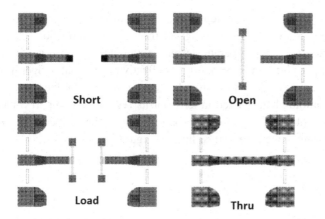

Figure 5.9 On-wafer patterns used to help characterize the FET device embedded in the layout of Figure 5.8.

for relative probe placement – the ports are physically closer together so the calibrations planes coincide – since this is a precise way to generate a good thru standard.

The S-parameters from the four patterns of Figure 5.9 are measured, and used together with the measured S-parameters of the test FET layout of Figure 5.8, to determine the S-parameters at the device reference planes. Algorithms are built into modern VNA software that use these four *SOLT* standards and the test structure measurements to return the S-parameters at the device planes. Alternative sets of on-wafer calibration patterns can be used with different calibration algorithms (e.g., LRRM or multiline TRL methods) to accomplish the same objectives [4].

5.4.1.3 Gate and Drain Manifolds

Before we can move the reference planes to the desired gate and drain boundaries indicated by the dotted lines in Figure 5.8, we must account for the behavior of the port manifolds.

The manifolds are the feed structures between the solid and dotted lines that couple the signals from the ends of the transmission lines of the test structure to the multiple gates and drain bars of the transistor. These structures are implemented very differently when a FET is physically realized as part of a MMIC design, for example. The idea is to model the manifolds of the test structure to move the reference planes and define the device at the dotted lines. The model we will ultimately build and extract, in this and the next chapter, will be for this device between the dotted lines of Figure 5.8. It will be incumbent on the MMIC designer to account for any parasitic effects introduced by the design-specific coupling of this device to others in the circuit, including the multiple gate and drain connections.

The gate manifold, in the layout of Figure 5.8, has source metal crossing over gate metal, with an interlayer dielectric in between. This could be modeled as a set of parasitic capacitors for each crossover and additional capacitors for fringing fields and the vertical bar of metal just to the right of the left solid line of Figure 5.8. A preferred modern approach, however, is to represent the manifold behavior by S-parameters computed directly from the layout using electromagnetic simulations. It is usually sufficient to simulate a 2-port S-parameter matrix with an input port at the calibration plane and an output port corresponding to an effective single gate finger at the inner (device) plane.[2] A similar simulation from layout can be done for the drain manifold.

With each manifold characterized in terms of its respective two-port S-parameters, simple linear de-embedding methods are used to move the reference planes to the desired device planes – the inner dotted lines in Figure 5.8.

5.5   The Device

The S-parameter data, now transformed to the device planes, are indicative of the transistor – not the test structure – and so the data are in as close as possible correspondence to the idealized structural representation of Figure 5.2.

5.5.1   Parasitic Shells of Series Impedances and Shunt Capacitances

The topology of Figure 5.2 can be abstracted to that of Figure 5.10, evidencing nested shells of fixed ECPs around the bias-dependent core of the intrinsic model. From a specific common terminal set of S-parameters, this suggests a conceptually simple method of parasitic stripping using linear algebra, by identifying generic Y- and Z- shells of parasitic elements of Figure 5.10 in the extrinsic topology of Figure 5.2 [6,7].

The common-source two-port S-parameters describing the device are a set of four complex numbers at each measured frequency. Certainly, at any fixed, single frequency, there are far too many ECPs in the device topology of Figure 5.2 and equivalent Figure 5.10 to be identified from these four numbers. Even using data over a wide range of frequencies, it is not practical to directly estimate the values of all extrinsic and intrinsic elements at a typical active bias condition where the linear device model is most likely to be useful. It is at this point we appeal to additional knowledge of the device behavior in certain distinct regions of operation in order to help us with the parameter extraction.

5.5.2   Parasitic Shell "Stripping"

5.5.2.1   Using Bias Conditions to Simplify the Equivalent Circuit for Parasitic Element Extraction from Measurements

Due to the significant dependence on bias of the intrinsic ECPs, the topology of the intrinsic device can be simplified in specific "cold FET" conditions, where the transistor is not active. The total topology of the device model in Figure 5.10 then becomes more

---

[2] That is, we do not simulate the 1-to-N terminal S-parameter network corresponding to the single extrinsic gate input signal port and N output ports, one output port per gate finger.

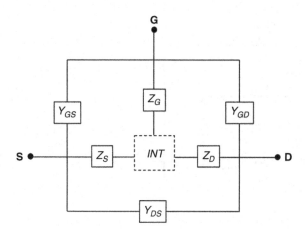

Figure 5.10 Device topology emphasizing Y–Z shells plus intrinsic device.

amenable to unambiguous extraction, where the parasitic elements can be more readily separated from the intrinsic elements. Under these conditions, the parasitic elements in the different shells of Figure 5.10 can be identified much more easily.

5.5.2.2   Identification of the Parasitic Capacitances

We consider one of the cold FET conditions where $V_{DS}^{DC} = 0$ and the device is reverse-biased to the point where the gate-source voltage is less than the threshold voltage, $V_{GS}^{DC} < V_T$. Under these conditions, where the device is not active, the device intrinsic transconductance, G_m must vanish. Under this strong reverse bias condition, we can also assume the intrinsic drain-source resistance, R_{DS}, is large enough to be neglected, which removes both of those branches from the equivalent circuit of the intrinsic device. If we further limit the characterization frequency range for the S-parameters to be far below the cutoff frequency of the device, we can also usually neglect the series inductances, and since we are ultimately interested in the imaginary part of the admittance elements in the outer shell, we can neglect the series resistances as well. The inductances and resistances will be dealt with by considering a different bias condition in Section 5.5.2.4. This assumes the value of intrinsic C_{DS} is zero at this particular bias condition.

We are left, therefore, with the combination of outer-shell parasitic capacitances in parallel with the intrinsic device junction capacitances at this reversed bias condition. This simple approximate equivalent circuit topology is shown in Figure 5.11.

Given the topology of the Figure 5.11, the correspondence between the 2-port linear data and the desired ECPs is most easily formulated in the admittance representation. We use the standard formula to convert the S-parameters to the Y-parameters given by (5.3), which follows from (3.36).

$$Y = \frac{1}{Z_0}(I - S)\cdot(I + S)^{-1} \tag{5.3}$$

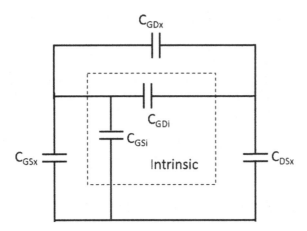

Figure 5.11 Simplified parasitic capacitance model of a FET under passive bias condition for parasitic capacitance identification.

The following simple relationships defining the element values in Figure 5.11 in terms of the common source Y-parameters are given in equations (5.4).

$$C_{GSx} + C_{GSi} = \frac{\text{Im}(Y_{11} + Y_{12})}{\omega}$$

$$C_{GDx} + C_{GDi} = \frac{-\text{Im}\,Y_{12}}{\omega} \tag{5.4}$$

$$C_{DSx} = \frac{\text{Im}(Y_{22} + Y_{12})}{\omega}$$

Here C_{GSi} and C_{GDi} are the intrinsic capacitances of the junction at the particular cold bias point, and C_{GSx}, C_{GDx}, and C_{DSx} are the parasitic values that, being assumed independent of bias, are to be identified.

There are still too many unknown parameters (five) on the left side of equations (5.4), to identify them all from one set of device admittances. Therefore, we characterize *an array of devices* of physically scaled layouts, shown in Figure 5.12, of various numbers of parallel gate fingers with various finger widths, and use the resulting geometrically dependent ECP values to separate parasitic from intrinsic element values.

Fitting the first equation of (5.4), from data for each set of devices with the same number of fingers separately, we get the plots of Figure 5.13 where the intercepts (extrapolation to zero width) provide the values for the parasitic capacitances of devices with different numbers of fingers. This interpretation assumes that the intrinsic capacitance values, extrapolated to zero gate width are zero. This is a physically reasonable approximation neglecting only fringing fields and perimeter effects that are usually insignificant for practical FET geometries.

It is a good sanity check that the slopes of the three independently fitted lines plotted in Figure 5.13, are so consistent.

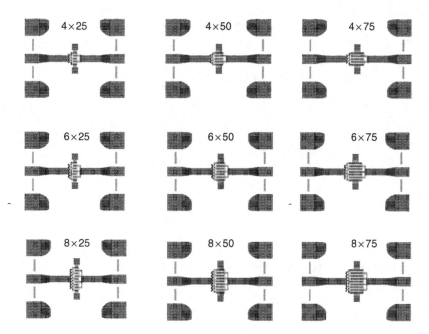

Figure 5.12 FET array of scaled layouts to identify parasitic elements.

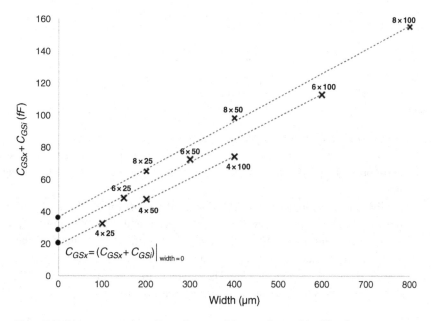

Figure 5.13 Linear regression allows for parasitic capacitance identification.

Plotting the values of C_{GSx} versus the number of fingers, we obtain the geometrical scaling rule from the linear fit shown in Figure 5.14.

A similar approach can be taken for the remaining parasitic elements in Figure 5.11.

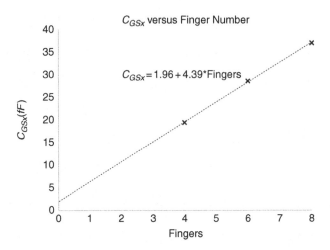

Figure 5.14 Scaling rule for shunt parasitic gate-source capacitance versus finger number.

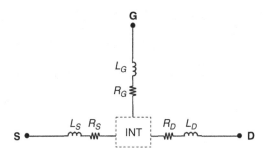

Figure 5.15 Structure after stripping parasitic capacitances using equation (5.5).

5.5.2.3   Stripping Off the Outer Admittance Shell

With the parasitic capacitance values of the outer parasitic shell now known, the two-port description at the device plane can be moved to the plane of the parasitic impedances.

Given the topological representation of these parasitics, the relationship between the common source two-port parameters at the Y and Z planes can be expressed most simply according to (5.5). The subtraction is element by element in the admittance representation, and it gives the common-source Y-parameters with the parasitic capacitances removed (Figure 5.15). The inverse of this difference is, therefore, the Z-parameter matrix, Z, at the plane of the series parasitics.

$$Z^{-1} = Y - Y^{para} = Y - j\omega \begin{pmatrix} C_{GSx} + C_{GDx} & -C_{GDx} \\ -C_{GDx} & C_{GDx} + C_{DSx} \end{pmatrix} \tag{5.5}$$

In (5.5), we have computed the common source parasitic admittance matrix, Y^{para}, in terms of the parasitic capacitance elements using the definition of the admittance matrix according to (5.6) and the structure of Figure 5.2.

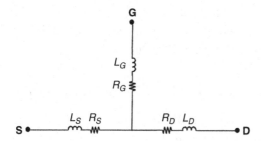

Figure 5.16 Idealized highly forward-biased FET equivalent circuit after the capacitance shell is stripped off. The intrinsic FET is assumed to be shorted out.

$$Y_{ij} \equiv \frac{\delta I_i}{\delta V_j}\Bigg|_{\delta V_{k \neq j}=0} \tag{5.6}$$

5.5.2.4   Parasitic Resistance and Inductance Extraction

We now move on to extracting the series parasitic shell containing the resistances and inductances. The strategy is similar to the capacitance parasitic extraction step of 5.5.2.2, in that we choose a bias condition to simplify the equivalent circuit, but this time the operating conditions are quite different. We will also do a matrix subtraction to remove the parasitic elements, but this time, due to the topology, we do so in the impedance domain.

Direct Method Based on Extreme Forward Bias Condition

A significantly forward-biased gate voltage is chosen such that there is substantial current flowing into the gate terminal across the Schottky barrier. The drain-source voltage is chosen to be zero, or, nearly equivalently, the drain current is biased to be equal and opposite to half the gate current. That is, $I_D^{DC} = \frac{-I_G^{DC}}{2}$. This condition forces the drain and source current to be equal.

This DC condition is assumed to be sufficiently extreme so as to effectively short out the Schottky barrier junction, interpreted as a strongly forward biased diode with, therefore, a very large conductance. The large conductance presents an effective electrical short between this intrinsic gate node and any other node to which it is connected in the intrinsic equivalent circuit. That is, the entire intrinsic FET can be collapsed to a single node, reducing the equivalent circuit topology within the Z-shell to a simple T-topology shown in Figure 5.16.

We remind the reader that gate current was not explicitly accounted for in the simple equivalent circuit of Figure 5.2. We assume a large conductance only at this stage in order to identify the series parasitic elements. The final linear model will be valid only for conditions where the actual gate leakage is negligible.

From Figure 5.16, we note only six equivalent circuit elements, all series parasitic elements, remain in the total model topology, now that the parasitic capacitances have been stripped off and the intrinsic device collapsed to a point at this bias condition. These EPCs can therefore be directly identified from the real and imaginary parts of the

Z-matrix of the data at this plane. This follows from the computation of the parasitic Z-matrix, given in (5.7) in common source configuration, which follows from the definition (5.8), applied to the circuit representation of Figure 5.16.

$$Z^{para} = \begin{bmatrix} R_G + R_S & R_S \\ R_S & R_D + R_S \end{bmatrix} + j\omega \begin{bmatrix} L_G + L_S & L_S \\ L_S & L_D + L_S \end{bmatrix} \tag{5.7}$$

$$Z_{ij}^{para} \equiv \frac{\delta V_i}{\delta I_j}\bigg|_{\delta I_{k \neq j} = 0} \tag{5.8}$$

While formally (5.7) can be solved for the six ECPs at any frequency by taking simple linear combinations of the real and imaginary parts of the Z-matrix, the frequency of the measurements must be high enough to measure the inductances without too much uncertainty. A fit to (5.7) from data over a range of moderate-to-high frequencies generally provides a more robust extraction than a direct solution from data at a single CW frequency.

This method has the following drawbacks, however. The extreme forward bias condition may be too stressful for modern short gate-length devices, causing damage during the measurements. Since R_D and R_S, are physically semiconductor access resistances, and as such can be current dependent, their extraction from data at an extreme bias condition may yield parameter values that might be different from their actual values under more normal operating conditions.

Modified Procedure at Less Extreme Forward Bias

A method that is potentially more accurate, but also more complicated, than that of 5.5.2.4.1, is presented in reference works [8–10]. In this case, the gate bias is only slightly forward biased, just enough to get the channel as open as possible but at much less extreme forward current conditions through the gate than 5.5.2.4.1. The intrinsic device is therefore not completely shorted out and an approximate modeling of its composition is required. The equivalent circuit of the series parasitics and the intrinsic device at this bias condition is shown in Figure 5.17 [1,10].

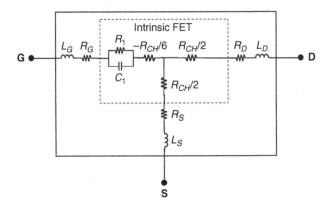

Figure 5.17 Forward bias FET equivalent circuit at $V_{DS} = 0$ V and $V_{GS} = 0.3$ V for parasitic resistance and inductance extraction.

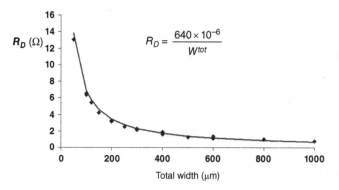

$$R_D = \frac{640 \times 10^{-6}}{W^{tot}}$$

Figure 5.18 Dependence of the parasitic drain resistance as a function of total gate width, as fitted to data from devices of multiple layouts.

The parameters R_1 and C_1 are associated with the total gate intrinsic resistance and capacitance, respectively, and R_{ch} represents the active channel resistance at this bias condition. The fractional factors of R_{ch} come from an analysis of the device as a distributed structure [11,12]. In this three-terminal schematic, the contribution of the seemingly strange $\frac{-R_{ch}}{6}$ factor at the gate with the $\frac{R_{ch}}{2}$ factors at the drain and source are equivalent to the more familiar common-source Z-parameters reported in the literature. The value of R_{ch} has to be estimated from data on other test structures, and its value depends on the material properties of the semiconductor. The authors use 1.55e-4 ohm-meter for GaAs and 6.75e-4 ohm-meter for GaN. Values of R_1 and C_1, along with the other six parameters in the impedance shell, are then optimized to fit the measured Z-parameters over a wide frequency range. More detailed considerations can be found in [1].

By applying either the methodology of Section 5.5.2.4.1 or that of 5.5.2.4.2, or equivalent, over the FET array of sizes, one can extract the values and the geometrical scaling rules for the parasitic elements. The data from an array of layouts and the resulting scaling rule for the drain resistance, R_D, of a GaAs FET is given in Figure 5.18.

Similar results are found for the source resistance.

A similar analysis leads to the gate resistance, R_G, but its scaling rule is quite different.

5.5.2.5 Gate Resistance Scaling

Geometrical Argument

From an examination of the common-source layout of Figure 5.3, given that the gate current flows along the fingers from left to right, we would expect, from simple geometrical arguments, the gate resistance to increase roughly linearly with the width per finger, w, for a fixed number, N, of parallel gate fingers.[3] For the same value of w, the resistance should scale inversely with N, because the fingers are arranged in parallel. The total gate width, W^{tot}, is just the product of the width per finger and the number of

---

[3] Neglecting distributed effects for simplicity.

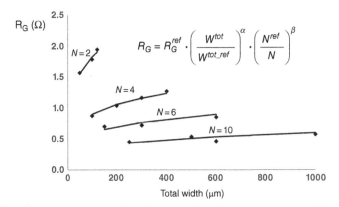

Figure 5.19 Empirical scaling rule for gate resistance of a GaAs pHEMT process. Measured (symbols) and best fit to formula (lines) with $\alpha = 0.23$, $\beta = 1$. An ideal lumped physical model predicts $\alpha = 1$, $\beta = 2$.

fingers, so we have $W^{tot} = N \cdot w$. From these considerations we derive a simple gate resistance scaling rule according to (5.9) with parameters $\alpha \approx 1$ and $\beta \approx 2$. Here the *ref* superscript refers to the values of a particular reference device. Note how different the rule for gate resistance scaling is compared to that of the source and drain resistance, the latter given by a simple expression such as shown in Figure 5.18. The rule (5.9) enables the estimation of the gate resistance for a range of geometries specified in terms of total gate width, W^{tot}, and number of parallel gate fingers, N. This can be simply re-expressed in terms of the width-per-finger, w, if desired (see Exercise 5.7). Obviously, care must be used to properly interpret the meaning of the geometrical parameters that define the device.

$$R_G = R_G^{ref} \cdot \left(\frac{W^{tot}}{W^{tot\_ref}}\right)^{\alpha} \cdot \left(\frac{N^{ref}}{N}\right)^{\beta} \qquad (5.9)$$

Empirical Results
Alternatively, an approach similar to that of 5.5.2.4 for the drain and source resistances can be followed for the gate resistance to arrive at an empirical expression for the gate resistance, R_G, as a function of the total gate width and the number of parallel gate fingers. The results are shown for a pHEMT process in Figure 5.19. While the scaled R_G values move in the same direction with W^{tot} and N as predicted by (5.9) for positive exponent values, the actual best fit values for the exponents are different. The data and parameter values are shown in Figure 5.19.

5.5.3   Stripping Off the Series Parasitics

The task is now to transform the two-port data through the parasitic Z-network to get to the intrinsic transistor.

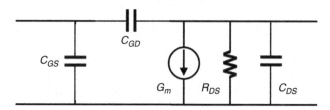

Figure 5.20 Intrinsic FET equivalent circuit after parasitic shell stripping.

The relationship of the intrinsic model linear two-port parameters (represented by the rectangular block within the parasitic impedances in Figure 5.11) to the two-port parameters after stripping the parasitics of the series impedances is most conveniently expressed by the simple relation (5.10), with Z^{para} as given by (5.7):

$$Y^{int} = [Z - Z^{para}]^{-1} \tag{5.10}$$

5.6 Intrinsic Linear Model

Now we are, finally, at the intrinsic equivalent circuit. The common source intrinsic linear equivalent circuit for a FET is shown in Figure 5.20. This neglects, as stated previously, leakage current through the gate as may happen at large reverse or forward bias conditions. It also neglects resistive elements, denoted R_{GS} (sometimes labeled R_i) and R_{GD}, elements in series with the junction capacitances, and also a phase associated with a time-delay of the transconductance. We will analyze an intrinsic model augmented with these elements in Section 5.6.5.

The topology of the intrinsic FET is seen from Figure 5.20 to be just a parallel combination of resistors and capacitors. We can easily compute the intrinsic model common source admittance matrix in terms of the individual ECPs in the equivalent circuit using the definition (5.6). The result is given in (5.11), with $G_{DS} = \frac{1}{R_{DS}}$. The hard part was all the work needed to move the reference planes for the various two-port description from the calibration plane all the way to the intrinsic device plane to arrive at the simple intrinsic model.

$$Y^{int}(\omega) = \begin{bmatrix} 0 & 0 \\ G_m & G_{DS} \end{bmatrix} + j\omega \begin{bmatrix} C_{GS} + C_{GD} & -C_{GD} \\ -C_{GD} & C_{GD} + C_{DS} \end{bmatrix} \tag{5.11}$$

5.6.1 Consistency Check of Linear Model Assumptions

Since conductances and capacitances from lumped elements are all real numbers, inspection of (5.11) reveals that the intrinsic FET model admittance matrix is a sum of a real part that is frequency independent and an imaginary part that is linearly dependent on frequency. That is, the intrinsic equivalent circuit topology predicts an especially simple explicit dependence on frequency. This prediction can be experimentally tested by de-embedding the measured (extrinsic) S-parameter data through the

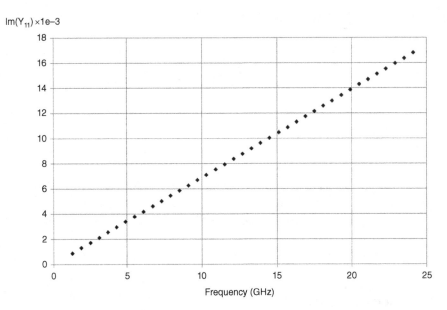

Figure 5.21 Measured imaginary part of intrinsic Y_{11} vs frequency of a GaAs FET.

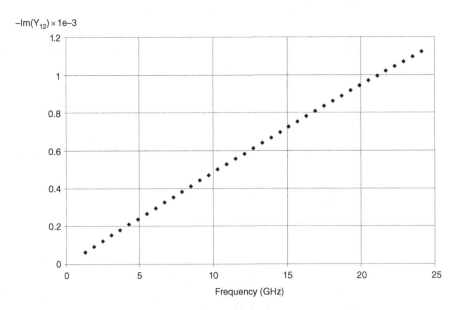

Figure 5.22 Measured imaginary part of intrinsic Y_{12} vs frequency of a GaAs FET.

parasitic network and plotting the real and imaginary parts of the intrinsic admittance parameters as functions of frequency. Some examples of intrinsic admittance parameters for a GaAs FET capacitances are shown in Figures 5.21 and 5.22.

The behavior of the intrinsic input admittance data versus frequency is a validation of the appropriateness of the equivalent circuit as a model of the transistor's linear behavior. It confirms that the reactive elements present in the gate-source and gate-drain branches are indeed two capacitances with different values.

5.6.2　Identification of Intrinsic Linear Equivalent Circuit Elements

Each of the elements of the common-source intrinsic admittance matrix (5.11) is a simple linear combination of resistive and capacitance terms. It is evidently possible to invert these equations and solve uniquely and explicitly for each of the equivalent circuit elements in terms of simple linear combinations of the intrinsic common-source admittance matrix elements. These equations are given in (5.12)–(5.16), where we drop the superscript *int* for intrinsic.

$$G_m = \text{Re}\,Y_{21}(\omega) \tag{5.12}$$

$$G_{DS} = \text{Re}\,Y_{22}(\omega) \tag{5.13}$$

$$C_{GD} = \frac{-\text{Im}\,Y_{12}(\omega)}{\omega} \tag{5.14}$$

$$C_{GS} = \frac{\text{Im}(Y_{11}(\omega) + Y_{12}(\omega))}{\omega} \tag{5.15}$$

$$C_{DS} = \frac{\text{Im}(Y_{22}(\omega) + Y_{12}(\omega))}{\omega} \tag{5.16}$$

5.6.3　Discussion and Implications

Model consistency means the right-hand sides (RHS) of (5.12)–(5.16) should end up being independent of frequency even though the numerators and denominators are themselves frequency dependent. This will never be exactly true in practice, but the validity is quite good for most FETs over much of the useful operating frequency range, provided the parasitic elements are properly identified and de-embedded from good measurements, as we demonstrate in the next section.

5.6.3.1　Problems Caused by Poor Parasitic Element Extraction

The following example illustrates what happens when there is a poor extraction of the parasitic element values. Broadband S-parameters are measured at a particular active bias condition for a GaN HFET. The device is a $0.15\ \mu m \times 6$ finger $\times 60\ \mu m$ GaN HFET with individual source vias from Raytheon Integrated Defense Systems [13]. The device is optimized for 1–40 GHz operation for applications including power amplifiers, low noise amplifiers, and switch applications. All further examples of GaN transistors in this chapter are based on this device.

The parasitic shells are stripped away using the methods of 5.5.2.4. The resulting intrinsic elements are computed, at each frequency, using the formulas (5.12)–(5.16). The resulting element values versus frequency are plotted for selected elements for two different values of the parasitic source inductance, L_G in Figure 5.23.

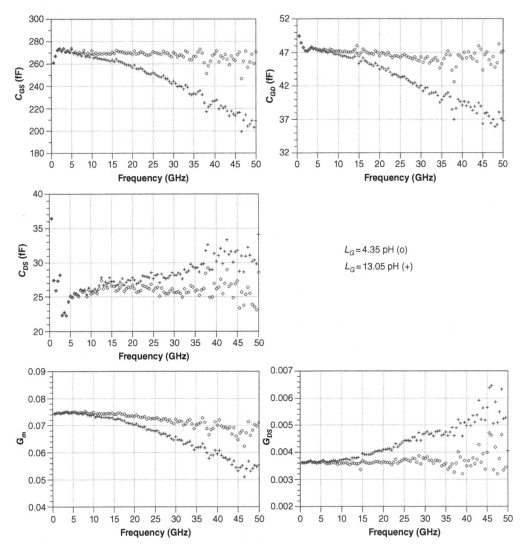

Figure 5.23 Flatness with frequency of intrinsic ECPs of a GaN HFET device for two values of parasitic inductance, L_G. The bias condition is $V_{GS} = -1.8$ V and $V_{DS} = 20$ V. The device is a 0.15 μm × 6 finger × 60 μm GaN HFET with individual source vias from Raytheon Integrated Defense Systems [13]. The device is optimized for 1–40 GHz operation for applications including power amplifiers, low noise amplifiers, and switch applications.

The frequency dependence of the intrinsic elements is much flatter over the wide range of the measurements with the (correct) smaller value of L_G. While this example may strike the reader as somewhat extreme, the gate inductance can be difficult to obtain without a large uncertainty. However, it evidently has a significant effect on the frequency flatness of the computed intrinsic elements. Flatness with frequency can be used as an optimization objective in more sophisticated (but more complicated) approaches to the total extraction problem. See for example [14].

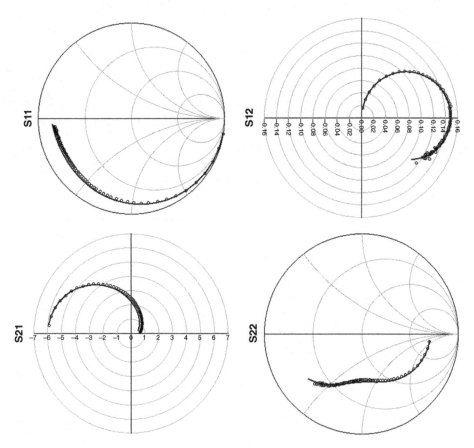

Figure 5.24 Broadband validation of the linear model of the $6 \times 60\,\mu m$ GaN HFET. Measured (symbols), modeled (lines). Frequency range is 0.5–50 GHz. DC bias conditions: $V_{GS} = -1.8\,V$, $V_{DS} = 20\,V$. The model ECPs were identified at 10 GHz.

With the correct values of the parasitic elements determined, the intrinsic ECPs of the model at each operating point can be determined from the small-signal intrinsic Y-parameters using (5.12)–(5.16) at a single frequency, provided the frequency is not so low that there is large uncertainty in elements, such as capacitances. Examining the capacitance values in Figure 5.23, we see large fluctuations in values below 5 GHz, indicating we should choose a higher frequency for the extraction point. Beyond about 30 GHz we observe an increasing spread in the values versus frequency. So in this case, we choose 10 GHz to be the extraction frequency.

The proof that the model topology and the parameter extraction methodology described here produce accurate linear models is shown in Figures 5.24 and 5.25. In these figures, the broadband measured S-parameters on the $6 \times 60\,\mu m$ GaN HFET are compared with the linear model simulations at two different bias conditions. The fit is quite good over the full range from 0.5 GHz to 50 GHz for both bias conditions, even though the identification of the ECPs for the intrinsic model were identified at only

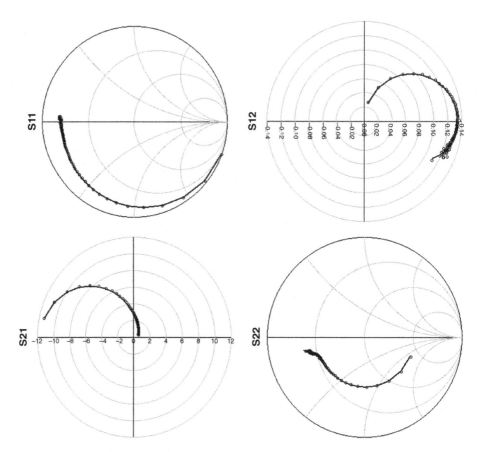

Figure 5.25 Broadband validation of the linear model of the $6 \times 60\,\mu m$ GaN HFET. Measured (symbols), modeled (lines). Frequency range is 0.5–50 GHz. DC bias conditions: $V_{GS} = -1.2\,V$, $V_{DS} = 6\,V$. The model ECPs were identified at 10 GHz.

the single CW frequency of 10 GHz. There is a slight discrepancy visible between the measurements and simulations for mid-frequency S_{11} and S_{22} and the mid-to-high-frequency gain, S_{21}, is slightly underestimated by the model. A more advanced model will be presented in Section 5.6.4 that improves upon these results.

An important implication of (5.12)–(5.16) is that the identification of the element values can be done at *any single CW measurement frequency*. This is the principle of "direct extraction" that has become standard industry practice since the late 1980s [6,7]. That is, once it is established that (5.16) is essentially independent of frequency, the particular fixed value of the ECPs can be obtained by a spot measurement at a single CW frequency.

Of course, in practice the extraction frequency must be high enough for the S-parameters at the calibration plane to register a capacitance contribution, but not so high that the data at the intrinsic transistor becomes overly sensitive to additional distributed effects or errors in the parasitic element value extraction (such as the port inductances).

In principle, for a given bias condition, there is an optimum value of frequency for identifying the element values from (5.12) to (5.16). Detailed considerations including uncertainty analysis are presented in [15,16].

The model of Figure 5.20 effectively *compresses* the independent measurement data from a great many frequencies, into only five real numbers, specifically the values of G_m, G_{DS}, C_{GS}, C_{GD}, and C_{DS}. These five numbers are *predictive* of the broadband frequency performance of the intrinsic device through the model. Even including the parasitic elements, the complete linear model is an extremely compact representation of the DUT frequency behavior at the given DC bias point.

Another consequence of the predictive power of the model is that the model can be used to *extrapolate* beyond the measurement data used to identify the element values. We can identify the element values using (5.12)–(5.16) at any one frequency, say 10 GHz, (or over a moderate range of frequencies, say 5–20 GHz) and then simulate with the model at a CW frequency of, say, 50 GHz. Recall the model frequency performance is determined completely from the element values and their arrangement in the circuit topology. So a well-extracted model will work well until the simulation frequency is so high the device topology is no longer valid. Predictions of DUT performance up to a large fraction of the transistor cutoff frequency, f_T, are reliable if the model parameter extraction has been done properly. Moreover, such simulations can be more accurate than an actual *measurement at the desired high frequency.* Of course this depends on the instrument measurement bandwidth and the quality of the test transistor layout and calibration patterns. Nevertheless, transistor figures of merit (FOM) such as f_T and f_{max}, can often be obtained more reliably and much less expensively by first extracting, from high-quality medium frequency data, a model of the device and then simulating with the model at frequencies higher than those used for the extraction to compute the FOM. In fact, this is often the way these device FOMs are defined.

5.6.4   A New Element Revealed from the Data

The above treatment followed from the *postulated linear equivalent circuit* model of Figure 5.20. There are five elements in this equivalent circuit, yet there are eight real numbers associated with the intrinsic small-signal parameters – the two-port matrix description in any representation – admittance, impedance, or scattering, etc., at any one frequency. An examination of the imaginary part of the intrinsic *model admittance* (5.11) shows that the off-diagonal matrix elements of the imaginary part are equal and proportional to frequency. But let's plot the *measured imaginary parts of the off-diagonal admittance elements* and compare them. This is shown in Figure 5.26 for the same 6×60 μm GaN HFET device as above at the DC bias of $V_{GS} = -1.8$ V and $V_{DS} = 20$ V.

The frequency dependence is (approximately) linear for each of the two imaginary terms, so they can be interpreted (modeled) as capacitances. But they are clearly not equal as the simple intrinsic model of Figure 5.20 predicts. So we must re-visit our equivalent circuit model and *add a new element!*

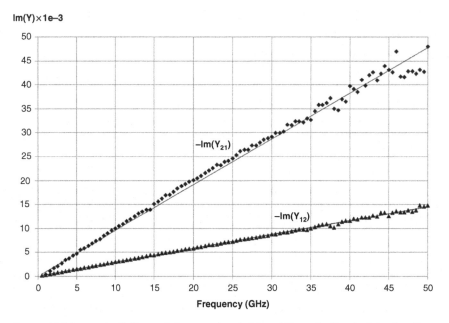

Figure 5.26 Measured off-diagonal elements (symbols) and linear fit (lines) of the 6×60 μm GaN HFET intrinsic input admittance in common source configuration.

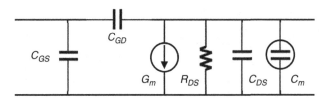

Figure 5.27 Equivalent circuit with the new transcapacitance element.

In formal analogy with the transconductance, we place the transcapacitance element in the drain-source branch, and represent it by a symbol of a capacitor surrounded by a circle. The augmented equivalent circuit diagram and the corresponding common-source intrinsic admittance are shown in Figure 5.27 and (5.17), respectively.

The corresponding identification formula for the new element is easily defined according to (5.17) in terms of the common-source intrinsic admittance matrix elements.

$$C_m = \frac{\text{Im}(Y_{21}(\omega) - Y_{12}(\omega))}{\omega} \tag{5.17}$$

The complete common-source intrinsic admittance matrix corresponding to the augmented equivalent circuit is given in (5.18). The transcapacitance element shows up in the second row of the first column in the second matrix of (5.18).

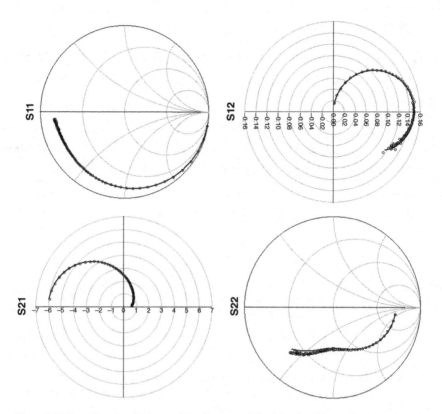

Figure 5.28 Broadband validation of the linear model of the $6 \times 60 \, \mu m$ GaN HFET with the transcapacitance element (model of Figure 5.27). Measured (symbols), modeled (lines). Frequency range is 0.5–50 GHz. DC bias: $V_{GS} = -1.8 \, V$, $V_{DS} = 20 \, V$. The ECPs of the model were identified at 10 GHz.

$$Y^{(CS)}(\omega) = \begin{bmatrix} 0 & 0 \\ G_m & G_{DS} \end{bmatrix} + j\omega \begin{bmatrix} C_{GS} + C_{GD} & -C_{GD} \\ C_m - C_{GD} & C_{GD} + C_{DS} \end{bmatrix} \tag{5.18}$$

It should be emphasized that this element, or an equivalent to be discussed presently, is *required* for improved model accuracy at (nearly) any frequency compared to the five-element model of Figure 5.20, because it was specifically identified, through (5.17), to account for the measured difference of the intrinsic off-diagonal matrix elements.

To prove this point, we show the result of the linear model based on the addition of the transcapacitance element, using Figure 5.27 as the augmented intrinsic model by adding the transcapacitance element. Comparing Figure 5.28 to that of Figure 5.24, we observe the augmented model fits S_{11} and S_{21} better, especially at mid-to-high frequencies, and fits mid-frequency S_{22} better as well. It does appear that S_{22} at the highest frequencies may agree slightly less well than the simpler model, but we must remember that the measurement uncertainty is also greatest at 50 GHz.

A more discriminating comparison between the models, with and without the transcapacitance element, can be observed by plotting separately the magnitudes and phases

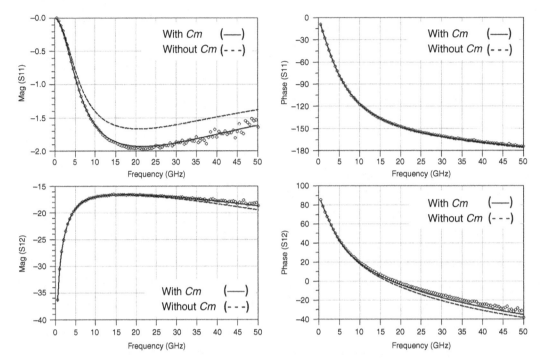

Figure 5.29 S_{11} and S_{12} versus frequency: Measured (symbols) versus model with the transcapacitance (solid line) and without the transcapacitance (dashed lines). Frequency range is 0.5–50 GHz. DC bias conditions: $V_{GS} = -1.8$ V, $V_{DS} = 20$ V. The ECPs of the model were identified at 10 GHz.

of the extrinsic S-parameters of each model and comparing them to the corresponding S-parameter data. This is done in Figures 5.29 and 5.30.

5.6.4.1   Discussion

The use of a transcapacitance element brings with it some complications, however, one of which will be described now. The magnitude of the transadmittance, (5.18), of the intrinsic model increases without bounds as the stimulus frequency increases due to the transcapacitance element. This follows from the small-signal relationship (5.19), deduced from (5.18) by calculating the magnitude of the drain current response to a perturbation in the gate-source voltage. The parameter τ is defined as the ratio of the magnitude of the transcapacitance to the transconductance. Fortunately, the terminal resistances R_G and R_S, in conjunction with the intrinsic gate capacitances, produce an input time constant that provides a frequency limit beyond which the intrinsic transadmittance won't increase. The overall model is, therefore, generally quite accurate up to a significant fraction of the transistor cutoff frequency. Nevertheless, the high-frequency behavior of (5.19), even for an intrinsic control frequency limited by input charging time constants, may not follow the measured and physically required decrease with frequency of the transadmittance at extremely high frequencies.

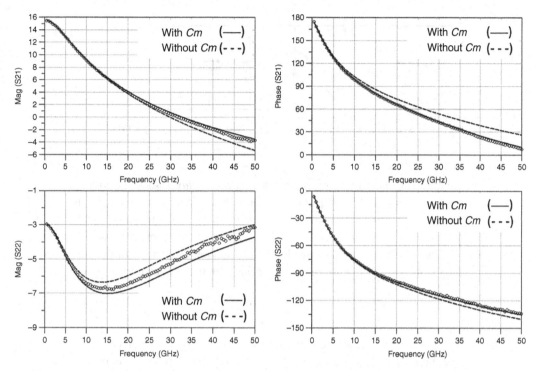

Figure 5.30 S_{21} and S_{22} versus frequency: Measured (symbols) versus model with the transcapacitance (solid line) and without the transcapacitance (dashed lines). Frequency range is 0.5–50 GHz. DC bias conditions: $V_{GS} = -1.8$ V, $V_{DS} = 20$ V. The ECPs of the model were identified at 10 GHz.

$$|\delta I_2| = \sqrt{G_m^2 + \omega^2 C_m^2} \cdot |\delta V_1| = G_m |\delta V_1| \cdot \sqrt{1 + (\omega \tau)^2}$$

$$\tau \equiv \frac{|C_m|}{G_m} \tag{5.19}$$

At the DC bias point $V_{GS} = -1.8$ V, $V_{DS} = 20$ V, the model parameter values are $C_m = -110$ fF and $G_m = 0.075$ S, so $\tau = 1.5$ ps. Even at 50 GHz, $\omega \tau = 0.47$, and the square root factor is only 1.1, so things still work well.

For linear models, there is a classical approach that associates a *time-delay* with the transconductance element that can account for the measureable behavior shown in Figure 5.26 without the need for a transcapacitance element. This will be described in Section 5.6.5.1. The time-delay approach is not as easily generalized to the large-signal case as the transcapacitance approach, however. We will return to the transcapacitance issues in Chapter 6 when we discuss the large-signal model.

5.6.5   Related Linear Models

5.6.5.1   Model with Time-Delay of the Transconductance

Historically, another parameter had long been introduced to deal with the imaginary part of the nonreciprocity of the off-diagonal intrinsic admittance parameters for linear

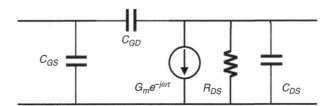

Figure 5.31 Variant of the intrinsic linear equivalent circuit model for a FET.

models. The most common approach is a *time-delay* associated with the transconductance element. The frequency-domain representation of a time-delay is a complex number on the unit circle with a phase proportional to the frequency, where the delay parameter, τ, is the coefficient of proportionality. The τ parameter varies with the DC bias just like the other intrinsic ECPs. The circuit diagram, with this term added, is shown in Figure 5.31 with the corresponding admittance matrix representation in (5.20).

$$Y^{(CS)} = \begin{bmatrix} 0 & 0 \\ G_m e^{-j\omega\tau} & G_{DS} \end{bmatrix} + j\omega \begin{bmatrix} C_{GS} + C_{GD} & -C_{GD} \\ -C_{GD} & C_{GD} + C_{DS} \end{bmatrix} \quad (5.20)$$

The new model parameter is now the time-delay, τ. The relationship between (5.18) and (5.20) can be understood by expanding the exponential of (5.20) in orders of $\omega\tau$.

$$G_m e^{-j\omega\tau} \approx G_m - j\omega G_m \tau + \cdots = G_m\left(1 - j\omega\tau + O\left((\omega\tau)^2\right)\right) \quad (5.21)$$

$$C_m = -G_m \tau$$

So the transcapacitance of (5.18) can be interpreted as a first order approximation to the delay of the transcondctance. Alternatively, the C_m value can be identified from data using the formula of (5.17) and used to solve for the parameter τ in (5.20) using the second equation of (5.21).

As discussed in Section 5.6.4.1, this is still a good approximation even at 50 GHz. Given the improved results using the transcapacitance demonstrated in Figures 5.29 and 5.30, it is clear the model is quite improved over this frequency range.

5.6.5.2   Adding R_{GS} and R_{GD} Elements

The intrinsic models of Figures 5.27 and 5.31 have six elements, each of which can be determined from a simple closed-form expression involving the intrinsic Y-parameters from data at any CW frequency. Since there are 8 (real) parameters in the 2×2 intrinsic Y-parameters, we might well ask if there are another two circuit elements that can be identified from the remaining 2 real parameters that could further improve the model. The answer is generally yes. Contrary to what is predicted by (5.18) and (5.20), the measured real parts of Y_{11} and Y_{12} are not exactly zero, and the measured imaginary parts are not exactly proportional to frequency. These facts suggest we might be able to add a resistor to each of the gate-source and gate-drain branches to improve the model agreement with the measurements.

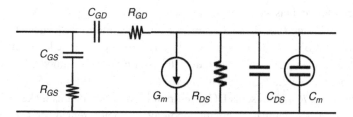

Figure 5.32 Eight-element intrinsic linear FET model.

From any single frequency measurement, it would not be possible to decide whether these new resistors should be placed in series with the input capacitances or in parallel. Over a broad frequency range, however, for bias conditions for which there is negligible gate leakage, the series combination fits the frequency-dependent data much better. The corresponding model appears in Figure 5.32. This choice can be seen as a natural consequence of good data analysis. There is also a physically based interpretation for these resistors, namely that the displacement current through the gate must go through some of the resistive semiconductor channel before appearing at either intrinsic source or drain terminal.

The two new elements require a modification of the formulas expressing the eight ECPs as explicit linear combinations of the intrinsic Y-parameters. We do this in a two-step process. We define the following linear combinations of common-source Y-parameters shown in (5.22). The expressions for each element, shown in (5.23), follow from simple algebra [6].

$$
\begin{aligned}
Y_{GD} &= -Y_{12} \\
Y_{GS} &= Y_{11} + Y_{12} \\
Y_{DS} &= Y_{12} + Y_{22} \\
Y_m &= Y_{21} - Y_{12}
\end{aligned}
\tag{5.22}
$$

$$
R_{GS} = \mathrm{Re}\left(\frac{1}{Y_{GS}}\right) \qquad R_{GD} = \mathrm{Re}\left(\frac{1}{Y_{GD}}\right)
$$

$$
C_{GS} = -\frac{1}{\mathrm{Im}\left(\dfrac{\omega}{Y_{GS}}\right)} \qquad C_{GD} = -\frac{1}{\mathrm{Im}\left(\dfrac{\omega}{Y_{GD}}\right)}
$$

$$
G_{DS} = \frac{1}{R_{DS}} = \mathrm{Re}(Y_{DS}) \qquad C_{DS} = \mathrm{Im}\left(\frac{Y_{DS}}{\omega}\right)
\tag{5.23}
$$

$$
G_m = \mathrm{Re}(Y_m) \qquad C_m = \mathrm{Im}\left(\frac{Y_m}{\omega}\right)
$$

The expressions for C_{GS} and C_{GD} in (5.23) reduce to those of (5.15) and (5.14), respectively, in the low-frequency limit. However, those of (5.23) are more sensitive to noisy data or if the data is affected by DC leakage at the measurement condition.

For some active bias conditions, the actual values of C_{GD} are so small that to accurately resolve the R_{GD} element in series with it, a stimulus frequency either beyond the measurement BW of conventional VNAs is required, or else the large uncertainty in the extracted R_{GD} element value can reduce the usefulness of this model. In such cases, it may be better to put the R_{GD} element *in parallel* with C_{GD}. (keeping R_{GS} in series with C_{GS}). In this case, R_{GD} represents DC gate-drain leakage. While this makes for a model with an asymmetric equivalent circuit topology, we must remember that the active bias condition itself "breaks the symmetry" of the device structure, so this presents no problem for a linear model where the element values are fixed.

5.7   Bias-Dependence of Linear Models

We now return to the DC bias dependence of the linear model. This will help with the transition to the large-signal model in Chapter 6. We assume the relations (5.12)–(5.16) hold for "all" bias conditions, making explicit the dependence not only on frequency but on bias. For example, the bias-dependence of the C_{GS} and C_{GD} elements are given by (5.24) (compare (5.15) and (5.14). A plot of these circuit element values as a function of the DC gate and drain bias voltages is given in Figure 5.33.

$$C_{GS}(V_{GS}, V_{DS}) \equiv \frac{\text{Im}(Y_{11}(V_{GS}, V_{DS}, \omega) + Y_{12}(V_{GS}, V_{DS}, \omega))}{\omega}$$

$$C_{GD}(V_{GS}, V_{DS}) \equiv \frac{-\text{Im}(Y_{12}(V_{GS}, V_{DS}, \omega))}{\omega}$$

(5.24)

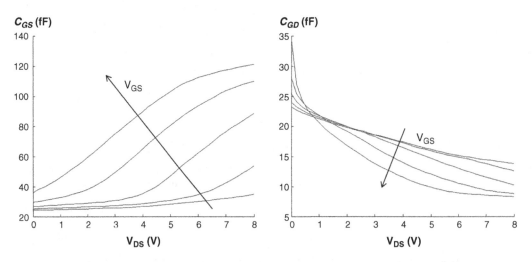

Figure 5.33 Bias-dependent junction capacitances of a GaAs pHEMT defined by (5.24). V_{GS} varies from −2 V to −1.2 V in 0.2 V steps in the direction of the arrows. The bias conditions are those applied to the extrinsic terminals.

We notice several interesting features of these characteristics. Each capacitance, C_{GS} and C_{GD}, depends in a complicated nonlinear way on *both* DC bias voltages, V_{GS} and V_{DS}. This is not so surprising in itself, based on the previous considerations articulated in this chapter. But the conventional definition of a two-terminal capacitor has a capacitance that depends only on the *single* voltage difference between the two terminals. We will have to come back to the circuit theory – and also to the physics – required to model this multiple voltage dependence. We will do this in Chapter 6.

From Figure 5.33, we see that for V_{DS} values greater than about 1.5 V, the actual measured value of the feedback capacitance C_{GD} *decreases* as the gate voltage increases. That is, the capacitance, C_{GD}, at pinchoff ($V_{GS} = -2$ V) is larger than when the channel is partially opened by increasing the gate voltage. It is interesting to note that C_{GD} exhibits exactly the *opposite* dependence on gate voltage, V_{GS}, compared to the behavior of capacitance C_{GS}, and indeed opposite to the conventional dependence of depletion capacitance on the applied voltage. Linear models with capacitance values that do not change with DC bias conditions consistent with these data will necessarily result in simulations that disagree with actual device behavior.

We also note that as V_{DS} approaches zero, the gate voltage dependence of both C_{GS} and C_{GD} becomes the same, and even the numerical values of the two capacitances become nearly the same. This is a manifestation of the drain-source symmetry of the intrinsic transistor. A more complete treatment of device symmetry, for all bias voltages, is presented in Chapter 6.

5.7.1   Bias-Dependent Linear Models

Simple bias-dependent linear models can be easily implemented in commercial simulators. The models are built as subcircuits using standard circuit elements configured in topologies like Figure 5.2 with intrinsic and parasitic elements. A typical symbol associated with such a model is given in Figure 5.34.

The model takes as input the two independent DC port voltages needed to determine the ECP values of the intrinsic model, and the two independent geometrical parameters such as the total gate width and number of parallel gate fingers for the scaling rules of both the parasitic elements and the intrinsic ECPs. In this case, a fifth input is the name

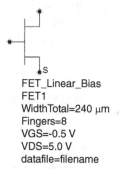

FET_Linear_Bias
FET1
WidthTotal=240 μm
Fingers=8
VGS=-0.5 V
VDS=5.0 V
datafile=filename

Figure 5.34 Symbol for scalable bias-dependent linear FET model.

of a file containing the bias-dependent intrinsic element values tabulated as a function of both bias voltages. Since the DC voltages correspond to the extrinsic terminals they remain on a grid if they were characterized on a grid. This means the data can be stored in standard tables readable by the simulator. The simulator also interpolates the values should the requested bias conditions not coincide with the discrete data points. For geometrical scaling the extracted algebraic rules for the parasitics are simply evaluated using the expressions developed and fit during the extraction process. The intrinsic elements scale proportionally to the ratio of the total gate width of the desired structure to that of the reference device.

5.7.1.1   Limitations of Simple Geometrical Scaling Rules

The geometrical scaling rules introduced above are only approximate. They hold over a certain range of layouts, or, equivalently, values of the scaling ratio, r. Simple scaling starts to break down at very high frequencies, where device distributed effects become important for large finger widths.[4]

Thermal Issues

Another limitation of simple geometrical scaling rules for the electrical component values of linear (and nonlinear) models is caused by thermal effects. Two identically biased FETs with the same DC power density (power per unit gate width) will differ in their "junction temperature" if the effective thermal resistance per unit gate width is different. This can be the case when one device uses more fingers in parallel compared to a second device that uses fewer but wider fingers. The multiple interior gate fingers see different local thermal boundary conditions, and therefore generally get hotter than the exterior fingers. Moreover, simple approximate analysis indicates that thermal resistance scales sublinearly with the inverse of the total gate width [17]. Since temperature affects the equivalent circuit element values – especially those of the intrinsic device, it is generally important to have at least a rough understanding of the geometrical scaling properties of the thermal resistance (and heat capacitance).

A practical approach for linear device models is to carefully model several layouts, say small, medium, and large, and "scale locally" around each layout. It is of course critical for first-time success that any FET actually used in a MMIC design be realized as a properly scaled version of the layout used for the device model! Otherwise, a unique layout may introduce an unanticipated – usually undesirable – performance characteristic when the circuit is actually fabricated.

5.7.1.2   DC Bias Conditions Are Not Computed

It is important to understand that the DC input parameters must be supplied by the user and are not related to the DC operating point computed by the simulator (unlike a general large-signal model). So if the transistor of interest is embedded in a circuit with a lossy path from the bias supplies, the designer must estimate the actual terminal

---

[4] Recall that for FETs the gate width is typically much greater than the gate length.

voltages the device will see and provide the correct, actual terminal voltage differences for the device through the model instance of Figure 5.34.

5.8   Summary

Linear equivalent circuit models of devices were introduced starting from physical considerations of the transistor structures. Microwave FETs were used as examples. The models are linear in the sense that the output RF signals are proportional to the input signals, and therefore they relate complex input phasors to complex output phasors at the same frequency. The ECPs of a linear model don't vary with the impressed RF signal. The frequency dependence of a complete linear model depends on the type of primitive electrical elements and their arrangement in the equivalent circuit topology. The models are predictive in the sense that the broadband frequency dependence, at a fixed DC operating point, is determined by only a few numbers, namely the value of the equivalent circuit elements.

Parameter extraction schemes were introduced for FET linear models, based on S-parameter data taken at different bias conditions in order to help separate the parasitic elements of the device from the bias-dependent elements associated with the intrinsic transistor. Measurements on an array of different scaled layouts were shown to also be necessary to discriminate all the ECPs based on their dependence on gate width and number of parallel gate fingers. This procedure also provided geometrical scaling rules enabling the resulting model to predict the performance of FETs with different number of fingers and total gate widths. The final model also provided more accurate simulated broadband linear frequency behavior because the parasitic elements were properly determined and, therefore, so were the intrinsic bias-dependent ECPs.

Explicit formulae were developed for the intrinsic ECPs of standard intrinsic models with 8 or fewer elements. For these models, the individual component element values can be identified in terms of measured, de-embedded single-frequency S-parameter data, once the parasitic element values in the topology are known.

The bias-dependence of intrinsic ECPs reveals interesting relationships among the elements that must be properly accounted for in nonlinear models. We will return to this in Chapter 6.

5.9   Exercises

Exercise 5.1   Derive the expression for the Y-shell parasitic capacitances from the Figure 5.2 using the definition (5.6).

Exercise 5.2   Show that the bias condition $I_D^{DC} = \frac{-I_G^{DC}}{2}$ forces the drain and source currents to be equal.

Exercise 5.3   Using the definition of the admittance matrix in equation (5.6), derive (5.11) from the equivalent circuit in Figure 5.20.

Exercise 5.4   Derive (5.12)–(5.16) from (5.11).

Exercise 5.5   Derive (5.2) from (5.11) and the appendix that relates common source and common gate admittances. Derive expressions for the other four element values in the linear equivalent circuit of Figure 5.20 in terms of their common-gate intrinsic admittance parameters.

Modify the formulas for the element values in terms of common-gate admittance parameters with the introduction of the transcapacitance element introduced in 5.6.4.

Exercise 5.6   Derive (5.22) and (5.23) for the 8-element FET model of Figure 5.32.

Exercise 5.7   Derive the ideal scaling rule (5.9) with $\alpha = 1$ and $\beta = 2$, for the gate resistance as a function of total gate width, W^{tot}, and the number of parallel gate fingers, N. Derive an equivalent expression in terms of the width-per-finger, w, and N.

References

[1] S. R. Nedeljkovic, W. J. Clausen, F. Kharabi, J. R. F. McMacken, and J. M. Gering, "Extrinsic parameter and parasitic elements in III-V HBT and HEMT modeling," chapter 3 in *Nonlinear Model Parameter Extraction Techniques*, M. Rudolph, C. Fager, and D. E. Root, editors, Cambridge University Press, 2012.

[2] D. E. Root, J. Xu, and M. Iwamoto, "Device modeling for FAB engineers: survey of selected foundations," 2015 CS-MANTECH Workshop: "RF for Device and Fab Engineers: Basic and Advanced Measurements, Device Modeling, Power Amplifier Design, and RF Packaging," Scottsdale, AZ, May 2015.

[3] G. Fisher, Cascade Microtech Europe Ltd, "A guide to successful on wafer millimeter wave RF characterization," www.keysight.com/upload/cmc_upload/All/OnWaferMillimeter.pdf

[4] A. Rumiantsev, "A Practical Guide for Accurate Broadband On-Wafer Calibration in RF Silicon Applications," *1st International MOS-AK Meeting*, Dec.13, 2008, San Francisco, CA, http://www.mos-ak.org/sanfrancisco/papers/09_Rumiantsev_MOS-AK_SF08.pdf

[5] A. Lord, Cascade Microtech Europe Ltd, *"Advanced RF Calibration Techniques,"* June 20–21, 2002, Wroclaw, Poland. http://ekv.epfl.ch/files/content/sites/ekv/files/mos-ak/wroc law/MOS-AK_AL.pdf

[6] B., Hughes; P. J. Tasker, "Bias dependence of the MODFET intrinsic model elements values at microwave frequencies," *IEEE Trans. Electron Devices*, vol. 36, no. 10, Oct. 1989, pp. 2267–2273.

[7] G. Dambrine, A. Cappy, F. Heliodore, and E. Playez, "A new method for determining the FET small-signal equivalent circuit," *IEEE Trans. Microw. Theory Techn.*, vol. 36, no. 7, Jul. 1988, pp. 1151–1159.

[8] J. Wood and D. E. Root, "Bias-dependent linear scalable millimeter-wave FET model," *IEEE Trans. Microw. Theory Techn.*, vol. 48, no. 12, Dec. 2000, pp. 2352–2360.

[9] A. Zarate de Landa, J. E. Zúñiga-Juárez, J. Loo-Yau, J. A. Reynoso-Hernández, M. C. Maya-Sánchez, and J. L. del Valle-Padilla, "Advances in linear modeling of microwave transistors," *IEEE Microw. Mag.*, April 2009, pp. 100–111.

[10] A. Zarate de Landa; J. E. Zúñiga-Juárez; J. Loo-Yau; J. A. Reynoso-Hernández; M. C. Maya-Sánchez; E. L. Piner, and K. J. Linthicum, "A new and better method for extracting

the parasitic elements of the on-wafer GaN transistors," *IEEE MTT-S International Symposium Digest*, Honolulu, Hawaii, June 3–8, 2007, pp. 791–794.

[11] R. Vogel, "Determination of the MESFET resistive parameters using RF-wafer probing," *17th European Microwave Conference*, September 1987 pp. 616–621.

[12] K. W. Lee, K. Lee, M. S. Shur, T. T. Vu, P. C. T. Roberts, and M. J. Helix, "Source, drain, and gate series resistances and electron saturation velocity in ion-implanted GaAs FETs," *IEEE Transactions on Electron Devices*, vol. ED-32, no. 5, May 1985, pp. 987–992.

[13] J. Xu, R. Jones, S. A. Harris, T. Nielsen, and D. E. Root, "Dynamic FET model – DynaFET – for GaN transistors from NVNA active source injection measurements," *International Microwave Symposium Digest*, Tampa, FL. June 2014.

[14] F. Lin and G. Kompa, "FET model parameter extraction based on optimization data-fitting and bi-directional search – a new concept," *IEEE Trans. Microw. Theory Techn.*, vol 42, no 7, 1994, pp. 1114–1121.

[15] C. Fager, K. Andersson, M. Ferndahl, "Uncertainties in small-signal equivalent circuit modeling," chapter 4 in *"Nonlinear Model Parameter Extraction Techniques,"* M. Rudolph, C. Fager, and D. E. Root, editors, Cambridge University Press, 2012.

[16] C. Fager, P. Linner, J. Pedro, "Optimal parameter extraction and uncertainty estimation in intrinsic FET small-signal models," *IEEE Trans. Microw. Theory Techn.*, vol. 50, December 2002, pp. 2797–2803.

[17] D. H. Smith, A. Fraser, and J. O'Neil, "Measurement and prediction of operating temperatures for GaAs ICs," *Proc. SEMITHERM*, Dec. 1986, pp. 1–20.

6 Nonlinear Device Modeling

6.1 Introduction

This chapter presents a survey of selected foundations and principles of large-signal device modeling for nonlinear circuit simulation. The chapter begins by recalling the early connection between simple transistor physics and nonlinear circuit theory. The evolution from simple physically based models to empirical models of various types is outlined, including table-based models and modern techniques based on artificial neural networks (ANNs). Formal as well as practical considerations related to ensuring robust model nonlinear constitutive relations for currents and charges are illustrated throughout. The relationship of large-signal models to linear data is treated extensively, for the purpose of proper parameter extraction and also for investigating the conditions under which bias-dependent small-signal data can be integrated to infer large-signal constitutive relations. Quasi-static models are analyzed in detail, and their consequences are deduced and checked against experimental device data.

Terminal charge conservation modeling concepts are considered in detail, including a treatment of charge modeling in terms of both depletion and "drift charge" components. Concepts of stored energy related to charge modeling are introduced including presently unresolved issues requiring future research.

Models for dynamic physical phenomena, especially self-heating and trapping mechanisms for III-V FETs, are introduced to explain the measured device behavior and to illustrate the reasons that quasi-static models are inadequate for most transistors. Symmetry principles are presented formally and as a practical tool to help the modeler ensure that the mathematical device model is consistent with the physical device properties.

Several modern modeling applications of nonlinear vector network analyzer (NVNA) data are presented. Even for simple models, the benefits of NVNA data, for parameter extraction and easy model validation at the time of extraction, offer large gains in modeling flow efficiency. For models incorporating multiple dynamic phenomena, such data can be used to efficiently separate and identify the independent effects. As an example, NVNA waveform data are used to generate a detailed nonlinear time-domain simulation model for III-V FETs, taking advantage of direct identification techniques for dynamical variables and the power of ANNs to fit scattered data in a multidimensional space. The model features dynamic self-heating and charge capture and emission mechanisms, and is extensively validated for several GaAs and GaN transistors.

The restriction to linear model behavior covered in Chapter 5 is removed, and therefore the economy of description afforded by the frequency domain treatment of Chapter 5 disappears. The models discussed in this chapter are lumped, meaning that their constitutive relations are defined via nonlinear current-voltage and charge-voltage relations, expressed in the time-domain. Lumped models are appropriate for transistors for excitation frequencies up to approximately the cutoff frequency, f_T, beyond which the devices become more and more distributed. Since the useful range of transistor applications is usually below f_T, we limit ourselves to the lumped nonlinear description.

6.2   Transistor Models: Types and Characteristics

6.2.1   Physically Based Transistor Models

The physics of semiconductor devices involves a complicated combination of quantum mechanics, solid state physics, material science, electromagnetism, and nonequilibrium thermodynamics. In the early 1950s, William Shockley derived the time-dependent terminal currents at the drain and gate of a field-effect transistor (FET) starting from the Poisson and current continuity partial differential equations [1]. The explicit solutions are shown in (6.1)–(6.4). Shockley made several simplifying assumptions along the way, such as field-independent charge carrier mobility (constant μ), the gradual channel approximation (the electric field is primarily perpendicular to the active channel), and neglecting the effects of self-heating. A simple derivation can be found in [2] (see also [3]).

$$I_{DS}(t) = I_{DS}^{DC}(V_{GS}(t), V_{DS}(t)) - \frac{dQ_{GD}(V_{GD}(t))}{dt} \tag{6.1}$$

$$I_G(t) = \frac{dQ_{GS}(V_{GS}(t))}{dt} + \frac{dQ_{GD}(V_{GD}(t))}{dt} \tag{6.2}$$

$$I_{DS}^{DC}(V_{GS}, V_{DS}) = \frac{W^{tot}\mu q N_D a}{\varepsilon L}\left(V_{DS} - \frac{2}{3}\left[\sqrt{\frac{2\varepsilon}{qN_D a^2}}\left((V_{DS} + \phi - V_{GS})^{3/2} - (\phi - V_{GS})^{3/2}\right)\right]\right) \tag{6.3}$$

$$Q_{GS}(V) = Q_{GD}(V) = Q(V) = -W^{tot}L\sqrt{2q\varepsilon N_D(\phi - V)} + const \tag{6.4}$$

Here W^{tot} is the total gate width, L is the gate length, a is the channel depth, q is the magnitude of the electron charge, ε is the dielectric constant of the semiconductor, N_D is the doping density of the semiconductor (assumed uniform), and ϕ is the built-in potential. The time-dependent drain and gate currents, (6.1) and (6.2), are evidently decomposable into distinct contributions from a voltage-controlled current source and two nonlinear two-terminal charge-based capacitors. That is, the solution of the physical partial differential equations leads to a rigorous representation, at the device terminals, in terms of standard lumped nonlinear circuit theory.

The nonlinear equivalent circuit corresponding to (6.1) and (6.2) is given in Figure 6.1. The functional forms (explicit nonlinear functions) of the

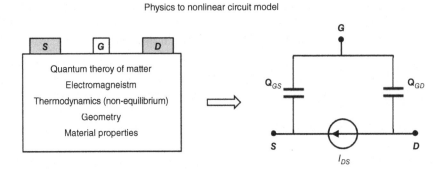

Figure 6.1 Physics to circuit representation.

constitutive relations for the current and charge sources are given by the expressions (6.3) and (6.4).

While the equivalent circuit of Figure 6.1 is usually justified based on its "resemblance" to the physical structure of the device (see also Figure 5.1), the above discussion provides a deeper explanation and justification. Moreover, in addition to the topology of the equivalent circuit, the details of the constitutive relations are obtained from this analysis.

The electrical performance of any circuit designed using the FET, in combination with other electrical components, can therefore be simulated by solving the set of ordinary nonlinear differential equations of circuit theory, using the methods of Chapter 2, rather than the partial differential equations of physics. In a very important sense, the nonlinear equivalent circuit representation of the transistor is a *behavioral model* suitable for efficient simulation of components and circuits at higher complexity levels of the design hierarchy [4]. This is quite fortunate, since it is not at all practical to solve the detailed partial differential equations of physics, at the transistor level, for a circuit composed of many transistors.

The constitutive relations, the particular functions for the I–V and Q–V relations (6.3) and (6.4), are closed-form nonlinear expressions involving the controlling terminal voltages, V_{GS} and V_{GD}, for the FET. We point out here that the domain of voltages over which these constitutive relations are defined is limited to $0 \le V_{DS} \le V_{DSsat}$ where $V_{DSsat} = \frac{qN_Da^2}{2\varepsilon} - (\phi - V_{GS})$. At $V_{DS} = V_{DSsat}$, the channel is pinched off and the slope of the I–V curve (the output conductance) is zero. This condition also determines the threshold voltage, V_T, as that value of V_{GS} required to pinch off the channel when the drain-source voltage is zero. We have $V_T = \phi - \frac{qN_Da^2}{2\varepsilon}$.

The domain of the model can be extended by defining the channel current to be constant, at the value $I_{DS}(V_{GS}, V_{DSsat})$, for values of $V_{DS} > V_{DSsat}$. The terminal voltages, V_{GS} and V_{GD}, are also limited to values less than the built-in potential, ϕ, at which the constitutive relations would become singular.

The notion that a model, well-defined initially in a limited domain, must have its domain properly extended for robust simulation, is a common and important consideration for successful nonlinear device modeling. We will discuss this in more detail in Section 6.2.5.3.

An important benefit of good physically based models is that the parameters entering the nonlinear constitutive relations tend to be meaningful. This is true for the Shockley model, where parameters include physical constants (e.g., q, the electron charge magnitude) and geometrical dimensions of the device (e.g., the gate width, W, the gate length, L, and the channel depth, a). There are material properties such as ε, the dielectric constant of the semiconductor, and the mobility μ, of the charge carriers. Also entering the constitutive relations is the device design parameter, N_D, the doping density of the semiconductor. Another important feature of the Shockley model, shared by many physically based models, is that the same parameters that enter the I–V constitutive relation (6.3) also enter the Q–V constitutive relations (6.4), coherently linking the resistive and reactive parts of the model through the common underlying physics.

A characteristic of the current-voltage constitutive relation (6.3) for the channel current is that it is a fully two-dimensional function that can't be de-composed into a product of two one-dimensional functions. That is, generally speaking, we have the condition expressed by (6.5).

$$I_{DS}(V_{GS}, V_{DS}) \neq g(V_{GS}) \cdot f(V_{DS}) \tag{6.5}$$

On the other hand, transistor technology has evolved rapidly since Shockley's analysis. The basic physical assumptions that led to (6.3) and (6.4) no longer apply, and it is generally not possible to arrive at simple closed form expressions for the current and charges as functions of the controlling voltages. Ironically, due to the phenomenon of charge-carrier velocity saturation (beyond the Shockley theory based on constant mobility), expressions like (6.5) can be more accurate for modern FET devices than the Shockley model suggests. Indeed, for simplicity of analysis, the approximation $I_{DS}(V_{GS}, V_{DS}) \approx g(V_{GS}) \cdot f(V_{DS})$ is used in the design example of Chapter 7.

Physically based models are predictive as long as the particular transistor technology is consistent with the physical assumptions. For example, where the Shockley model applies, the doping density could be changed and the resulting device characteristics would follow from (6.3) and (6.4) just by changing the numerical value for N_D. Of course this also means device technologies based on different principles of operation require distinct physical models, even within the same class of transistors (e.g. FETs). For example, a Si MOSFET and a GaAs pHEMT will have different physical models, despite some semi-quantitative similarity in their measured performance characteristics.

6.2.1.1 Limitations of Physically Based Models

Classical physically based models may neglect (intentionally or otherwise) certain physics that can influence the actual DUT performance. A closed-form expression may not exist for the constitutive relations without further approximations that may render the model insufficiently accurate for circuit design. The detailed physics of new technologies may not be fully known when the technologies come on line. This means physically based models, robustly implemented in commercial simulators, may not be available for the latest device technologies when the transistors themselves become useful for practical designs and products. It often takes years for the development and implementation of a good physically based nonlinear simulation model for a new semiconductor technology that is itself continuing to evolve. Some physical models may require knowledge of parameters, such as trap energy distributions, that may not be knowable or extractable from simple measurements, thus reducing much of the practical utility for these types of models.

6.2.1.2   Brief Note on Modern Physically Based Transistor Models

Over the past decade, physically based transistor models have made a comeback with the advent of surface potential models, primarily for MOSFET technologies [5–7]. These models define nonlinear constitutive relations for the terminal voltages in terms of the surface potential and also the currents as functions of the surface potential. Together, these equations define, implicitly, the current-voltage and charge-voltage relationships. This is in contrast to defining the currents and charges explicitly as functions of the voltages. More recently, the surface potential approach has been applied to III-V FET technologies (e.g. GaN) [8]. Other modern physically based III-V FET models, such as that proposed in [9], are based on different approaches. We will not delve into the important topic of modern physically based RF and microwave transistor models any further here but mention that this could be a re-emerging and important future trend.

6.2.2   Empirical Models

To deal with the limitations of physically based models discussed in Section 6.2.1.1, so-called empirical models were introduced [10–17] and still play a dominant role in microwave and RF nonlinear design applications, especially in the III-V semiconductor systems (e.g., GaAs, GaN, and InP).

Empirical models assume that the same equivalent circuit topology, suggested by basic physical considerations, is capable of representing the device dynamic characteristics, but that the I–V and Q–V constitutive relations can be treated as generic and independent functions to be chosen based on simple measurements and ease of parameter extraction.

For example, a classical empirical FET model due to Curtice [17] defines constitutive relations by equations (6.6)–(6.9) for the same equivalent circuit as shown in Figure 6.1.

$$I_{DS}(V_1, V_2) = \left(A_0 + A_1 V_1 + A_2 V_1^2 + A_3 V_1^3\right) \tanh{(\gamma V_2)} \tag{6.6}$$

$$Q_{GS}(V) = -\frac{C_{GS0}\phi}{\eta + 1}\left(1 - \frac{V}{\phi}\right)^{\eta+1} \tag{6.7}$$

$$C_{GS}(V) = \frac{C_{GS0}}{\left(1 - \dfrac{V}{\phi}\right)^{-\eta}} \tag{6.8}$$

$$Q_{GD}(V) = C_{GD0}V \tag{6.9}$$

For our purposes, V_1, in (6.6), can be taken to be the gate-source voltage, V_{GS}, and V_2 the drain-source voltage, V_{DS}.[1] Contrary to the general case, (6.5), and neglecting the V_{DS}-dependence of V_1, the channel current (6.6) takes the form of a product of two univariate functions. The polynomial dependence on V_1 is chosen for convenience. The corresponding parameters, $A_0 - A_3$ have no physical significance. The parameter, γ, is related to the voltage at which velocity saturation becomes important – a phenomenon not taken into account by the Shockley theory.

To simulate with the model, numerical values must be specified for each of the model parameters in all of the constitutive relations (6.6)–(6.9), typically by relating them to measurements on a reference device. This is the goal of parameter extraction. Obviously, the specific form of the intrinsic model constitutive relations can influence the parameter extraction strategy – the methodology by which the parameters are assigned numerical values. For the above model, $\{A_n\}$ define the gate-voltage dependence of the channel current, γ determines where the I–V curves saturate with drain bias, and C_{GS0}, C_{GD0}, ϕ, and η are parameters determining the charge variation of the two capacitors as functions of their respective controlling voltages.

It is also clear from (6.6)–(6.9), that the channel current and charge storage (nonlinear capacitance) model are less strongly coupled to one another in this model than in the Shockley model. In particular, the built-in potential, ϕ, does not even enter the channel current constitutive relation, while it plays a determinant role in the input capacitance model for C_{GS}.

Equation (6.7) is a standard physically based bias-dependent junction model, with corresponding capacitance value given by (6.8) with typical parameter values in the range $0 < \phi < 1$, $0 < \eta \leq 0.5$. Equation (6.9) describes a capacitance with a fixed value independent of bias. Notice that despite having a symmetric equivalent circuit topology, Figure 6.1, the different functional forms for $Q_{GS}(V)$ and $Q_{GD}(V)$ makes the Curtice model quite asymmetric with respect to the source and drain. Compare this to the fully drain-source symmetric Shockley charge model where the two nonlinear capacitors have the same Q–V functional form given by (6.4). The significance of these statements will be discussed in the context of symmetry principles and their consequences, presented in Section 6.5.

6.2.3   Quasi-Static Models

Both the Shockley model of (6.1)–(6.4) and Curtice model of (6.6)–(6.9) are quasi-static models. That is, the constitutive relations depend only on the instantaneous values of the

---

[1] In [17], V_1 was allowed to have a weak dependence on V_{DS}.

controlling variables (voltages in this case). There is no explicit time or frequency dependence in the relationships. This means, for example, that the I_{DS} current source will respond, instantaneously, to the applied intrinsic voltages, no matter how quickly the voltages may change with time. Similarly, the charge functions will respond, instantaneously, to the time-dependent applied intrinsic voltages.

An important consequence of a true quasi-static model, be it empirical or physically based, is that the resistive part of the intrinsic high-frequency RF and microwave model is fully determined and follows completely from the DC model characteristics. However, in real RF and microwave applications, the physical transistor is almost never operated in a true quasi-static condition. The RF signal is almost always much faster than the important thermal dynamic response, something we have not yet considered. For FETs, we will examine electro-thermal modeling in Section 6.6.

6.2.4   Empirical Models and Parameter Extraction

Since the constitutive relations of empirical models are chosen to fit the data, any deficiency of fit – at least for the I–V model fit to DC data – can be rectified by increasing the complexity of the function. For example, higher order polynomial terms can be added to the Curtice model I_{DS} constitutive relation (6.6). A problem with this approach, however, is that whenever the constitutive relations are changed, the parameter extraction process may also need to be modified.

Additionally, errors in the measurements will generally affect the model, since the parameter extraction process usually finds the best fit to the experimental data. The reader may wish to review, at this point, the more detailed discussion in Chapter 1, Section 1.5.1 about fitting static transfer functions.

For microwave FET devices of various types, several empirical models, notably the EEFET model family [14] and the Angelov (Chalmers) model family [15] and [16] are popular. The latter has evolved over 20 years to include many important dynamic effects such as self-heating and other dispersive or non-quasi static effects.

6.2.5   Nonlinear Model Constitutive Relations

6.2.5.1   **Good Parameter Extraction Requires Proper Constitutive Relations**

Even a simple intrinsic model constitutive relation, like that of Eq. (6.6), can be extracted improperly leading to disastrous results. The straightforward way to extract parameters in (6.6) is to measure I_{DS} versus V_1, say at fixed V_2, for which the tanh(.) term is effectively unity, and solve for the $\{A_n\}$ coefficients using a robust least-squares fitting process. Negative consequences of this direct approach appear when the constitutive relation (6.6) is evaluated outside the range of bias used to extract the coefficients. The resulting channel current may take physically unreasonable values for reasonable values of V_1. This is illustrated in Figure 6.2 [18–21]. The polynomial

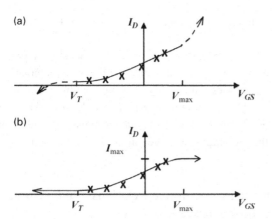

Figure 6.2 Problems caused by naïve parameter extraction methods for poorly defined constitutive relations. Data (x), models (lines). Top: Evaluation of polynomial beyond domain of validity (dashed lines). Bottom: proper extension of constitutive relations.

model may never pinch off (even if the device does), or the current can become large and negative and therefore unphysical in other cases.[2]

6.2.5.2 Properties of Well-Defined Constitutive Relations

The root cause of these problems is the formulation of the model constitutive relations themselves. Care must be used to define the domain of voltages where Eq. (6.6) applies and then to appropriately extend this domain to define the constitutive relationship properly for all values of V_1, while imposing reasonable conditions on the global I–V relationship. We encountered a similar issue with the Shockley physical model in Section 6.2.1. Model constitutive relations must have certain mathematical properties for a robust device model. They must be well-defined for all voltages, even at values far outside the range over which a real device might operate in any application. They must be continuous functions of their arguments. Moreover, the first order partial derivatives of the constitutive relations must be continuous everywhere. Usually the second partial derivatives should be bounded as well. These conditions are imposed mainly by the underlying Newton-type algorithms used by the simulator to converge to a solution of the circuit equations, something discussed in Chapter 2.

Even for solutions known to be within a subdomain of intrinsic terminal voltages where the constitutive relations are well-behaved, the process of iterating to convergence may require the evaluation of the constitutive relations at values of the controlling variables far outside this region. If, in these extreme regions, a singularity in the function evaluation of the constitutive relation is encountered, or if the corresponding derivative values point in the wrong direction, the simulator can be led astray for the next iteration, and convergence may fail altogether. It may also happen that the

---

[2] Channel current should be positive for positive V_{DS} assuming gate current is modeled by other current sources.

simulation converges to a nonphysical solution that happens to satisfy the circuit equations.

Further constraints on constitutive relations are induced by accuracy requirements for certain types of simulations. Accuracy requirements for distortion simulation, such as IM3 or IM5 at low signal amplitudes, impose higher order continuity constraints on the model constitutive relations. That is, constitutive relations should have nonvanishing partial derivatives of sufficiently high orders. Note that the constitutive relations defined by (6.6) do not satisfy this requirement since all fourth order and higher partial derivatives are identically zero.

6.2.5.3   Regularizing Poorly Defined Constitutive Relations: An Example

The solution to making (6.6) well-defined can be obtained by enforcing additional constraints on the model – namely that it pinches off and attains a maximum value – to make it physically reasonable for all values of the independent variables. Specifically, the model channel current should be constrained to be zero for all values of voltages at or below a value of $V_1 = V_T$, the threshold voltage. Continuity of the channel current and its first derivative at V_T means $I_{DS}(V_1 = V_T, V_2) = \frac{\partial I_{DS}(V_1 = V_T, V_2)}{\partial V_1} = 0$. Therefore, the cubic polynomial factor in (6.6) has a double root at $V_1 = V_T$, allowing us to factor out $(V_1 - V_T)^2$ from the general cubic expression, making it easier to obtain the remaining polynomial coefficients. We can also assert that there is a value, $V_1 = V_{max}$, where the current attains its maximum value and is constant for all higher values of V_1. These conditions enable us to reformulate (6.6) in terms of the three new parameters V_T, V_{max}, I_{max} as given in (6.10).

$$I_D(V_1, V_2)$$

$$= \begin{cases} 0 & V_1 < V_T \\ I_{max} \dfrac{(V_1 - V_T)^2}{(V_{max} - V_T)^3}(V_{max} - V_T + 2(V_{max} - V_1))\tanh(\gamma V_2) & V_T \leq V_1 \leq V_{max} \\ I_{max}\tanh(\gamma V_2) & V_1 > V_{max} \end{cases}$$

$$(6.10)$$

Equation (6.10) satisfies all the constraints required for a well-defined and reasonable constitutive relation [18–21],[3] provided only $V_T < V_{max}$ and I_{max} and γ are positive. Moreover, the new parameters now have a clear interpretation in terms of minimum and maximum values of the model current (see Exercise 6.1). However, this formulation of the model with corresponding parameter extraction methodology has used up all of the original fitting degrees of freedom in Eqn. (6.6). Distortion figures of merit, such as IM3, are completely determined by the three parameters, V_T, V_{max}, I_{max}. For real transistors, variations in the doping density can result in devices with identical values of V_T, V_{max}, I_{max} but with different shapes of the I–V curves between zero and

---

[3] This discussion neglects the general consideration of adding some residual very small positive conductance as a further aide to convergence.

I_{max}. Therefore, more general and flexible constitutive relations than those of (6.10) are required to independently and accurately model intermodulation distortion. See also [22].

It should be clear from the above discussion of the channel current constitutive relation that the charge constitutive relation (6.7) also needs to be extended for values of V_{GS} approaching ϕ and beyond. This is easily accomplished by linearizing (6.7) at some fixed voltage, $V_0 < \phi$. The result is given by Eq. (6.11).

$$Q_{GS}(V) = \begin{cases} -\dfrac{C_{GS0} \cdot \phi}{\eta + 1}\left(1 - \dfrac{V}{\phi}\right)^{\eta+1} & V < V_0 \\[4mm] -\dfrac{C_{GS0} \cdot \phi}{\eta + 1}\left(1 - \dfrac{V_0}{\phi}\right)^{\eta+1} + C_{GS0} \cdot \left(1 - \dfrac{V_0}{\phi}\right)^{\eta}(V - V_0) & V \geq V_0 \end{cases}$$

$$(6.11)$$

6.2.5.4 Comment on Polynomials for Model Constitutive Relations

Polynomials are fast to evaluate – so models using polynomial constitutive relations tend to simulate quickly. Polynomials are linear in the coefficients so their numerical values can be extracted from data efficiently and without the need for nonlinear optimization (e.g. by using least squares or pseudo-inverse methods). However, polynomials diverge for very large magnitudes of their arguments. They have only finite orders of nonvanishing derivatives, so they can cause discontinuities of simulated distortion at low signal levels. Extending the domain of polynomial constitutive relations beyond the boundary over which they are used for extraction becomes much more difficult when the expressions depend on more than one variable (e.g. both V_1 and V_2), unlike the simple case discussed above where the polynomial part of (6.6) depends only on V_1. In general, therefore, polynomial constitutive relationships should be used with great care or avoided altogether if possible.

6.2.5.5 Comments on Optimization-Based Parameter Extraction

For constitutive relations more complicated than (6.10), parameter extraction generally involves a simulation-optimization loop. An example of the general flow is given in Figure 6.3. However, such direct approaches can be slow. The model must be evaluated and parameters updated many times before a good result is obtained. Gradient-based optimization schemes may be sensitive to the initial parameter values or get stuck in local minima in the cost function (the error function between the desired value and actual value of the simulation at particular values of the model parameters). There are other techniques, such as simulated annealing [23] and genetic algorithms [24] that can help find a global solution to the nonlinear optimization problem, but these techniques are usually much slower and more complex.

The parameters in (6.11) must be constrained not to take specific values during optimization where the constitutive relations might become singular (e.g., for $\eta = -1$).

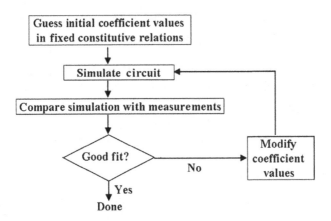

Figure 6.3 Optimization-based parameter extraction flow.

Modern parameter extraction software usually allows the user to restrict the parameter values to specific ranges during the iterative optimization process.

Advanced nonlinear models (see, e.g., [25]) have complicated nonlinear electrical constitutive relations. Most of these electrical nonlinear constitutive relations have nonlinear thermal dependences as well. It is usually best to extract parameters of such models by using an iterative scheme where dominant electrical parameters are extracted from specific subsets of data to which those parameters exhibit high sensitivities. A good flow is necessary for good, global fits to the data and also to get physically reasonable parameter values, which can then be scaled to model devices of other sizes without the need for an additional comprehensive extraction.

Of course, a given model with fixed, *a priori* closed form constitutive relations may *never* give sufficiently accurate results, for any possible set of parameter values. The model may be flawed, or just too simple to represent the actual behavior of the device. We turn now to other approaches that are more flexible.

6.2.6   Table-Based Models

Table models are usually classified as extreme forms of empirical models since the constitutive relations are not physically based but depend directly on measured data. In fact, there are no fixed, *a priori* model constitutive relations with parameters to be extracted at all. Table models are examples of "nonparametric" models. The *data are the constitutive relations*. The idea is simple enough, at least for the I–V constitutive relations of a quasi-static model. For the resistive model, just measure the DC I–V curves, tabulate the results, and interpolate as needed during simulation to evaluate the constitutive relations and their derivatives during the solution of the circuit equations. For the nonlinear reactive model, much more analysis needs to be done. We will address this in Section 6.2.9.

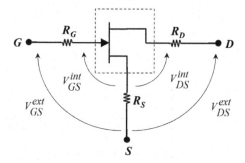

Figure 6.4 Extrinsic and Intrinsic device planes for nonlinear constitutive relations re-referencing

6.2.6.1 Nonlinear Rereferencing: Extrinsic-Intrinsic Mapping between Device Planes[4]

The intrinsic model nonlinear constitutive relations are defined on the set of intrinsic voltages, V_{GS}^{int} and V_{DS}^{int}, after accounting for the voltage drop across the parasitic resistances when voltages V_{GS}^{ext} and V_{DS}^{ext} are applied at the extrinsic device terminals (see Figure 6.4). Measured I–V data, on the other hand, are defined on the extrinsic voltages that correspond to the independent variables of the characterization. The relationship between extrinsic and intrinsic DC voltages is simple, given the resistive parasitic element values, previously extracted, as explained in Chapter 5, and the simple equivalent circuit topology. The equations are given in (6.12) [26]. An important issue for table models, however, is that the extrinsic voltages at which the measurements are taken are usually defined on a grid, but the resulting intrinsic voltages, explicitly computed by substitution using (6.12), do not fall on a grid, and therefore cannot often be directly tabulated. This is shown in Figure 6.5.

$$\begin{bmatrix} V_{GS}^{int} \\ V_{DS}^{int} \end{bmatrix} = \begin{bmatrix} V_{GS}^{ext} \\ V_{DS}^{ext} \end{bmatrix} - \begin{bmatrix} R_g + R_s & R_s \\ R_s & R_d + R \end{bmatrix} \cdot \begin{bmatrix} I_G^{DC} \\ I_D^{DC} \end{bmatrix} \tag{6.12}$$

If the measured extrinsic I–V data is fit or interpolated, equation (6.12) can be interpreted as a set of implicit nonlinear equations for the extrinsic voltages, V_{GS}^{ext} and V_{DS}^{ext}, given specified intrinsic voltages V_{GS}^{int} and V_{DS}^{int}, [18,26]. Solving (6.12) in this sense enables the data to be re-gridded on the intrinsic space so that the terminal currents can be tabulated as functions of the intrinsic voltages.

Modeling the measured I–V data as functions of the intrinsic voltages reveals characteristics quite different from the model expressed in terms of extrinsic data. This is shown in Figure 6.6. In part (a) of the figure, the modeled I–V curves as functions of the applied (extrinsic) voltages V_{GS}^{ext} and V_{DS}^{ext} are plotted. In part (b), intrinsic I–V modeled constitutive relations, defined on V_{GS}^{int} and V_{DS}^{int} are plotted. There is a big difference between Figure 6a and b, especially around the knee of the curves. This process also makes clear that errors in parasitic extraction – where the values of the

---

[4] In this Section, we associate V_{GS} with V_1 and V_{DS} with V_2, respectively.

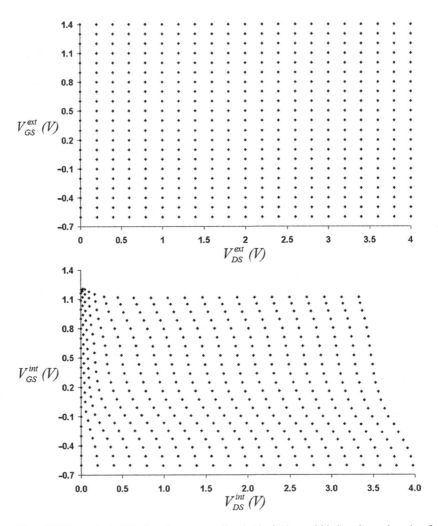

Figure 6.5 Extrinsic (gridded) and corresponding intrinsic (nongridded) voltage domain of a FET.

terminal resistances are determined – can distort the characteristics that we would otherwise attribute to the intrinsic model.

Alternatively, one can tabulate the extrinsic I–V data, as measured on the original grid, and treat the coupled equations (6.12) as additional model equations to be solved dynamically during simulation. This allows the simulator to sense the intrinsic voltages and look up the associated interpolated values of the measured I–V curves consistent with the solution of (6.12). This saves the post-processing step of regridding during the parameter extraction but adds the two nonlinear equations (6.12) to the model, thus increasing simulation time.

Table-based models can be both accurate and general. The same procedure and modeling infrastructure can be used to model devices in very different material systems (e.g. Si and GaAs) and manufacturing processes [26]. An example of the same

(a)

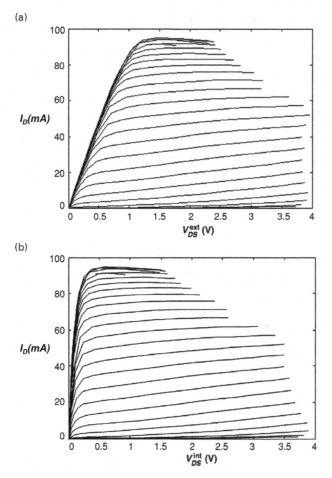

(b)

Figure 6.6 FET model I–V constitutive relations expressed as functions of (a) extrinsic and (b) intrinsic voltages.

table-based model applied to Si and GaAs transistors is given in Figure 6.7. For physically based models, each transistor would have very different constitutive relations requiring different parameter extraction procedures.

6.2.6.2   Issues with Table-Based Models

A critical issue with table-based models is the nature of the interpolation algorithms that define the constitutive relations as differentiable functions on the continuous domain containing all the discrete data points stored in the tables. The interpolator needs to define the partial derivatives continuously, and extrapolate appropriately – the same conditions that apply to all constitutive relations. At relatively small signal levels, it has been shown that simulations of harmonic distortion can become inaccurate when the amplitude (in volts) of the signal is comparable to or smaller than the distance between voltage data-points at which the constitutive relations are sampled [27]. This is

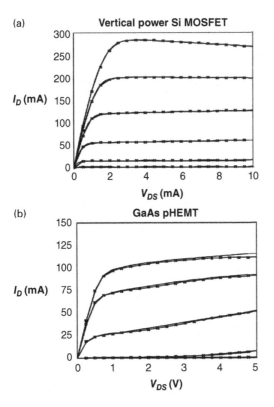

(a)

(b)

Figure 6.7 Table-based I–V models (−) and measurements (×) for Si MOSFET and GaAs pHEMT transistors.

demonstrated in Figure 6.8. At small voltage swings associated with low power signals, simulation results depend on the mathematical properties of the interpolant between data points rather than the underlying data itself. For larger signals, the corresponding applied voltage swings average out the local characteristics of the interpolant, and the table-based model simulations become quite accurate. It can sometimes help to increase the density of the data points. For some spline schemes, however, this can make the interpolant oscillate nonphysically between data points. Ultimately, there is a practical limit on taking too many data points leading to increased measurement time, file size, and interpolation of noise [27].

Various types of spline-based models have been explored in the literature. Methods involving B-splines have been applied in [28], but unphysical behavior due to spline oscillations between knots can still sometimes occur. Variation-diminishing splines [29] can tame oscillations, but their low polynomial order precludes their use for intermodulation simulation. "Smoothing splines" [30] can have a variable spline order and trade off accuracy for smoothing noise.

For good large-signal simulation results with table-based models, it is necessary to acquire data over the widest possible range of device operating conditions. This range

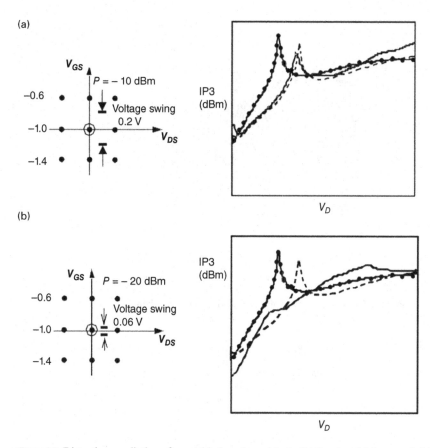

Figure 6.8 Distortion predictions from table-based models (solid lines) at (a) large and (b) small-signal amplitudes compared to empirical model (black dots) and measured data (dashed lines).

should include regions up to or just into breakdown, high power dissipation, and forward gate conduction, since these phenomena are critical to limiting the large-signal RF device performance. A portion of the domain of static measurements is shown in Figure 6.9 for a GaAs FET. The boundary illustrates the major mechanisms that constrain the data to the interior of the boundary. The precise shape depends on the detailed device-specific characteristics and the compliance limits on currents, set by the user, on the measurement equipment [31,32]. Covering a wide range of device operation during characterization reduces the likelihood of uncontrolled extrapolation during simulation and possible poor convergence as a consequence. Unfortunately, these extreme operating conditions can stress the device to the point of changing its characteristic during characterization [27]. This is especially true for static operating conditions at which DC I–V and S-parameters are measured. An excellent model of a degraded device can be obtained unless care is used. A delicate balance must be maintained between complete characterization and device safety. It is therefore very important to defer stressful static measurements until as late as possible in the characterization process [27].

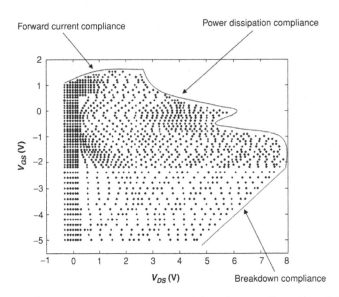

Figure 6.9 Data domain for pHEMT device with compliance boundaries.

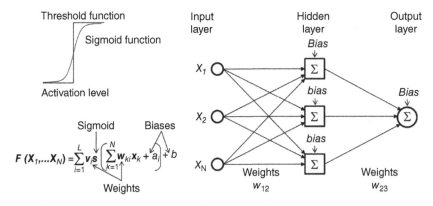

Figure 6.10 ANN illustrating sigmoid function, layer structure, and mathematical formula.

6.2.7   Models Based on Artificial Neural Networks (ANNs)

Many of the problems of table models, including issues of gridding, ragged boundaries, and poor interpolation properties, can be obviated by replacing the tables with artificial neural networks (ANNs) [33–36]. An ANN is a parallel processor made up of simple, interconnected processing units, called *neurons*, with weighted connections and biases that constitute the parameters [37,38]. A schematic of an ANN is given in Figure 6.10. Each neuron represents a simple univariate nonlinear "sigmoid" function, with range zero to unity, monotonically increasing, and infinitely differentiable with respect to its argument. The layer structure and interconnectedness of the neurons – specified by the weights – endows the overall network with powerful mathematical properties. The

Universal Approximation Theorem states that any nonlinear function, in an arbitrary number of variables, can be approximated arbitrarily well by such a network [37].

ANNs provide a powerful and flexible way to approximate the required model multivariate constitutive relations by smooth nonlinear functions from discretely sampled scattered data. ANNs provide an alternative to using multivariate polynomials, rational functions, or other more conventional basis sets to approximate the data. There are now powerful third party software tools [38,39] available to train the networks, that is, extract the weights and biases that form the parameters of the resulting function, such that the network approximates the measured nonlinear constitutive relationship well.

A key benefit of ANNs is the infinite differentiability of the resulting constitutive relations, providing a smooth approximation also for all the partial derivatives necessary for good low-level distortion simulation.[5] Another key benefit is that ANNs can be trained on scattered data. In particular, they can be trained directly on the scattered intrinsic $I - V^{int}$ data without the need for regridding. An example of an ANN I–V constitutive relation trained on the nongridded intrinsic voltage space of a pHEMT device is presented in Figure 6.11. Hard constraints on model constitutive relations, such as required by discrete symmetry properties, can also be accommodated by ANN technology. Symmetry conditions are discussed in Section 6.5. An example of an ANN-based FET model with drain-source exchange symmetry is presented in [34].

The mathematical form of an ANN nonlinear constitutive relation is a very complicated expression, typically involving many transcendental functions – even nested transcendental functions if there are multiple hidden layers. The expression can take many lines of mathematical symbols just to write out explicitly. However, from the point of view of the simulator, it is just a closed-form nonlinear expression like that of (6.10). The implementation of an ANN-based model in the simulator requires the values of the neural-based constitutive relations and their partial derivatives at all values of the independent variables, just like any conventional compact model. The parameters (weights and biases) can be placed in a datafile and read by the simulator for each model instance. The partial derivatives can be efficiently computed by evaluating a related neural network, called the adjoint network, obtained from the original network and weights [40].

6.2.8   Extrapolation of Measurement-Based Models

Conventional parametric empirical models, when properly formulated, are well-defined everywhere, even far outside the training data used to extract the parameters. As we have discussed, this can be achieved by defining the constitutive relations first in a bounded domain and then appropriately extending this domain, often by linearization.

More of a challenge is how to systematically extend the domain of table-based or ANN models. Table models based on polynomial splines can extrapolate very poorly, causing failures in convergence of simulation. An example of a table-model

---

[5] Well-modeled partial derivatives of constitutive relations are necessary, but not sufficient, for good low-level distortion simulations. The correct dynamical model equations are also needed.

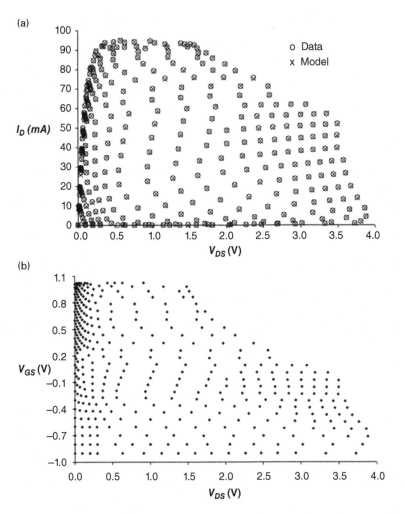

Figure 6.11 (a) ANN FET I–V model (x) and data (circles). (b) Nongridded intrinsic voltage domain.

extrapolation is shown in Figure 6.12a. The symbols indicate the actual data points. The solid lines correspond to the table-based model. Within the region of actual data, the model fits extremely well. At high drain voltages, the model extrapolates and the model curves cross in a nonphysical way. Eventually, the drain current of the model becomes negative (nonphysical) and the model does not converge robustly.

ANN models don't diverge as rapidly as polynomials, but their extrapolation properties are also poor from the perspective of simulation robustness. Successful deployment of table-based or ANN models can depend on a good "guided extrapolation" to help the simulator find its way back into or near the training region should it stray far from the region at any particular iteration. The method reported in [41] defines a compact domain containing the training region in terms of the convex hull constructed from the data points themselves [42]. Inside this region, the table or ANN is evaluated. Outside the boundary, an algorithm is applied to smoothly extend the current constitutive relation in

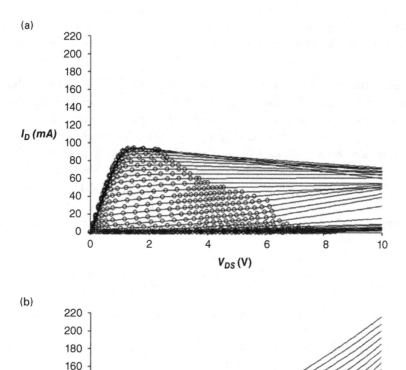

Figure 6.12 Extrapolation of measurement-based FET models. Data (+), model (lines): (a) without guided extrapolation, (b) with guided extrapolation.

a way that sharply increases the model branch conductances outside the training range. An example is shown in Figure 6.12b. This method increases the robustness of DC convergence and also the maximum power levels at which the model converges in large-signal harmonic balance analysis. Transient analysis becomes more robust as well. Other methods of extrapolation of ANN models can be found in [35,43].

6.2.9   Large-Signal Model Connection to Small-Signal Data

This section shows how to establish a connection between a large-signal model and small-signal measurements, in particular, linear S-parameter measurements. This is important for a variety of reasons, but primarily for parameter extraction based on readily available, calibrated data obtained from vector network analyzer instruments.

6.2.9.1   Linearization of Large-Signal Models around a DC Operating Point

We already encountered this procedure in Chapter 3 where we derived an S-parameter model from a simple nonlinear model of a FET. Here we derive small-signal models from large-signal models more generally, but still assuming a simple lumped description. We assume the following model for the intrinsic nonlinear two-port given in (6.13). Here the index, i, goes from 1 to 2, labeling the port currents and port voltages. Equation (6.13) allows gate leakage terms in parallel to the junction capacitances at the gate, terms that can become important as the controlling voltage values approach breakdown and forward conduction during the RF time-dependent large-signal excitation.[6]

$$I_i(t) = I_i^{vccs}(V_1(t), V_2(t)) + \frac{d}{dt} Q_i(V_1(t), V_2(t)) \tag{6.13}$$

We now assume the time-varying port voltages have a DC component, V_i^{DC}, and a small RF sinusoidal component, V_i^{rf}, a complex phasor at the angular frequency ω. We write the excitations as (6.14), substitute them into (6.13), compute the port current response to first order in the RF amplitudes, and finally evaluate the result in the frequency domain.

$$V_i(t) = V_i^{DC} + V_i^{rf} e^{j\omega t} + V_i^{rf*} e^{-j\omega t} \tag{6.14}$$

The result is (6.15), describing the port current complex phasors, I_i, associated with frequency ω, as a linear combination of the port voltage phasors, V_j^{rf}, at the same frequency. The small-signal model admittance matrix, $Y_{i,j}$, is simply expressed in terms of the partial derivatives of the port current and charge constitutive relations from the linearization of (6.13), according to (6.16). Here we introduce the real conductance and capacitance matrices, G and C, respectively, into which the complex admittance naturally separates.

$$I_i = \sum_j Y_{ij}(V_1^{DC}, V_2^{DC}, \omega) V_j^{rf} \tag{6.15}$$

$$Y_{ij}(V_1^{DC}, V_2^{DC}, \omega) = \frac{\partial I_i}{\partial V_j}\bigg|_{V_1^{DC}, V_2^{DC}} + j\omega \frac{\partial Q_i}{\partial V_j}\bigg|_{V_1^{DC}, V_2^{DC}} \equiv G_{ij} + j\omega C_{ij} \tag{6.16}$$

If we identify ports 1 and 2 with the gate and drain (common source configuration with the specified ordering), then we can write the matrix (6.16) in terms of elements indexed with the letters corresponding to gate and drain, as in (6.17).

---

[6] We neglected gate leakage in FETs for linear models in Chapter 5 because the small-amplitude RF signals could be assumed not to come close to these extreme conditions starting from a normal DC bias condition.

$$
Y_{ij}^{(CS)}\left(V_1^{DC}, V_2^{DC}, \omega\right) = \begin{bmatrix} G_{11} & G_{12} \\ G_{21} & G_{22} \end{bmatrix} + j\omega \begin{bmatrix} C_{11} & C_{12} \\ C_{21} & C_{22} \end{bmatrix}
$$

$$
= \begin{bmatrix} \dfrac{\partial I_G^{vccs}\left(V_{GS}^{DC}, V_{DS}^{DC}\right)}{\partial V_{GS}} & \dfrac{\partial I_G^{vccs}\left(V_{GS}^{DC}, V_{DS}^{DC}\right)}{\partial V_{DS}} \\[3ex] \dfrac{\partial I_D^{vccs}\left(V_{GS}^{DC}, V_{DS}^{DC}\right)}{\partial V_{GS}} & \dfrac{\partial I_D^{vccs}\left(V_{GS}^{DC}, V_{DS}^{DC}\right)}{\partial V_{DS}} \end{bmatrix} + j\omega \begin{bmatrix} \dfrac{\partial Q_G\left(V_{GS}^{DC}, V_{DS}^{DC}\right)}{\partial V_{GS}} & \dfrac{\partial Q_G\left(V_{GS}^{DC}, V_{DS}^{DC}\right)}{\partial V_{DS}} \\[3ex] \dfrac{\partial Q_D\left(V_{GS}^{DC}, V_{DS}^{DC}\right)}{\partial V_{GS}} & \dfrac{\partial Q_D\left(V_{GS}^{DC}, V_{DS}^{DC}\right)}{\partial V_{DS}} \end{bmatrix}
$$

$$(6.17)$$

From the analysis of Chapter 5, we had an independently derived expression for the *measured* intrinsic admittance matrix, expressed in terms of the ECPs based on the specific linear model topology considered. We reproduce equation 5.18, here for convenience as (6.18).

$$
Y_{ij}^{(CS)}\left(V_1^{DC}, V_2^{DC}, \omega\right) = \begin{bmatrix} G_{11} & G_{12} \\ G_{21} & G_{22} \end{bmatrix} + j\omega \begin{bmatrix} C_{11} & C_{12} \\ C_{21} & C_{22} \end{bmatrix}
$$

$$
= \begin{bmatrix} 0 & 0 \\ G_m & G_{DS} \end{bmatrix} + j\omega \begin{bmatrix} C_{GS} + C_{GD} & -C_{GD} \\ C_m - C_{GD} & C_{DS} + C_{GD} \end{bmatrix}
$$

$$(6.18)$$

Equating (6.17) and (6.18), matrix element by matrix element, *we can identify simple linear combinations of the measured linear* ECPs *at any particular bias point with specific partial derivatives of the large-signal model constitutive relations* evaluated at the corresponding bias. Even more generally, without any reference to the linear model equivalent circuit topology, the correspondence of (6.17) and (6.18), predicts that the intrinsic admittance matrix element values, obtained from the measured S-parameter data, should equal the partial derivatives of the large-signal model constitutive relations at the corresponding DC bias condition. This conclusion follows quite generally from the quasi-static large-signal modeling assumption that the intrinsic device can be modeled as a parallel combination of lumped nonlinear current sources and nonlinear charge sources. This assumption will be tested and analyzed extensively throughout the rest of this section.

6.3 Charge Modeling

6.3.1 Introduction to Charge Modeling

Nonlinear charge modeling has been shown to be critical for accurate simulation of bias-dependent high-frequency S-parameters [44], intermodulation distortion and ACPR in FETs [45,46], and harmonic and intermodulation distortion in III–V HBTs [47–50]. FET models with identical equivalent circuit topologies and identical I–V constitutive relations, differing from one another only by the form of their Q–V constitutive relations, can result in differences of 5–10 dB or more in simulated IM3 and differences of more than 5 dB in simulations of ACPR [45]. Moreover, conventional nonlinear charge models, based on textbook junction formulae, tend to show significant discrepancy compared to actual measured device characteristics.

The charge modeling problem can be simply stated as the specification of the non-linear constitutive relations defining the independent charges at the (intrinsic) terminals of the circuit model, as functions of the relevant independent controlling variables, usually voltages. The charge-based contribution to the current at the ith terminal is then the total time derivative of the charge function, Q_i. This is expressed in (6.19). If we choose one of the three terminals as a reference to define a particular 2-port description, we can interpret Q_i as the charge function at the ith independent port.

$$I_i(t) = \frac{dQ_i(V_1(t), V_2(t))}{dt} \tag{6.19}$$

Here V_1 and V_2 are the two independent intrinsic port voltages, which, for a FET, can be taken to be V_{GS} and V_{DS}. For an HBT, V_1 and V_2 can be taken to be V_{BE} and V_{CE}. I_1 is the gate (base) current and I_2 is the drain (collector) current for the FET (HBT) cases, respectively. Charge constitutive relations contribute to the current model through the time derivative operator in (6.19). This makes it apparent that charge plays an increasingly important role as the stimulus frequency increases.

The Curtice charge model with branch elements given by (6.7) and (6.9) can be cast in terms of terminal charges at the gate and drain, respectively, according to (6.20). Included in (6.20) is an additional nonlinear charge-based capacitor to model the drain-source capacitive coupling [11]. We see that the total terminal, or port charge function, is just the sum of the charges associated with all branch charges attached to the terminal with an appropriate reference direction. The arguments of (6.20) are time-varying intrinsic voltage differences.

$$\begin{aligned} Q_G(V_{GS}, V_{DS}) &= Q_{GS}(V_{GS}) + Q_{GD}(V_{GD}) \\ Q_D(V_{GS}, V_{DS}) &= -Q_{GD}(V_{GD}) + Q_{DS}(V_{DS}) \end{aligned} \tag{6.20}$$

For a three-terminal intrinsic model with gate and drain charges, Kirchhoff's Current Law (KCL), constrains the model to also have a charge function associated with the source node equal and opposite of the sum of the drain and gate charges. That is, we must have (6.21). Strictly speaking, the right-hand-side of (6.21) can be any constant, but we can set the constant to zero since the charges only enter the circuit equations in terms of a total time derivative. In the port description, we only need to define two independent port charge functions, Q_1 and Q_2.

$$\sum_{i=G,D,S} Q_i = 0 \tag{6.21}$$

We note that (6.21) is a circuit-level expression of *physical charge conservation*, a consequence of KCL and is to be carefully distinguished from the independent modeling concept that we will refer to by *terminal charge conservation*, to be introduced in Section 6.3.4.

The Shockley charge model (6.4) and the Curtice charge models (6.7) and (6.8), evidently fit into the form (6.20). In both the Curtice and Shockley models, the gate charge expressions separate into the sum of two functions, each defined on one variable only. For the Curtice model, this is also the case for the drain charge as evident from (6.20). It is a consequence of the idealized simplicity of these models that there are only two one-dimensional functions, $Q_{GS}(V)$ and $Q_{GD}(V)$, defining the entire two-port Shockley charge model, and three one-dimensional functions, $Q_{GS}(V)$, $Q_{GD}(V)$, and $Q_{DS}(V)$, defining the entire two-port Curtice charge model. The more general equation (6.19) is defined in terms of two arbitrary functions of two variables, $Q_G(V_1, V_2)$ and $Q_D(V_1, V_2)$, neither of which needs to be separable in terms of sums of univariate functions.

6.3.2   Measurement-Based Approach to Charge Modeling

Unlike DC I–V curves, charge constitutive relations cannot be measured directly. To specify the charge model, it is most convenient to establish a relationship between the model nonlinear charge constitutive relations and simpler quantities that can be directly obtained from bias-dependent S-parameter data. In fact, these relationships have been established by equating the imaginary parts of (6.17) and (6.18).

Assuming a particular two-port configuration, we obtain the following matrix equation, which relates the four nonlinear functions of bias constituting the imaginary parts of the admittance matrix, to the partial derivatives of the two model nonlinear charge functions, Q_i, evaluated at the operating point. Here indices i and j range from 1 to 2, the number of ports. For devices like MOSFETs, with additional terminals (e.g. bulk), i and j range from 1 to 3. The right-hand equality defines a capacitance matrix, C_{ij}, used for notational simplicity.

$$\frac{\text{Im}\left[Y_{ij}(V_1, V_2, \omega)\right]}{\omega} = \frac{\partial Q_i(V_1, V_2)}{\partial V_j} \equiv C_{ij}(V_1, V_2) \tag{6.22}$$

The assumption that the middle and right-hand side of (6.22) are independent of frequency is necessary for consistency with (6.19). In Chapter 5 we showed examples of measured device data consistent with the left-hand side of (6.22) being independent of frequency. In practice, with a good intrinsic and extrinsic equivalent circuit topology, and good parasitic extraction, (6.22) is approximately true for frequencies approaching the cutoff frequency of the device. For higher frequencies, the intrinsic model topology needs to be augmented, such as adding additional elements in series with the nonlinear capacitors, to deal with "non quasi-static effects." We described such R_{GS} and R_{GD} elements, for example, in Chapter 5, Section 5.6.5.2 for the small-signal model. For the large-signal model, however, this is a more difficult problem, and we won't deal with it further here.

The measured admittance parameters can be obtained by simple linear transformations of the (properly de-embedded) S-parameters according to (6.23), which is equivalent to 3.36 (see also [51]). The measured common source capacitance matrix then follows by taking the imaginary part of (6.23) and dividing by angular frequency.

$$Y_{ij} = \frac{1}{Z_0}\left[(I - S)(I + S)^{-1}\right]_{ij} \tag{6.23}$$

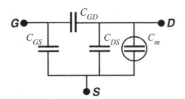

Figure 6.13 Linear equivalent circuit model of capacitance part of intrinsic FET model. See also Figure 5.27.

At this point, modeled and measured intrinsic imaginary admittance functions, or equivalently, the capacitance matrix elements, C_{ij}, can be directly compared.

However, it is more customary to compare modeled and measured small-signal responses in terms of linear equivalent circuit elements. There are many different intrinsic linear equivalent circuit representations that lead to the same intrinsic capacitance matrix. An alternative will be presented in Section 6.3.3. Defining equivalent circuit elements therefore requires a specific choice of equivalent circuit topology.

The linear equivalent circuit shown in Figure 5.27 maps, via the imaginary part of 5.18, the elements into the four independently measured capacitance matrix elements defined by 5.18. We solve these equations for the ECP values in (6.24). When applied to the transformed S-parameter measurements using (6.23), equations (6.24) define "measured" equivalent circuit elements. When applied to the linearized model admittances, (6.24) results in "modeled" equivalent circuit elements.

$$C_{GS} = C_{11} + C_{12}$$
$$C_{GD} = -C_{12}$$
$$C_{DS} = C_{22} + C_{12}$$
$$C_m = C_{21} - C_{12}$$

(6.24)

We note that there are four capacitance-equivalent circuit elements defined by Eq. (6.24). This is not surprising given that there are four functions, C_{ij}, corresponding to the four imaginary parts of the two-port admittance matrix. However, there are only three nodes in the equivalent circuit diagram of Figure 6.13. Historically, it was customary to place one capacitance between each pair of nodes. This procedure neglected, entirely, the fourth element, C_m, called the transcapacitance. As shown in Chapter 5, its contribution to the small-signal data is very significant.

A three-terminal (two-port) intrinsic equivalent circuit with two independent terminal charges generally leads to (at least) one transcapacitance. The equivalent circuit of Figure 6.13 or equivalently (6.24), places the transcapacitance in the device channel branch connecting drain and source, (parallel to the transconductance as is shown in Figure 5.27). It is important to note that the relationship of terminal charge partial derivatives to admittance matrices given in (6.22) is unique and is more fundamental than the set of transformations that define linear equivalent circuit elements in Eq. (6.24).

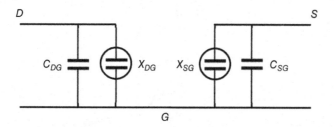

Figure 6.14 Alternative linear equivalent circuit for capacitance model of a FET.

6.3.3   Nonuniqueness of Equivalent Circuits

Here we demonstrate another way of looking at transcapacitances, starting from an equivalent circuit that looks a little different from that of Figure 6.13.

We consider the topology given by Figure 6.14. Another possibility is given in [52].

The elements in this circuit are defined by how they produce port currents, which is given by (6.25). Note that there are two capacitances and two transcapacitances in this description corresponding to the equivalent circuit of Figure 6.14, whereas the equivalent circuit of Figure 6.13 had three capacitances and one transcapacitance.[7]

$$I_{DG} = C_{DG}\frac{dV_{DG}}{dt} + X_{DG}\frac{dV_{SG}}{dt}$$
$$I_{SG} = C_{SG}\frac{dV_{SG}}{dt} + X_{SG}\frac{dV_{DG}}{dt}$$
(6.25)

The common gate capacitance matrix corresponding to (6.25) is given in (6.26).

$$\frac{\mathrm{Im}Y_{ij}^{CG}}{\omega} = C_{ij}^{CG} = \begin{bmatrix} C_{DG} & X_{DG} \\ X_{SG} & C_{SG} \end{bmatrix}$$
(6.26)

The forms of (6.25) and (6.26) are very simple and symmetric, and lend themselves to easier expressions for drain-source exchange symmetry. There is also a unique identification of the ECPs of Figure 6.14 and the common-gate imaginary part of the admittance parameters. This is even simpler than the linear combinations of common source Y-parameters needed to identify the ECPs of Figure 6.13 using (6.24).

Again we note that at a general bias condition, $X_{SG} \neq X_{DG}$ and so (6.26) is still not reciprocal, just like (6.24). If the device model has a nonreciprocal capacitance matrix in any port configuration, the conclusion is that the two-port model admits a transcapacitance.

From the point of view of the terminal charges, the results work out exactly the same, independent of where the transcapactiance element is placed in the equivalent circuit. The gate, drain, and source charges will have precisely the same functional forms on a fixed set of port voltages whether recovered from common gate Y-parameters using the

---

[7] In Figure 6.14 the transcapacitances are labeled with "X" while in Figure 6.13 the transcapacitance is labeled C_m.

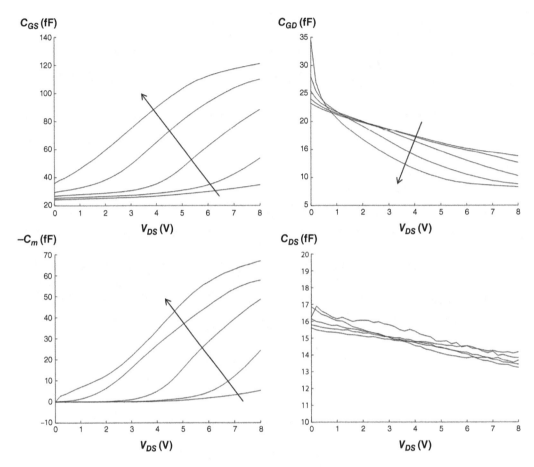

Figure 6.15 Bias-dependence of GaAs pHEMT capacitance matrix elements. V_{GS} ranges from -2 V to -1.2 V in 0.2 V steps. The arrows point in the direction of increasing V_{GS}.

topology of Figure 6.14 or common source Y-parameters using the topology of Figure 6.13.

Using (6.24), the measured capacitances can be compared to theory. An example of measured capacitances, as functions of the bias conditions, is given by Figure 6.15 for a GaAs pHEMT. Several facts are immediately apparent from the figures. The value of the transcapacitance, C_m, is generally zero or negative. This follows from the discussion in Chapter 5 related to equation (5.21). C_{GS} depends not just on V_{GS}, the voltage across the element, but also on the other independent voltage, V_{DS}. This is qualitatively different from the Shockley and Curtice models, where the model C_{GS} is completely independent of V_{DS}. This also means that C_{GS} cannot be modeled by a standard two-terminal nonlinear capacitor,[8] for any functional dependence on the (single) voltage, V_{GS}, across the element. The feedback capacitance, C_{GD}, depends on both

---

[8] A two-terminal lumped capacitor is defined by its branch relationship, Q(V), where V is the voltage difference across the two terminals [68].

V_{GS} and V_{DS}, in a more complicated way than $V_{GD} = V_{GS} - V_{DS}$. That is, C_{GD} also cannot be modeled by a standard two-terminal nonlinear capacitor, despite the familiar looking symbol in the linear equivalent circuit. Moreover, the V_{GS} dependence of C_{GD}, for large V_{DS}, when the device is in the saturation region of operation, is exactly the opposite of the Shockley model's prediction. That is, the feedback capacitance is actually much larger when the device is pinched off ($V_{GS} = -2\,\text{V}$) than when the channel is open ($V_{GS} = -1.2\,\text{V}$) and conducting current. More elaborate physical theories, which yield results closer to measured characteristics of modern FETs, lead to equations sufficiently complicated that they can be expressed, usually, only in approximate form [53]. A more recent approach, based on a decomposition of the charge model into simple one-dimensional depletion charges and a two-dimensional drift charge defined in terms of voltage and current, has been proposed in [54] and is presented in Section 6.3.9.

6.3.4   Terminal Charge Conservation

The above development, beginning with (6.19), started from large-signal model equations and then computed the small-signal responses that can be compared easily to measured S-parameter (Y-parameter) data. In what follows, we present a treatment based on trying to reverse the above flow. That is, we seek to solve the inverse problem to determine the functional form of the large-signal model constitutive relations, Q_i, that enter (6.19), directly from the measured bias-dependence of the small-signal characteristics. Unfortunately, but not surprisingly, this inverse process is generally ill-posed. However, under certain specific and verifiable conditions, a practical inverse modeling process can be constructed for transistors manufactured in a variety of material systems. This enables the generation of accurate nonlinear circuit simulation models for devices of great practical utility, from simple DC and *linear* (S-parameter) measurements.

Equation (6.22), with the left-hand side referring to measured data, can be interpreted as the mathematical statement of the inverse problem. That is, the *measured* bias-dependences of the intrinsic admittance elements are equated to the partial derivatives of the respective *model* terminal charges. The mathematical problem then becomes determining the conditions under which (6.22) can be solved for the model terminal charge functions, Q_i.

The necessary and sufficient conditions for the terminal charges to be recovered from bias-dependent capacitance matrix elements defined from measured data using Equations (6.22), are succinctly expressed by (6.27) [26,31,55].

$$\frac{\partial C_{ij}(V_1, V_2)}{\partial V_k} = \frac{\partial C_{ik}(V_1, V_2)}{\partial V_j} \tag{6.27}$$

Through definitions (6.22), equations (6.27) are constraints on pairs of bias-dependent measured admittances, one pair per row of the admittance matrix labeled by the index i. If (6.27) is satisfied, then the terminal charges can be constructed directly from the measured capacitance matrix elements by a path-independent contour (line) integration expressed by (6.28).

$$Q_i = \oint C_{i1}dV_1 + C_{i2}dV_2 \tag{6.28}$$

This result is unique up to an arbitrary constant that has no observable consequences and so the constant can be set equal to zero. Moreover, the partial derivatives of this charge reduce exactly to the measured bias-dependent capacitance measurements. That is,

$$\frac{\partial Q_i^{model}}{\partial V_j} = C_{ij}^{meas} \tag{6.29}$$

If (6.27) is not exactly satisfied for fixed index, i, then, strictly speaking, there is no function, Q_i^{model}, that is consistent with (6.29). In this case, the line integral using measured capacitance functions in Eq. (6.28) produces charge functions that depend on the path chosen for the contour. Different contours produce models that fit some capacitances versus bias better than others, with no perfect fits of all capacitances possible.

Equations (6.27), expressing the equality of mixed partial derivatives with respect to voltages of different capacitance functions with the same first index, can be interpreted as meaning that pairs of capacitance functions attached to the ith node form a conservative vector field in voltage space [31]. An alternate but mathematically equivalent representation of this concept is presented in [56]. Capacitance functions that obey (6.27) are said to obey "terminal charge conservation at the ith node."[9] We use the nomenclature *terminal charge conservation* to distinguish it from the fundamental physical law of charge conservation that is embodied in circuit theory by Kirchoff's Current Law (KCL) [compare (6.21) that embodies KCL]. In contrast, terminal charge conservation is a constraint that can, but needs not, be imposed by the modeler to approximate the behavior of a device. Physical charge conservation is a fundamental physical law and a requirement of any circuit model that is consistent with KCL. An example of a nonlinear model not consistent with terminal charge conservation, and its consequences, is presented in Section 6.3.7.

Any *model* starting from Eq. (6.19) has model capacitance functions that conserve terminal charge at each node. This is true because the model capacitances are derived in (6.22) starting from model charges and then (6.27) follows from the derivative properties of smooth functions. However, starting from independent *measurements* and trying to go back to *model* charges via (6.29) requires the constraints (6.27) to be satisfied by the measured C_{ij}^{meas} data.

The degree to which actual bias-dependent admittance data is consistent with the modeling principle of terminal charge conservation was investigated in [44] and [57]. For small GaAs FETs, it was found to hold extremely well at the gate, and slightly less well at the drain. For larger, high-power GaN HEMTs, these relationships don't seem to

---

[9] Early treatments did not use the "terminal" prefix and denoted this concept by "conservation of charge," leading to the confusion between the distinct concepts.

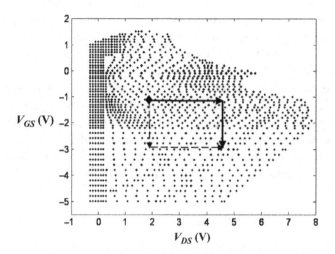

Figure 6.16 Two different paths in voltage space for line-integral calculation of terminal charges.

hold quite as well, and more elaborate models are required (see Section 6.6). The applicability of terminal charge conservation to III-V HBTs was discussed in [58] and [60] and found to be extremely useful in accurate modeling of delays and "capacitance cancellation" effects in such transistors.

6.3.5   Practical Considerations for Nonlinear Charge Modeling

The parameterization of line-integral (6.28) for two distinct paths, shown in Figure 6.16, are written explicitly in (6.30) and (6.31), respectively. Path independence means the same charge function can be computed from completely independent sets of bias-dependent data along the two paths of Figure 6.16.

$$Q_G(V_g, V_d) = \int_{V_{g0}}^{V_g} \left(C_{11}(\overline{V}_g, V_{d0})d\overline{V}_g + \int_{V_{d0}}^{V_d} \left(C_{12}(V_g, \overline{V}_d)d\overline{V}_d \right. \tag{6.30}$$

$$Q_G(V_g, V_d) = \int_{V_{g0}}^{V_g} \left(C_{11}(\overline{V}_g, V_d)d\overline{V}_g + \int_{V_{d0}}^{V_d} \left(C_{12}(V_{g0}, \overline{V}_d)d\overline{V}_d \right. \tag{6.31}$$

There are several issues with respect to implementing (6.30) and (6.31) directly on measured data. The measured capacitance data is defined only at the discrete voltages (points of in Figure 6.16) so the integrals have to be done numerically. If the data is not on a rectangular grid, interpolation along some of the paths may be required (as in in Figure 6.16 along the V_{GS} direction). Fundamentally, if (6.27) is not exactly satisfied due to measurement errors or the neglect of effects like temperature and traps (to be considered later), the different paths effectively trade-off model fidelity of $\text{Im}Y_{12}$

and $\mathrm{Im}Y_{11}$ versus bias, respectively. Integration error accumulates for large paths so the charge value at points far away from the starting point of integration (V_{G0} and V_{D0}) will be less accurate. The integration is also difficult to perform along paths to or near the ragged boundary of the data domain (see Figure 6.16).

Despite these practical difficulties, table-based models using full two-dimensional nonlinear gate charge functions constructed directly from small-signal device data have found their way into practical commercial tools [59]. Table-based charge models can be much more accurate than closed form empirical models, where the complex two-dimensional nature of the Q-V constitutive relations have not received as much attention as have I–V relations which are directly measureable.

6.3.6   Charge Functions from Adjoint ANN Training.

There are now robust methodologies to train ANNs to construct Q-V constitutive relations directly from knowledge of the desired function's partial derivatives as represented by measured capacitances [33,40]. This has been a major breakthrough for practical measurement-based charge modeling of transistors.

All the practical problems described above of computing multidimensional charge functions by line integration of suitably decomposed small-signal data are ameliorated by using the adjoint ANN training approach [40]. This method directly results in a neural network that represents the $Q_i(V_{GS}, V_{DS})$ functions from information only about their partial derivatives, as represented by the bias-dependent measured capacitances defined by (6.29). A diagram of the training method is given in Figure 6.17. If (6.27) is not exactly satisfied by the device data, the adjoint method still returns a charge function that generally gives a much better global compromise

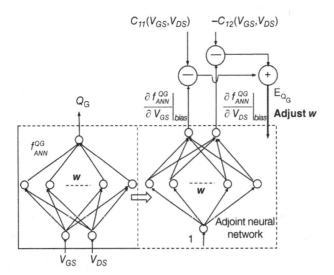

Figure 6.17 Adjoint ANN training of model gate terminal charge function from C_{11} and C_{12} data.

(a)

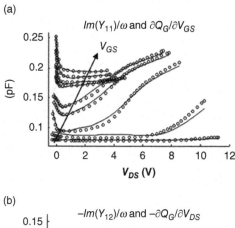

(b)

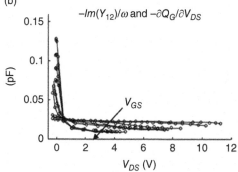

(c)

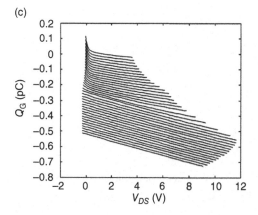

Figure 6.18 Validation of ANN-based gate charge model fitting, (a) C_{11}, (b) C_{12}. Data (symbols), model (lines); (c) gate charge function, Q_G.

between the capacitances than the typical line-integral methods. The training can take place directly on the intrinsic nongridded intrinsic bias data, and the ragged boundary presents no difficulty. Validation of the adjoint ANN approach to simultaneously fit the detailed two-dimensional FET input capacitance behavior with bias is shown in Figure 6.18. The validation for the independent fit for the drain capacitances is shown in Figure 6.19.

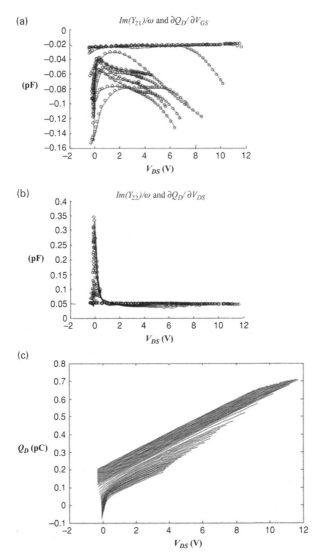

Figure 6.19 Validation of ANN-based drain charge model fitting, (a) C_{21}, (b) C_{22}. Data (symbols), model (lines); (c) gate charge function, Q_D.

With current-voltage and charge-voltage nonlinear constitutive relations modeled by ANNs, the improvement in simulation accuracy over spline-based table models can be demonstrated. A comparison is shown for the case of a GaAs pHEMT device in Figure 6.20. At moderate to high power levels, where the voltage swings are comparable or greater than the distance between discrete data points, the ANN and table-based models are nearly identical and compare well with measurements. At low power levels, the distortion simulation of the table-based model is determined by the numerical properties of the interpolating functions. Piecewise-cubic splines, used in this case, don't do a good job for high-order distortion at low signal levels, hence the

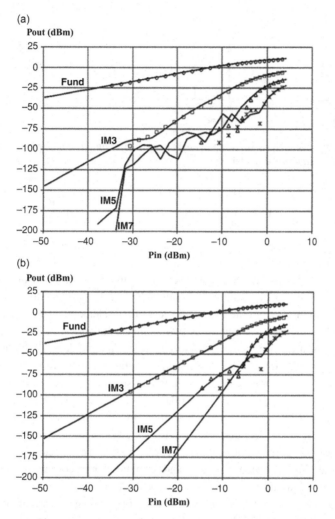

Figure 6.20 Distortion simulation results from (a) table-based model and (b) ANN-based model constructed from the same DC and small-signal data. Measured nonlinear validation data (symbols) and model predictions (lines).

ragged variation of distortion with power. However, the ANN model is very well-behaved at all power levels, and has the correct asymptotic dependence as the power decreases.

6.3.7 Capacitance-Based Nonlinear Models and Their Consequences

Strictly speaking, if (6.29) is not exactly satisfied, the assumption that (6.19) models the non-current-source terms in the intrinsic device is not consistent with the data. An alternative is to write the time-dependent port currents directly in terms of the measured two-port capacitance matrix elements [26,60]. That is, one can propose to replace (6.19) with more general equations according to (6.32).

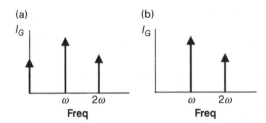

Figure 6.21 Spectra generated by (a) non terminal-charge-conserving gate capacitances and (b) gate terminal charge based model.

$$I_i(t) = C_{i1}(V_1(t), V_2(t))\frac{dV_1(t)}{dt} + C_{i2}(V_1(t), V_2(t))\frac{dV_2(t)}{dt} \tag{6.32}$$

In (6.32) the model functions C_{i1} and C_{i2} can be completely independent of one another, without the constraints of (6.27). It is quite easy to implement models like (6.32) for non terminal-charge-conserving capacitance-based large-signal equations in nonlinear circuit simulators. Equivalently, using the definitions (6.24), it is possible to rewrite (6.32) in terms of contributions from the four equivalent circuit elements of Figure 6.13 [26]. Compared to (6.19), which is specified by two nonlinear functions, (6.32) is defined by four model nonlinear functions (for a two-port device), which, if desired, can be taken to be precisely the measured (or independently fit) relations from equation (6.24). Such a model will exactly fit the measured bias-dependent small-signal dependence, by construction. However, as proved in [19,52,60,61] such models will generally lead to spectra containing a DC component, proportional to the stimulus frequency and the square of the signal amplitude, generated from such capacitance elements when simulated by large signals at the device terminals [18,26,55,62]. The spectra of non terminal-charge-conserving capacitance models and terminal charge conserving models are shown in Figure 6.21a and b, respectively. A spectrum with a DC component cannot result from true displacement current, the physical origin of gate current (neglecting leakage) in reverse biased FETs due to modulated stored charge. Things are less clear in the channel of a FET, where current arises by a combination of charge transport and time varying electric field [63]. Nevertheless, enforcing terminal charge conservation on large-signal intrinsic models results in a simpler model, with no "strange" consequences in large-signal analysis. Finally, it is possible to model the gate current using Equation (6.19) (for $i = 1$) and the drain current using (6.32) (for $i = 2$). That is, terminal charge conservation can be enforced at the gate terminal but not at the drain terminal, if desired.

6.3.8   Transcapactiances and Energy Conservation[10]

The transcapacitance, C_m, clearly shows up in device small-signal data as evidenced in Figure 6.15 (see also [64]). However, attempts to calculate stored charge from simple

---

[10] This section corresponds to an advanced topic and may be skipped on a first reading.

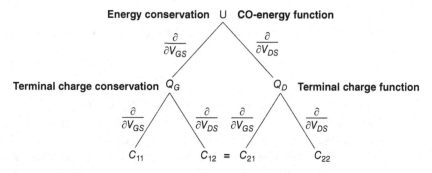

Figure 6.22 Computation of terminal charges and capacitance matrix elements from co-energy function.

theoretical conditions including energy conservation leads to the conclusion that $C_m = 0$, which is inconsistent with the small-signal data [26,65,66]. The magnitude of the channel current can increase with signal frequency in models that have large-signal terminal charges that admit transcapacitcances (this is the general case unless the mixed partial derivatives of the terminal charge functions are equal). This can cause an anomalously large simulated gain at high frequencies. Fortunately, the model parasitic network limits the rate of intrinsic voltage variation to a maximum frequency determined by the total input resistance and input capacitance product, partially mitigating the undesirable consequences.

The modeling principle of energy conservation for charge modeling begins with the assumption that all the electrical stored energy of an active device is computable only from the terminal charges [61]. The mathematical embodiment of stored energy-based terminal charges is perfectly analogous to that of the principle of terminal charge-based capacitance functions. That is, we would like to compute the function, U, called the co-energy, such that, for arbitrary terminal charge functions, Q_i, (6.33) holds for $i=1,2$ in the case of the two-port considered here [66]. It is important to emphasize that the co-energy, U, is generally not the same as the stored energy, W, even though they have the same units. There is a relationship between these functions that will be presented in equation 6.39.

$$\frac{\partial U(V_1, V_2)}{\partial V_i} = Q_i(V_1, V_2) \tag{6.33}$$

The formal relationship between (6.22) and its consequences, and (6.33), implies that for given terminal charge functions, Q_i, there is a unique (up to a constant) solution of (6.33) for U if and only if the terminal charge functions satisfy (6.34). If (6.34) is not satisfied, there is no solution to (6.33).

$$\frac{\partial Q_i}{\partial V_j} = \frac{\partial Q_j}{\partial V_i} \tag{6.34}$$

Assuming for the moment that (6.34) is satisfied, we can write a hierarchy of conservation principles according to Figure 6.22.

Assuming (6.33) admits a solution for U, we can formally assume a general co-energy function and try to fit the resulting capacitance functions. For example, we can write

$$U = \sum A_{nm} V_1^n V_2^m \qquad (6.35)$$

Then using (6.33) we can derive the forms for Q_1 and Q_2. Finally, taking partial derivatives of the charges gives us expressions for the four elements of the capacitance matrix, C_{ij}. We can try to determine the set of coefficients, A_{nm}, to best fit the measured capacitance matrix elements. However, this methodology raises an interesting question. We will find that we get only three unique expressions for the four capacitance matrix elements. In fact, we recognize equation (6.34) is a constraint on the bias-dependence of the terminal charge functions. Is this constraint satisfied by the data? We already know the answer. In fact, assuming only that the terminal charge functions are smooth [such as example (6.35)], (6.34) immediately leads to a reciprocal capacitance matrix (6.36).

$$C_{ij} = C_{ji} \qquad (6.36)$$

That means the transcapacitance, derived from a co-energy function, must vanish at all bias conditions [see the last of equations (6.24)]. The off-diagonal elements of the capacitance matrix are equal, since they are the mixed partial derivatives of the same co-energy function, U, assumed to be smooth. This is summarized in (6.37). Models that start only with the presumption of terminal charge conservation, but without the additional constraint of energy conservation, are less restrictive in terms of the resulting capacitance matrix elements, which can generally admit nonreciprocal capacitance matrices, as summarized in (6.38). A hierarchy of conservation laws is shown in Figure 6.22.

$$\begin{bmatrix} \dfrac{\partial^2 U}{\partial V_{GS}^2} & \dfrac{\partial^2 U}{\partial V_{DS}\partial V_{GS}} \\ \dfrac{\partial^2 U}{\partial V_{GS}\partial V_{DS}} & \dfrac{\partial^2 U}{\partial V_{DS}^2} \end{bmatrix} = \begin{bmatrix} C_{GS} + C_{GD} & -C_{GD} \\ -C_{GD} & C_{DS} + C_{GD} \end{bmatrix} \qquad (6.37)$$

$$\begin{bmatrix} \dfrac{\partial Q_G}{\partial V_{GS}} & \dfrac{\partial Q_G}{\partial V_{DS}} \\ \dfrac{\partial Q_D}{\partial V_{GS}} & \dfrac{\partial Q_D}{\partial V_{DS}} \end{bmatrix} = \begin{bmatrix} C_{GS} + C_{GD} & -C_{GD} \\ C_m - C_{GD} & C_{DS} + C_{GD} \end{bmatrix} \qquad (6.38)$$

However, such a conclusion is clearly at odds with the experimental data as shown in Figure 6.15. So we are now in a quandary. The clear evidence for a nonzero bias-dependent transcapacitance element is seemingly inconsistent with a simple embodiment of conservation of stored energy. Of course the basic physical law of energy conservation can't be violated, so we will need to expand the model to be compatible with physics on the one hand and the data on the other.

Energy conservation constrains the model terminal charge constitutive relations in precisely the same mathematical way that terminal charge conservation constrains the capacitance functions attached to a charge-based node. Since the mixed partial derivatives of U are equal, capacitance matrices are necessarily reciprocal and there is no model transcapacitance.

The energy function, W, can be derived from the co-energy function, U according to (6.39).

$$W(Q_1, Q_2) = \sum_i Q_i V_i - U(V_1, V_2) \tag{6.39}$$

This is consistent with the definition of stored energy in circuit theory given by (6.40) (see [66,67]).

$$W(Q_1, Q_2) = \oint V_1(Q_1, Q_2) dQ_1 + V_2(Q_1, Q_2) dQ_2 \tag{6.40}$$

6.3.9 FET Charge Modeling in Terms of Depletion and Drift Charges[11]

We saw in Section 6.3.3 that the bias-dependence of the measured FET capacitances, shown in Figure 6.15, could not be accurately modeled by simple lumped nonlinear two-terminal capacitors controlled by the single voltage difference across the respective branch elements. The multivariate approach of Sections 6.3.4 through 6.3.6, based on terminal-charge conservation principles, was able to produce models capable of accurately reproducing the complicated two-dimensional characteristics, as demonstrated in Figures 6.18 and 6.19. But this procedure does not give much insight into the origins of this bias-dependent behavior. That is, the complexity of the mathematical formalism obscures the simple physical relationship between carrier transport and charge storage mechanisms responsible for the behavior exhibited in Figure 6.15.

An approach that suggests the origins of this behavior was proposed in [54] and will be reviewed here, in a slightly simplified form. It is based on a previous application of terminal charge conservation modeling principles to large-signal HBT models [58,60,68].

We postulate here that the total FET charge model can be constructed from a simple depletion model of separable two-terminal nonlinear charge-based capacitors, with the addition of a new voltage and current-dependent "drift charge" term. The depletion capacitors have the simple one-dimensional dependence on the single voltage difference across the elements, just like the Shockley and Curtice models, expressed by (6.20). The drift charge is a two-dimensional function, depending on one voltage, which we take as V_{GD} (based on V_{BC} used in the HBT case), and the channel current, I_D. The charge model is given by (6.41). We assume for convenience

---

[11] This section corresponds to an advanced topic and may be skipped on a first reading.

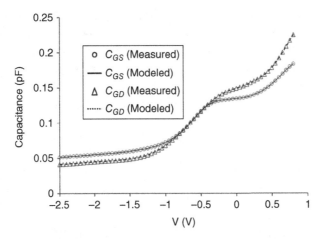

Figure 6.23 Depletion capacitances for a pHEMT defined according to (6.43).

the model is defined only for $V_{DS}, I_D \geq 0$; it can be globally extended using the methods of Section 6.5.

$$Q_G = Q_{GS}(V_{GS}) + Q_{GD}(V_{GD}) + Q^{drift}(V_{GD}, I_D)$$
$$Q_D = -Q_{GD}(V_{GD}) + Q_{DS}(V_{DS}) - Q^{drift}(V_{GD}, I_D) \tag{6.41}$$

The drift charge of a unipolar FET device plays the role of the diffusion charge of a bipolar HBT transistor. The idea is that the physical charge carriers responsible for the transport current, moving at the saturated velocity, modify the otherwise simple bias-dependent contribution to the imaginary part of the intrinsic model admittance from the depletion capacitances. The velocity is inversely proportional to a transit delay, τ, that itself depends on bias and can be extracted from simple S-parameter measurements.

The interpretation of Q^{drift} as a drift charge requires that it vanish when no current flows. This is expressed by equation (6.42), and this property will be confirmed by the final expression to be derived.

$$Q^{drift}(V_{GD}, I_D = 0) = 0 \tag{6.42}$$

At $V_{DS} = 0$ we can assume $I_D = 0$ and hence $Q^{drift} = 0$, simplifying (6.41) for the identification of the depletion charges. We identify the simple univariate nonlinear depletion capacitances, $C_{GS}(V)$ and $C_{GD}(V)$ from the (intrinsic) admittances, (6.43), evaluated at $V_{DS} = 0$, as functions of the reverse bias voltage, $V = V_{GS}$. The depletion charges, $Q_{GS}(V)$ and $Q_{GD}(V)$, follow from (6.43) by simple one-dimensional integration with respect to V. Measured and modeled gate depletion capacitances for a GaAs pHEMT are shown in Figure 6.23. The shapes are consistent with simple one-dimensional depletion physics, as expected. This means simple physically based constitutive relations can be used for this part of the nonlinear charge model, rather than just numerical calculations from measured data.

$$C_{GS}(V) = \frac{\text{Im}(Y_{11}(V_{GS} = V, V_{DS} = 0, \omega) + Y_{12}(V_{GS} = V, V_{DS} = 0, \omega))}{\omega}$$

$$C_{GD}(V) = \frac{-\text{Im}(Y_{12}(V_{GS} = V, V_{DS} = 0, \omega))}{\omega}$$

(6.43)

We now define a *transit delay* function, $\tau(V_{GD}, I_D)$, in terms of the partial derivative of the model drift charge function with respect to the drain current. With this definition, and using (6.41), we can identify $\tau(V_{GD}, I_D)$ from small-signal measurements by subtracting the depletion capacitance contribution from the imaginary part of the intrinsic admittance after changing variables to the pair V_{GD} and I_D. The result is (6.44).

$$\tau(V_{GD}, I_D) \equiv \frac{\partial Q^{drift}}{\partial I_D}$$

$$= \frac{\text{Im}(Y_{11}(V_{GD}, I_D, \omega)) + \text{Im}(Y_{12}(V_{GD}, I_D, \omega)) - \omega C_{GS}(V_{GS}(V_{GD}, I_D))}{\omega(G_m(V_{GD}, I_D) + G_{DS}(V_{GD}, I_D))}$$

(6.44)

Since we started from a charge-based model, (6.41), we know that there is a corresponding partial derivative of the drift charge with respect to the voltage, V_{GD}, which defines an effective capacitance according to (6.45)[12].

$$C_{GD}^{drift}(V_{GD}, I_D) \equiv \frac{\partial Q^{drift}}{\partial V_{GD}}$$

$$= \frac{1}{2} \left[\begin{array}{l} \frac{\text{Im}(Y_{11}(V_{GD}, I_D, \omega)) - \text{Im}(Y_{12}(V_{GD}, I_D, \omega))}{\omega} \\ -C_{GS}(V_{GS}(V_{GD}, I_D)) - 2C_{GD}(V_{GD}) - \tau(V_{GD}, I_D) \cdot (G_m(V_{GD}, I_D) - G_{DS}(V_{GD}, I_D)) \end{array} \right]$$

(6.45)

We know from the discussion in previous sections that terminal charge is conserved at the gate, and given the results of Exercise 6.4, we can use the adjoint training procedure of 6.3.6, in V_{GD}, I_D coordinates, to compute the model function, Q^{drift}, where the right-hand sides of (6.44) and (6.45) are provided from bias-dependent RF data.

The insight comes from realizing that Q^{drift} can be evaluated by a path-independent contour integral involving $\tau(V_{GD}, I_D)$ and $C_{GD}^{drift}(V_{GD}, I_D)$, as shown in (6.46). But since we must have $C_{GD}^{drift}(V_{GD}, I_D = 0) = 0$ from (6.42), we can choose a convenient contour where only one leg is nonzero, and we have the very simple result (6.47), which vanishes at $I_D = 0$ verifying (6.42). Note that result (6.47) means that we don't even need (6.45) to compute Q^{drift}.

$$Q^{drift} = \oint \left(\tau dI + C_{GD}^{drift} dV_{GD} \right)$$

(6.46)

$$Q^{drift}(V_{GD}, I_D) = \int_0^{I_D} \tau(V_{GD}, I) dI$$

(6.47)

---

[12] Equation (6.45) corrects a misprint in Equation 4 of [54].

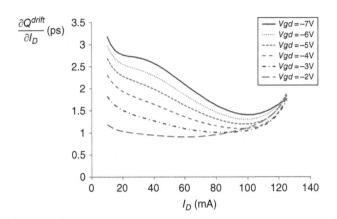

Figure 6.24 Bias-dependent delay from GaAs FET drift charge model.

The bias-dependent delay, recoverable from the modeled drift charge computed from data on a GaAs pHEMT, is shown in Figure 6.24. The delay function can be directly related to the velocity-field characteristics of the III-V semiconductor transport curves, enabling modeling from a physical perspective.

The charge model (6.41) fits the bias-dependent admittance data quite well [54]. It provides nearly the same accuracy as that of the general charge-based model (6.28), but with the benefit of separately identifying the depletion and transport mechanisms that, in combination, account for the overall behavior exhibited in Figures 6.15, 6.18, and 6.19. Moreover, if the drain current model includes thermal effects, then incorporating the drain current dependence in the drift charge automatically includes the temperature dependence into at least part of the reactive model, as emphasized in [69].

This approach has the potential for enabling intuitive and accurate bottoms-up analytical expressions for empirical and physically based FET models, while preserving the simple principles of device operation.

6.4   Inadequacy of Quasi-Static Large-Signal Models

This section uses the connection between partial derivatives of large-signal constitutive relations and linear data derived in 6.2.9.1 to demonstrate that the fundamental quasi-static assumption is not satisfied.

Specifically, the equality of real matrix elements on the bottom rows of (6.17) and (6.18) constitutes a prediction that the derivatives with respect to V_{GS} and V_{DS}, of the DC drain current constitutive relation should numerically equal the transconductance and output conductance, respectively, as measured at RF and microwave frequencies. That is, the DC and RF conductances should be equal.

These *predictions* beg to be tested experimentally. To do so we have to account for parasitic resistances consistently between DC and RF cases, to properly take the partial derivatives of the currents with respect to the intrinsic voltages, and to de-embed the S-parameter data.

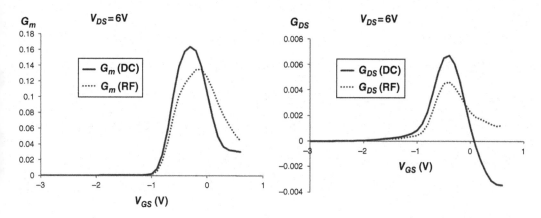

Figure 6.25 Measured conductances from DC data and RF data versus extrinsic bias for a GaAs pHEMT.

6.4.1   Non-Quasi-Static Conductances

Figure 6.25 shows direct comparisons of measured data versus bias for a GaAs pHEMT transistor. There is a significant difference between the DC and RF characteristics over most of the bias range. We note that the value of the DC output conductance, G_{DS}, can even become negative at some biases. The data in Figure 6.25 is typical of real microwave transistors and therefore proves that the quasi-static modeling assumption is not justified. This confirms the statement made in Section 6.2.3. We can express the failure of the quasi-static approach through (6.48). The simple textbook relations don't hold! The discrepancy between device properties at DC versus RF conditions is sometimes referred to as *frequency dispersion*. It can have multiple causes, some of which we will examine in more detail, in particular in Section 6.6.

$$G_m^{DC} \equiv \frac{\partial I_D^{DC}}{\partial V_{GS}^{DC}} \neq \mathrm{Re} Y_{21}(\omega) \equiv G_m^{RF}$$

$$(6.48)$$

$$G_{DS}^{DC} \equiv \frac{\partial I_D^{DC}}{\partial V_{DS}^{DC}} \neq \mathrm{Re} Y_{22}(\omega) \equiv G_{DS}^{RF}$$

The inadequacy of a quasi-static model is further supported by the analysis of the tradeoffs shown in Figure 6.26. The top two plots show (left to right) a quasi-static model for the channel current constitutive relation that fits the DC I–V curves very well but predicts poorly the RF conductances. The bottom two plots show (right to left) a quasi-static model for the channel current constitutive relation that fits well the bias-dependent RF data but poorly predicts the DC I–V curves. This is a fundamental limitation of the quasi-static approach, independent of the details of the constitutive relations for modeling real devices.

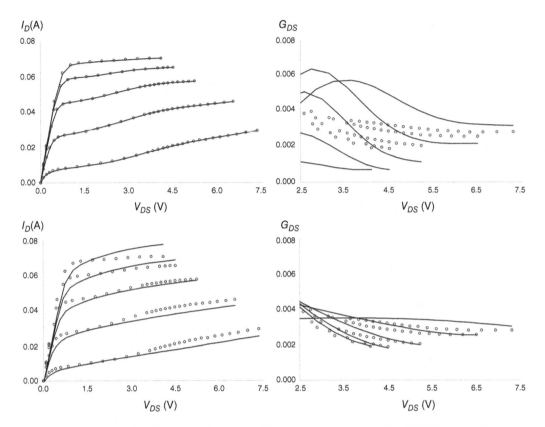

Figure 6.26 Inability of quasi-static models to simultaneously predict DC I–V characteristics and bias-dependent RF data.

6.4.2   High-Frequency Current Model

Just as we did to try to compute terminal charges from pairs of bias-dependent capacitance functions, we can try to find a single model function, $I_D^{RF}(V_{GS}, V_{DS})$, such that its distinct partial derivatives fit, simultaneously, the measured RF conductances, $G_m^{RF}(V_{GS}, V_{DS})$ and $G_{DS}^{RF}(V_{GS}, V_{DS})$. The mathematical conditions for a solution of this problem are the same as discussed in Section 6.3.4, relating charge functions to pairs of capacitance functions. We know therefore, there is a unique solution provided the *measured* RF conductances satisfy the condition (6.49). This implies there is a relationship between the bias-dependent data, through (6.48), that can be directly tested [57].

$$\frac{\partial G_m^{RF}}{\partial V_{DS}} = \frac{\partial G_{DS}^{RF}}{\partial V_{GS}} \tag{6.49}$$

If and only if (6.49) is satisfied we can construct the model function $I_D^{RF}(V_{GS}, V_{DS})$ uniquely up to a constant [usually fixed by requiring $I_D^{RF}(V_{GS} = 0, V_{DS} = 0) = 0$], using the relationship (6.50). The contour integral expressed by (6.50) is path-independent if (6.49) is satisfied.

$$I_D^{RF} = \oint G_m^{RF} dV_{GS} + G_{DS}^{RF} dV_{DS} \tag{6.50}$$

$$\frac{\partial I_D^{RF}}{\partial V_{GS}} = G_m^{RF}$$

$$\frac{\partial I_D^{RF}}{\partial V_{DS}} = G_{DS}^{RF} \tag{6.51}$$

Both small-signal *measured* conductances are then perfectly recoverable from the single *model* function, $I_D^{RF}(V_{GS}, V_{DS})$ according to (6.51).

Condition (6.49) is approximately satisfied for many devices over a reasonable range of bias conditions, but not perfectly. It does not hold as well for large power transistors over the full bias range. The expression (6.50) can be used to construct the model function, $I_D^{RF}(V_{GS}, V_{DS})$, but the result will have some residual path-dependence since (6.49) is not exactly satisfied. Alternatively, the adjoint neural network method can be used to compute an optimal solution to (6.51), that trades off errors in fitting the RF transconductance with that of the RF output conductance.

The adjoint training method was in fact used to produce the fit in the lower right plot of Figure 6.26 to G_{DS}^{RF} and also G_m^{RF} (not shown). The constructed $I_D^{RF}(V_{GS}, V_{DS})$ function is depicted by the set of solid lines in the lower left plot of Figure 6.26. It is evident from this plot, as expected, that $I_D^{RF}(V_{GS}, V_{DS}) \neq I_D^{DC}(V_{GS}, V_{DS})$.

Some quasi-static models are based on *substituting* constitutive relation $I_D^{RF}(V_{GS}, V_{DS})$ for the constitutive relation $I_D^{DC}(V_{GS}, V_{DS})$ and ignoring the error at DC. This will certainly improve the prediction of the RF model behavior but, as we have seen, will degrade the DC fit thus causing problems with biasing of multiple transistor models with DC coupling. For some calculations, such as power added efficiency, both the RF large-signal behavior and the DC behavior must be predicted simultaneously.

Is it possible to somehow use both constitutive relations, $I_D^{RF}(V_{GS}, V_{DS})$ and $I_D^{DC}(V_{GS}, V_{DS})$, in a non-quasi-static model together so as to be correct at DC and also at RF? One such phenomenological approach was taken in [31] and later [33], which combined the two constitutive relations to define an effective channel current in terms of the solution of the linear differential equation given by (6.52). Here $I_D^{RF}(V_{GS}(t), V_{DS}(t))$ and $I_D^{DC}(V_{GS}(t), V_{DS}(t))$ are understood to mean the constitutive relations constructed from the bias-dependent RF and DC data, respectively, evaluated at the time-dependent intrinsic voltages during the simulation. The frequency-dependent small-signal model resulting from linearizing (6.52) smoothly transitions from the DC conductances to the RF conductances, at a characteristic frequency given by τ^{-1}. This parameter is estimated as the time constant in a one-pole approximation to the thermal or trapping response.

$$\tau \frac{dI(t)}{dt} + I = \tau \frac{dI_D^{RF}(V_{GS}(t), V_{DS}(t))}{dt} + I_D^{DC}(V_{GS}(t), V_{DS}(t)) \tag{6.52}$$

Analytical models, specifically the EEHEMT model [14], also use both $I_D^{RF}(V_{GS}(t), V_{DS}(t))$ and $I_D^{DC}(V_{GS}(t), V_{DS}(t))$ constitutive relations and combine them to produce a non-quasi-static model. In this case, the two functions $I_D^{RF}(V_{GS}(t), V_{DS}(t))$ and $I_D^{DC}(V_{GS}(t), V_{DS}(t))$ have the same functional form but are independently extracted to RF and DC data, respectively, so they have different numerical values for the parameters.

Such phenomenological approaches to non-quasi-static models have had considerable success over the past 25 years. However, they may result in unphysical results when used in certain large pulsed transient applications. With the power presently available in standard engineering workstations, and the evolution of fast, reliable non-linear simulators, it is preferable to instead model the basic phenomena, namely dynamic self-heating and trapping effects, that are known to be responsible for the non-quasi-static effects in the first place. These topics will be addressed in Sections 6.6 and 6.7.2.

6.5   Symmetry

When a device or system is physically transformed in some way, but an observable property doesn't change, we say the property is invariant with respect to that transformation. For such a case we say the device or system has a particular *symmetry*. A simple example is afforded by a circle. If a circle is rotated by an arbitrary angle around its center its shape is invariant. It has a (continuous) symmetry of rotational invariance. One cannot tell after the rotation that the circle had been changed in any way from its original state. A rectangle does not have this symmetry property. A rectangle does have discrete symmetry properties, with respect to rotations of 180 degrees, and reflections about diagonals and lines bisecting the sides.

Any mathematical model that is not endowed with the symmetry property of the object or system it is describing *cannot be globally valid* and will, under certain large-signal conditions, predict results inconsistent with data. Models with incorrectly implemented symmetric constitutive relations have been known to cause serious problems of unphysical behavior [70]. It is therefore a requirement of correct models that they be implemented with the necessary mathematical symmetry[13]. This is quite analogous to the principle that led to building time-invariance directly into the nonlinear spectral maps in our development of X-parameters in Chapter 4.

6.5.1   Drain-Source Exchange Symmetry

In the case of an ideal single-finger FET, such as shown in Figure 5.1, the physical device structure is clearly unchanged if we interchange the source and drain terminals.

---

[13] An exception to the requirement is if it can be somehow guaranteed that the model will be exercised only within a local region of operation that never crosses into the corresponding symmetric region of operation under large-signal stimulus.

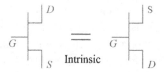

Figure 6.27 Drain-source exchange symmetry of an ideal intrinsic FET

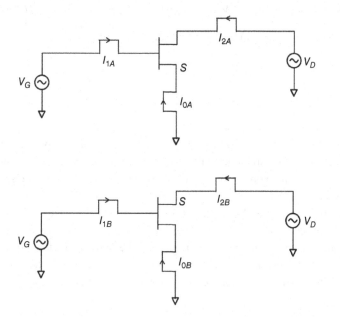

Figure 6.28 Connections of device and its symmetrical counterpart to deduce symmetry relations.

While the specific multigate layouts with feed structures and parasitic elements of Figures 5.3 and 5.5 clearly violate this symmetry, we can assume, if the gate is deposited exactly between the source and drain terminals, at least that the intrinsic device possesses drain-source exchange symmetry. The principle of drain-source exchange symmetry is expressed schematically in Figure 6.27.

A consequence of this symmetry is that if we imagine applying the same stimulus conditions to the device, wired in each of the two configurations above, identical responses must result from the device in both cases. We illustrate the configuration in Figure 6.28.

The symmetry condition requires that the currents measured into each of the device terminals, including the ground terminal, must be identical in the top and bottom configurations of Figure 6.28.

Starting with two of the current meters in the corresponding cases, we must have, for a symmetric device, relations (6.53).

$$I_{1B} = I_{1A}$$
$$I_{0B} = I_{0A}$$

(6.53)

We now express the currents in (6.53) by the values of the terminal current constitutive relations of the FET in the respective configurations. This results in (6.54), where in the second line we use KCL to re-express the source current in terms of the drain and gate currents of the upper configuration. The final results are summarized in (6.55).

$$I_{1B} = I_G(V_{GS} - V_{DS}, -V_{DS}) = I_G(V_{GS}, V_{DS}) = I_{1A}$$

$$I_{0B} = I_D(V_{GS} - V_{DS}, -V_{DS}) = I_S(V_{GS}, V_{DS}) = I_{0A} = -I_{1A} - I_{2A} \qquad (6.54)$$

$$= -I_G(V_{GS}, V_{DS}) - I_D(V_{GS}, V_{DS})$$

$$I_G(V_{GS} - V_{DS}, -V_{DS}) = I_G(V_{GS}, V_{DS})$$
$$\qquad\qquad\qquad\qquad\qquad\qquad\qquad\qquad (6.55)$$
$$I_D(V_{GS} - V_{DS}, -V_{DS}) = -I_G(V_{GS}, V_{DS}) - I_D(V_{GS}, V_{DS})$$

Physically, (6.55) means, that for any value of the pair of independent voltages, $\{V_{GS}, V_{DS}\}$ there is a corresponding pair, $\{V_{GS} - V_{DS}, -V_{DS}\}$ that defines an equivalent (symmetric) operating condition, where the currents in the two cases are specifically and simply related.

We express these properties here in terms of DC currents at related DC voltages, for simplicity, but similar relationships hold for general stimulus signals and responses.

If we imagine that the symmetry extends beyond the intrinsic device to also include access parasitics (e.g., R_D and R_S) then we can logically conclude the full parasitic topology must be symmetric and the numerical values of all corresponding ECPs are equal for (6.55) to be satisfied.

6.5.2   Testing Constitutive Relations for Proper Symmetry

Figure 6.28 and expressions (6.55) provide unambiguous tests of a model claimed to be symmetric with respect to source-drain exchange. A device model from a simulator palette can be wired in each of the two configurations of Figure 6.28. If the terminal currents through the corresponding voltage sources are not identical for all applied voltages, then the model is not symmetric.

Algebraically, if (6.55) is not exactly satisfied, the model is not symmetric. We can use (6.55) as a test of proposed global constitutive relations for models of symmetric devices.

As an example, we test (6.56) to see if it is suitable as a proposed global constitutive relation for a model of a symmetric transistor. We assume we always have $I_D = 0$ for $V_{GS} < V_T$.

$$I_G = 0$$
$$\qquad\qquad\qquad\qquad\qquad\qquad\qquad\qquad (6.56)$$
$$I_D(V_{GS}, V_{DS}) = \beta \cdot (V_{GS} - V_T)^2 \tanh(\gamma V_{DS})$$

To test (6.56), we evaluate it at the arguments $V'_{GS} = V_{GS} - V_{DS}$ and $V'_{DS} = -V_{DS}$. This result is then compared to that obtained by applying transformations (6.55) to (6.56).

From (6.56) we obtain (6.57).

$$
\begin{aligned}
I_D(V_{GS} - V_{DS}, -V_{DS}) &= \beta(V_{GS} - V_{DS} - V_T)^2 \tanh\left(-\gamma V_{DS}\right) \\
&= -\beta(V_{GS} - V_{DS} - V_T)^2 \tanh\left(\gamma V_{DS}\right)
\end{aligned}
\tag{6.57}
$$

Applying transformation (6.55) to (6.56) we obtain (6.58).

$$
I_D(V_{GS} - V_{DS}, -V_{DS}) = -I_D(V_{GS}, V_{DS}) = -\beta(V_{GS} - V_T)^2 \tanh\left(\gamma V_{DS}\right)
\tag{6.58}
$$

Since the two expressions, (6.57) and (6.58) are not identical, we conclude (6.56) *is not an admissible global constitutive relation* for a symmetric device.

However, we can restrict the domain of (6.56) to be valid for $V_{DS} \geq 0$ only, and then extend the drain current expression for values of $V_{DS} < 0$ by using (6.55).

In this case, we have the piecewise definition, valid for all bias conditions.

$$
\begin{aligned}
&\text{If } V_{DS} \geq 0 \\
&\quad \text{Then } I_D = \beta(V_{GS} - V_T)^2 \tanh\left(\gamma V_{DS}\right) \\
&\quad \text{Else } I_D = \beta(V_{GS} - V_{DS} - V_T)^2 \tanh\left(\gamma V_{DS}\right) \\
&\text{EndIf}
\end{aligned}
\tag{6.59}
$$

An equivalent set of expressions to (6.55) can be formulated using a more symmetric coordinate system for the terminal currents and voltage difference, and appears in (6.60).

$$
\begin{aligned}
I_D(V_{GS}, V_{GD}) &= I_S(V_{GD}, V_{GS}) \\
I_S(V_{GS}, V_{GD}) &= I_D(V_{GD}, V_{GS})
\end{aligned}
\tag{6.60}
$$

Expressions (6.60) make the symmetry condition manifest in a "covariant" way. That is, just exchanging the labels "S" and "D" in the left-hand side of (6.60) for terminal currents and voltage differences give the correct results on the right-hand side. The drain and source terminal currents swap when the gate-drain and gate-source voltages swap. From (6.60) and KCL it follows that the gate current is invariant when the gate-source and gate-drain voltages are interchanged.

It should be emphasized that (6.55) and (6.60) are completely equivalent descriptions of the same symmetry property. It is not a requirement that a properly symmetric mathematical model be written in the manifestly covariant form of (6.60). It may actually be more convenient to model and validate the device in the space defined for all V_{GS} and for the half-space $V_{DS} \geq 0$ and use (6.55) to extend the domain of the model to the full space.

We can write the correspondence between the related bias conditions in matrix form, where the variables with the *prime* symbols are the transformed variables, that is the corresponding voltages in the other half-space.

$$
\begin{bmatrix} V'_{GS} \\ V'_{DS} \end{bmatrix} = \begin{bmatrix} 1 & -1 \\ 0 & -1 \end{bmatrix} \cdot \begin{bmatrix} V_{GS} \\ V_{DS} \end{bmatrix}
\tag{6.61}
$$

For any point $\{V_{GS}, V_{DS}\}$ where $V_{DS} \geq 0$, (6.61) gives the corresponding value of the bias conditions for the equivalent operating point in the other half-space

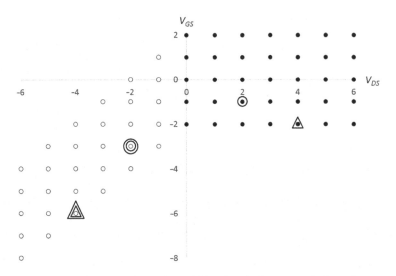

Figure 6.29 Half-space and transformed bias domain for source-drain exchange symmetry. The points for $V_{DS} > 0$ (solid points) are mapped into their image (open circles) using transformation (6.61). Two particular points are selected to show how each is specifically mapped to the other half-space.

(where $V_{DS} < 0$. As an example, for the choice $V_{GS} = -1$ and $V_{DS} = 2$, the symmetrical operating condition, using (6.61), is just $V'_{GS} = -3$ and $V'_{DS} = -2$. A plot showing this case and another related pair of operating points is shown in Figure 6.29.

Together with the independent voltages transforming according to (6.61), the dependent currents transform according to (6.62), which follows from (6.55).

$$\begin{bmatrix} I'_G \\ I'_D \end{bmatrix} = \begin{bmatrix} 1 & 0 \\ -1 & -1 \end{bmatrix} \cdot \begin{bmatrix} I_G \\ I_D \end{bmatrix} \tag{6.62}$$

Both transformations, (6.61) and (6.62) are invertible; in fact, they are their own inverses, as can easily be checked.

Another convenient form for the symmetry relations is given by (6.63).

$$I_G(V_{GS}, V_{DS}) = I_G(V_{GS} - V_{DS}, -V_{DS})$$
$$I_D(V_{GS}, V_{DS}) = -I_G(V_{GS} - V_{DS}, -V_{DS}) - I_D(V_{GS} - V_{DS}, -V_{DS}) \tag{6.63}$$

This give us rules for deducing the terminal currents at particular bias conditions whenever $V_{DS} \leq 0$ from known terminal current values associated with bias conditions for which $V_{DS} \geq 0$.

For example, suppose we are given the point $(V_{GS} = -6, V_{DS} = -4)$ and want to know the corresponding drain current when we have only modeled the device for values of $V_{DS} \geq 0$. We just plug $(V_{GS} = -6, V_{DS} = -4)$ into the right side of (6.63) and evaluate the gate and drain currents of the model at the bias condition$(V_{GS} = -2, V_{DS} = 4)$. In fact, this example corresponds to the mapping from

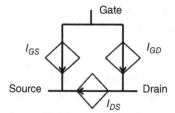

Figure 6.30 Simple intrinsic equivalent circuit model in terms of branch currents to illustrate symmetry transformations.

the point labeled with double triangle to the point labeled by single triangle in Figure 6.29.

6.5.3   Symmetry in Terms of Branch Elements

If we postulate a large-signal equivalent circuit, with current sources given in Figure 6.30, we can conclude from the above discussion that the transformations (6.64) must hold.

$$I_{GS}(V_{GS} - V_{DS}, -V_{DS}) = I_{GD}(V_{GS}, V_{DS})$$
$$I_{GD}(V_{GS} - V_{DS}, -V_{DS}) = I_{GS}(V_{GS}, V_{DS}) \qquad (6.64)$$
$$I_{DS}(V_{GS} - V_{DS}, -V_{DS}) = -I_{DS}(V_{GS}, V_{DS})$$

It follows from the equivalent circuit of Figure 6.30 and the transformation properties (6.64) that the functional dependence of the gate-source and gate-drain branches must be identical for a symmetric device. In particular, symmetry is violated even if the I_{GS} and I_{GD} current sources use the same diode-like nonlinearities but with different numerical values for the parameters. That is, not only must the equivalent circuit topology of a symmetric device be symmetric, but the functional forms and even the parameter values of the corresponding elements that are mapped into one another must be identical.

6.5.4   Common Model Failures for Symmetric Devices

To summarize, there are several reasons a model may not be symmetric. The equivalent circuit may not be properly symmetric. For example, there may be an R_{GS} element (see Chapter 5) in series with C_{GS} but not a corresponding R_{GD} element in series with C_{GD}. Alternatively, the constitutive relations of symmetrically placed elements in the topology may not be the same. Finally, the parameter values of the corresponding constitutive relations may not be identical.

6.5.5   Asymmetric Devices

No actual device is exactly symmetric, even if it was designed to be. There can be great value in modeling the full device, including its asymmetry, to be able to simulate the consequences of asymmetry in circuits that may rely on presumed symmetry for some performance properties (e.g., like double-balanced FET mixers). Also, of course, many transistors are specifically designed to be asymmetric. An example is a transistor for high-power RF amplifier applications. But such devices are rarely used for signals that take it from one half-space to the other during an RF cycle. So if it can be established – and verified – that for all applications, the device model will always satisfy $V_{DS}(t) > 0$, there is no need to consider issues of drain-source exchange symmetry. The device model symbol should distinguish the source and drain nodes whenever the model is not symmetric.

6.5.6   Terminal Charges and Drain-Source Exchange Symmetry

Symmetry constraints induce requirements also on the terminal charges (assuming we have a terminal charge-conserving model). The transformation rules (6.65) apply to the independent terminal charges. Note the formal equivalence to (6.63).

$$Q_G(V_{GS}, V_{DS}) = Q_G(V_{GS} - V_{DS}, -V_{DS})$$
$$Q_D(V_{GS}, V_{DS}) = -[Q_G(V_{GS} - V_{DS}, -V_{DS}) + Q_D(V_{GS} - V_{DS}, -V_{DS})]$$

(6.65)

6.5.6.1   Drain-Source Exchange Symmetry Is a Necessary but Not Sufficient Condition

The discrete symmetry relations (6.63) and (6.65) of drain-source exchange are necessary conditions for a proper nonlinear model of a symmetric FET. However, that is not sufficient. As discussed in Section 6.2.5.2, the constitutive relations must be continuous and differentiable everywhere, in particular at the boundary of distinct subdomain definitions.

Clearly (6.59) is continuous at the domain boundary, $V_{DS} = 0$, since the current evaluates to zero when V_{DS} reaches zero from above and from below. It is a simple exercise to show that (6.59) is also differentiable at $V_{DS} = 0$. However, higher order (partial) derivatives of (6.59) are not defined at $V_{DS} = 0$. For this reason, simple, piecewise constitutive relations are not often good choices where high orders of continuity are required at special operating points, especially $V_{DS} = 0$, and other more global formulations should be used [71]. This is most critical for switch and resistive mixer applications. Another approach to discrete symmetry with high smoothness is presented in [34]. Although the context of [34] is artificial neural networks, the decomposition in terms of symmetric and antisymmetric functions, each of which can be infinitely differentiable, is generally applicable and is a powerful technique for advanced applications.

6.5.7   Symmetry Simplifies Characterization and Parameter Extraction

If a model is known to be drain-source symmetric, then it is only necessary to characterize the device over the half-space (e.g. $V_{DS} \geq 0$) and extract parameters based on this reduced set of data. This saves measurement time.

Equations (6.63) and (6.65) will then guarantee that if the model fits well for $V_{DS} \geq 0$, it will fit just as well for operating conditions for which $V_{DS} < 0$, provided only that the device is indeed symmetric.

6.5.8   Final Comments on Symmetry

Symmetry principles are far broader and more powerful than might be apparent from the level at which we have presented them here. We have focused in this chapter on a discrete symmetry of lumped nonlinear models. But symmetry arguments are actually independent of the particular dynamics and even the physics assumed for the model. That is, symmetry arguments apply at whatever level of description (physical partial differential equations, distributed nonlinear circuit theory, or simple lumped nonlinear circuit theory) is chosen to describe the device. In fact, we can generally deduce that for any signals incident into the two independent ports of a symmetric device, the corresponding time-dependent terminal currents must still be related through transformations like (6.55), but now with time-dependent arguments that might also require more dynamical independent variables.

6.6   Self-Heating and Trapping: Additional Dynamical Phenomena

6.6.1   Static vs Dynamic Self-Heating

Figure 6.31 shows the static I–V characteristics of a GaAs FET where the computed junction temperature, T_j, for each curve of constant gate voltage is superimposed and can be read on the scale of the right y-axis. There is a particular DC operating point at $V_{DS} = 6\ V$ and $I_D = 60\ mA$, where the slope of the DC curve is negative. That is, the DC output conductance is less than zero; $G_{DS}^{DC} \equiv \frac{\partial I_D^{DC}}{\partial V_{DS}} < 0$. However, at RF and microwave frequencies, the linear model element, G_{DS}^{RF}, extracted from the linear model analysis of Chapter 5, is positive. That is, $G_{DS}^{RF} \equiv \mathrm{Re} Y_{22}(0.5V, 6V, \omega) > 0$.

This is another manifestation of the results of (6.48) and precisely corresponds to the right plot of Figure 6.25. How do we understand and resolve this dilemma?

6.6.2   Modeling Dynamic Self-Heating

We model the electro-thermal effect by coupling the electrical large-signal model and a thermal equivalent circuit that computes, approximately, the temperature rise above ambient due to the power dissipated in the device. The simple equivalent circuit is given in Figure 6.32.

The electrical constitutive relations for current and charge now are taken to depend on the junction temperature, $T_j(t)$, as well as on the instantaneous intrinsic terminal voltages. This is indicated, for the drain current, by (6.66).

$$I_D(t) = I_D\big(V_{GS}(t), V_{DS}(t), T_j(t)\big) \tag{6.66}$$

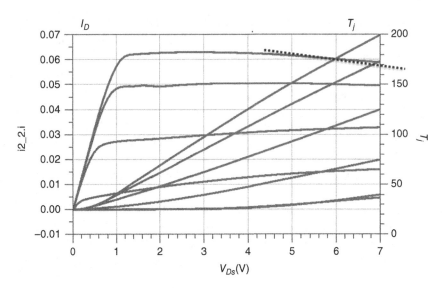

Figure 6.31 Static I–V characteristics of a GaAs FET, $V_{GS} = -1.5$ V to 0.5 V, in 0.5 V steps. For each gate-source voltage, the junction temperature is plotted as a function of drain-source voltage. The dashed line emphasizes the negative slope of the top I–V curve at $V_{DS} = 6$ V, corresponding to a junction temperature of 175°C. Ambient temperature is 25°C.

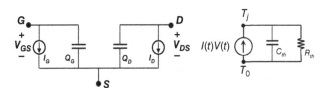

Figure 6.32 Simple coupled electrical and thermal circuit for FET self-heating.

The simple thermal equivalent circuit provides a single-pole approximation for the heat diffusion equation that is actually a partial differential equation. The first-order ordinary differential equation that can be used as a simple approximation is given in (6.67). This can be easily implemented as an equivalent circuit. Here R_{th} is the thermal resistance, $\tau_{th} = R_{th}C_{th}$ is the thermal time constant, and T_0 is the ambient temperature. More poles can be used to get a better approximation to the actual distributed thermal response. In MMIC design, especially with very temperature-sensitive components, as many as seven poles or even more may be needed [72], or a *fractional pole* representation can be used if the nonlinearity of the thermal parameters is neglected [52].

$$\tau_{th}\frac{dT_j}{dt} + T_j - T_0 = R_{th}(I_D(t)V_{DS}(t) + I_G(t)V_{GS}(t)) \tag{6.67}$$

With a suitable electrical model and reasonable values for the thermal parameters, we can begin to understand the failure of the non-quasi-static model. Figure 6.33 shows the

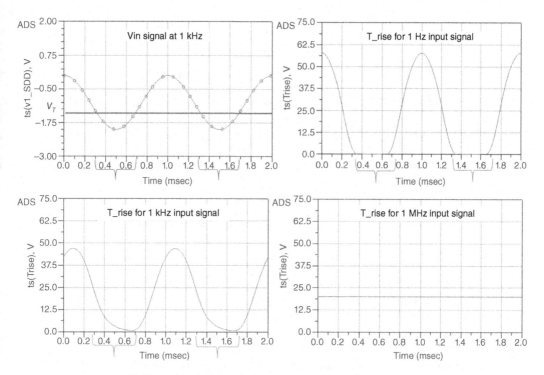

Figure 6.33 Incident electrical signal and thermal response at various time-scales.

simulated results of a transistor's time-dependent junction temperature in response to incident sinusoidal signals at three different frequencies, 1 Hz, 1 kHz, and 1 MHz. The incident waveform at 1 kHz is shown in the upper left position.

From the upper left plot of the incident waveform, it can be seen that the gate-source voltage drops below the threshold voltage, so the device is actually off during a fraction of the stimulus period. The time-dependent temperature rise above ambient is shown in the lower left plot corresponding to the 1 kHz input signal. We note the temperature rise lags the voltage stimulus but still varies by nearly 50 degrees over the cycle. Corresponding to the electrical signal at only 1 Hz, the device temperature reported in the upper right plot shows that now $T_j(t)$ perfectly tracks the voltage, and it peaks at over 55 degrees. In this case the device is operating quasi-statically. Finally, in response to the electrical signal at 1 MHz, the junction temperature $T_j(t)$ no longer varies in time, but instead is seen to assume a constant value, of about 18 degrees above ambient, as shown in the lower right plot. The temperature can no longer follow the electrical signal. The thermal time-constant used for the simulation was one ms.

We can compute the DC and RF conductances from (6.66) and (6.67) in the high and low frequency limits. At RF conditions, the temperature is constant as we have seen from Figure 6.33, independent of the voltage, so we have the expression in (6.68).

$$G_{DS}^{RF} = \frac{\partial I_D(V_{GS}, V_{DS}, T)}{\partial V_{DS}}\bigg|_{V_{GS},T} \tag{6.68}$$

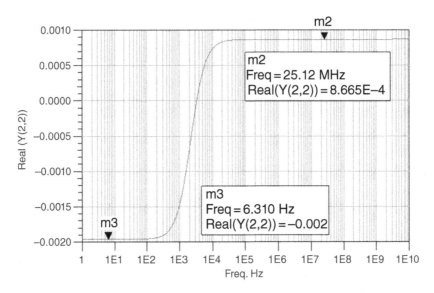

Figure 6.34 Frequency dependence of ReY22 at low frequencies.

Under DC conditions, the junction temperature is itself a static function of the voltage because the time-derivative in (6.67) vanishes. We find the result (6.69).

$$
\begin{aligned}
G_{DS}^{DC} &= \left. \frac{\partial I_D(V_{GS}, V_{DS}, T)}{\partial V_{DS}} \right|_{V_{GS}, T} + \frac{\partial I_D}{\partial T} \cdot \frac{\partial T}{\partial V_{DS}} \\
&= \qquad\qquad G_{DS}^{RF} \qquad\qquad + \frac{\partial I_D}{\partial T} \cdot \frac{\partial T}{\partial V_{DS}}
\end{aligned}
\tag{6.69}
$$

Since usually $\frac{\partial I_D}{\partial T} < 0$ due to mobility degradation[14], while $\frac{\partial T}{\partial V_{DS}}$ is positive, we have the result that $G_{DS}^{DC} < G_{DS}^{RF}$. Usually G_{DS}^{RF} is so small that the negative contribution from the product terms can be significant enough, at certain bias conditions, to make the G_{DS}^{DC} less than zero.

These effects can be observed by doing a simulation of $Y_{22}(\omega)$ at very low frequencies. This is shown in Figure 6.34. At 1 Hz the frequency dependent admittance is real, negative, and equal to the DC output conductance – the small negative slope in Figure 6.34. By 10 kHz, the conductance has become positive, and saturates beyond 100 kHz.

If we plot S_{22} versus frequency over this range, we observe, at low frequencies, that the results are outside the Smith Chart (Figure 6.35). As the stimulus frequency increases, the trajectory moves back inside the Smith Chart, as expected from RF and microwave measurements.

---

[14] This means essentially that charge carriers move more slowly in semiconductors at higher temperatures.

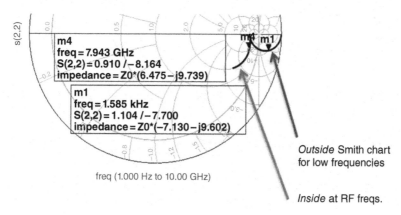

Figure 6.35 Frequency dependent S22 showing low-frequency behavior outside the Smith Chart due to dynamic thermal effects.

6.6.3   Temperature under Large-Signal Conditions

The temperature of a transistor under large-signal conditions depends on the self-consistent calculation of energy conversion to heat that is dissipated in the device. This depends on details of the current and voltage waveforms generated in response to the input signals. These waveforms depend strongly on the incident signal, the complex loads at the output – at the fundamental frequency and harmonics – and the bias conditions.

It is well-known that a class A power amplifier actually gets cooler as the incident RF power increases. The reason for this is that, for no applied RF, all the DC power at the class A bias point is being dissipated in the device, causing a temperature rise over ambient. As the RF power increases, up to half the DC battery power previously being dissipated as heat is instead up-converted to RF and delivered to the load outside the device. The device junction temperature is therefore reduced at high RF incident power compared to low RF incident power. However, the junction temperature is still hotter than ambient since there is always some energy dissipated as heat.

For a class C amplifier, however, there is no dissipation without incident RF power because the device is biased in an "off" state. So with no RF, the device junction temperature is the ambient temperature. As the RF signal increases, the transistor is turned on for more and more of the RF cycle. So there is now some power dissipation, increasing the junction temperature over ambient.

A plot of measured and modeled drain current (left axis) and calculated junction temperature (right axis) versus output power for a class A and a deep class AB amplifier is shown in Figure 6.36.

6.6.4   Trapping Phenomena in III-V FETs

Detailed characterization of III-V FETs in GaAs and, more recently, GaN material systems, show evidence of additional dynamic effects for which the model represented by the intrinsic equivalent circuit in Figure 6.32 is not sufficient. Frequency dispersion

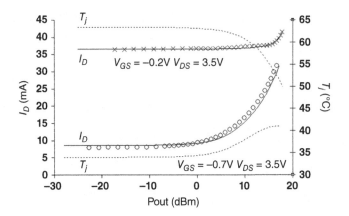

Figure 6.36 Drain current and computed junction temperature versus output power for class A (top) and deep class AB (bottom) amplifiers. Model (lines) and measurements (symbols).

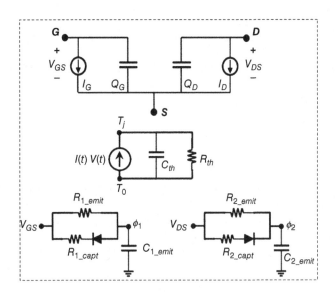

Figure 6.37 Equivalent circuit of advanced FET model with dynamic self-heating and two trap capture and emission processes.

of small signal behavior – differences in the equivalent circuit conductances at high frequencies compared to their values at DC – can only partially be accounted for by self-heating mechanisms. Pulsed transient phenomena like gate-lag and drain-lag, and related effects of "power slump" and knee walk-out with increasing RF power, require more complex explanations [73] and correspondingly more complicated dynamical models [74,36]. Important performance measures, such as bias current and power-added efficiency, as functions of input RF power, were notoriously difficult to simulate accurately without including mechanisms for trapping [75].

A more complete intrinsic equivalent circuit model for III-V FETs is presented in Figure 6.37 [74]. The top subcircuit is the conventional lumped electrical topology, in common source configuration, for the currents and charges. The middle subcircuit is a simple one-pole thermal equivalent circuit that models the self-heating. The two remaining subcircuits model dynamic charge capture and emission controlled by the gate and drain potentials, respectively. The electrical constitutive relations for current and charge now depend on five state variables: the instantaneous gate and drain voltages, V_{GS} and V_{DS}, the time-varying junction temperature, T_j, and two time-varying voltages, ϕ_1 and ϕ_2, associated with trap mechanisms [74,36]. More formally, we can write the dynamic equations of this intrinsic model as

$$I_G(t) = I_G\big(V_{GS}(t), V_{DS}(t), T_j(t), \phi_1(t), \phi_2(t)\big)$$
$$+ \frac{d}{dt}Q_G\big(V_{GS}(t), V_{DS}(t), T_j(t), \phi_1(t), \phi_2(t)\big) \tag{6.70}$$

$$I_D(t) = I_D\big(V_{GS}(t), V_{DS}(t), T_j(t), \phi_1(t), \phi_2(t)\big)$$
$$+ \frac{d}{dt}Q_D\big(V_{GS}(t), V_{DS}(t), T_j(t), \phi_1(t), \phi_2(t)\big) \tag{6.71}$$

$$\frac{dT_j}{dt} = \frac{T_0 - T_j(t)}{\tau_{th}} + \frac{1}{C_{th}}\big(I_D(t)V_{DS}(t) + I_G(t)V_{GS}(t)\big) \tag{6.72}$$

$$\frac{d\phi_1}{dt} = f(\phi_1(t) - V_{GS}(t)) + \frac{V_{GS}(t) - \phi_1(t)}{\tau_{1\_emit}} \tag{6.73}$$

$$\frac{d\phi_2}{dt} = f(V_{DS}(t) - \phi_2(t)) + \frac{V_{DS}(t) - \phi_2(t)}{\tau_{2\_emit}} \tag{6.74}$$

whose corresponding equivalent circuit representation is shown in Figure 6.37. Equations (6.72)–(6.74) are state equations – first order differential equations for the evolution of key dynamical (state) variables that are arguments of the electrical constitutive relations appearing in equations (6.70) and (6.71). The function f that appears in (6.73) and (6.74) is a diode-like nonlinearity that accounts for preferential trapping rates when the instantaneous gate (drain) voltage becomes more negative (positive) than the values of ϕ_1 (ϕ_2). The parameters, $\tau_{n\_emit} = R_{n\_emit}C_{n\_emit}$ for $n = 1,2$ in Figure 6.37 are characteristic emission times, typically assumed to be in the millisecond range or longer. Typical capture times in Figure 6.37 are in the picosecond range.

The primary method for characterization of trapping phenomena for modeling has been the use of pulsed I–V and also pulsed S-parameter measurements [52,75–78]. As these techniques are well-described in the references, we do not present them here. Instead, in Section 6.7.2, we present an alternative method based on recently introduced instruments for large-signal continuous wave (CW) waveform measurements.

6.7   NVNA-Enabled Characterization and Nonlinear Device Modeling Flows

A nonlinear vector network analyzer (NVNA), or large-signal network analyzer (LSNA), measures magnitudes and relative phases of incident and scattered RF or

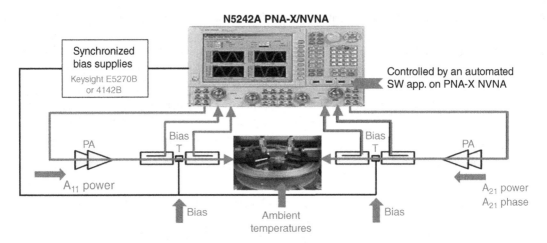

Figure 6.38 NVNA configured for active large-signal waveform measurements. The measurements are performed at various A_{11} powers, A_{21} powers, A_{21} phases, DC bias conditions, and ambient temperatures. In this setup, external couplers replace the built-in PNA-X couplers for high-power application.

microwave frequency tones and their harmonics, at each DUT port, which can then be converted into the corresponding time-domain voltage or current waveforms [79], [80], [81], and [82]. An illustration of a typical hardware configuration for NVNA transistor measurements is shown in Figure 6.38. While the PNA-X NVNA instrument has internal couplers, Figure 6.38 shows external couplers that can be necessary for high power applications. For transistor characterization, it is most useful to stimulate both ports of the device, simultaneously, with large-amplitude incident waves, at the fundamental applied frequency. Varying the relative phase between the incident waves at ports 1 and 2 is then equivalent to active time-domain load-pull. At each port, the NVNA measures the scattered waves at the fundamental frequency, DC, and the magnitudes and relative phases of all harmonic signal components generated by the DUT in response to the stimuli. Since the stimuli signals are commensurate, the measured complex spectral response of the transistor can be transformed to the time domain as periodic current and voltage waveforms under significant degrees of compression. Samples of NVNA waveforms are shown in Figures 6.39 and 6.40. Notice that the drain-source voltage and drain current waveforms of Figure 6.39 take negative values during part of the periodic cycle.

Over the past many years, there has been significant research applying such large-signal measurement systems to the field of nonlinear device modeling. Much of this work has focused initially on the validation of compact models under realistic large-signal operating conditions. But it has also included the waveforms as target data for parameter extraction, and for obtaining insight into the limitations of existing models under conditions closer to those that the physical device is expected to experience in the actual application [83–86].

An example of the extent of the characterization range compared to static DC measurements is shown in Figure 6.41 for the case of a GaN device. The gray cloud of thousands of measured load-lines extends well beyond the range over which DC and

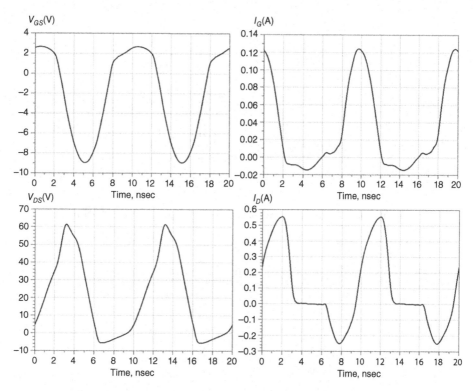

Figure 6.39 Large-signal voltage and current periodic waveforms of a GaN FET measured with an NVNA system like that of Figure 6.38.

S-parameter data can be taken. This is especially important for characterizing high-power devices in regions where the device will be operating during large-signal applications, and to obtain, benignly, information in limiting regions of operation such as breakdown and high instantaneous power dissipation regions.

This generic capability enables several distinct innovative characterization and modeling flows for transistors. Figure 6.42 shows a schematic representation of three of these flows. The bottom flow is of course the X-parameter behavioral modeling approach discussed in Chapter 4 that can be applied to transistors given systems like Figure 6.38. We present two additional flows here, the top and middle flows in Figure 6.42, related to compact models in the time domain.

6.7.1 Parameter Extraction of Conventional Empirical Models to Waveforms

It is ironic that it is still common practice for nonlinear transistor models to have their parameter values extracted from DC and linear S-parameter data. A model can fit DC characteristics and S-parameters perfectly but still not give accurate results under large-signal conditions. Extractions based on such simple data are therefore not reliable indicators of nonlinear model performance. For some models, generally those that include dynamic self-heating and trapping phenomena, it is impossible to properly

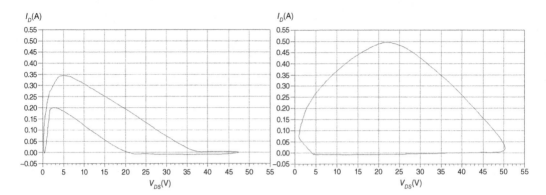

Figure 6.40 Measured dynamic load-lines at the output port of a GaN FET.

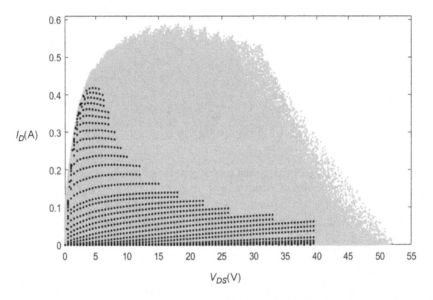

Figure 6.41 Measured DC I–V characteristics (black symbols) of a GaN HFET and ensemble of measured dynamic load-lines (gray region) using an NVNA system like that of Figure 6.38. The device is described in more detail in 6.7.4.1. In this case the measurements were done at 100 MHz.

extract the parameters from data limited to DC and S-parameters. Without large-signal data, there is no direct evidence that the model will perform properly under the large-signal conditions for which nonlinear models are designed and required.

The most obvious modeling flow based on NVNA/LSNA data is therefore to extract and tune model parameters by directly using the NVNA large-signal measured waveforms of the device as optimization targets. That is, model parameter values are adjusted until the measured and simulated large-signal waveforms agree [87–95].

There are certain parameters, such as related to breakdown voltage in FETs, or Kirk effect in HBTs, that can be better extracted from such data. The data is closer to the device conditions of actual use, so model parameters obtained under such conditions are

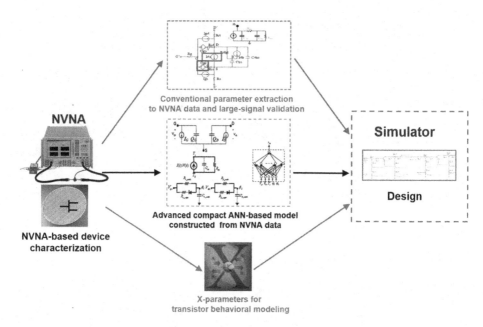

NVNA

NVNA-based device
characterization

Conventional parameter extraction
to NVNA data and large-signal validation

Advanced compact ANN-based model
constructed from NVNA data

X-parameters for
transistor behavioral modeling

Simulator

Design

Figure 6.42 Multiple characterization and modeling flows enabled by NVNA data.

likely to be more useful for such applications. Since no model is perfect, the large-signal data provides an ability to tune the model parameters for better results in certain applications. In any case, supplementing conventional DC and linear data with NVNA large-signal data ultimately leads to better sets of extracted parameter values for models. In general, better (richer) data lead to better models.

Due to inherent limitations of any model, optimal parameter extraction may still result in discrepancies between simulated and measured performance under the desired large-signal conditions. But this information would usually not be available until the design validation stage much later on in the modeling flow, if at all. If parameters have to be re-extracted at this late point, it becomes a cumbersome, time-consuming, and expensive process to iterate between parameter extraction and design validation. Using NVNA data, it is possible to combine the parameter extraction and design validation phases into one step, a significant simplification of the modeling flow. NVNA data provides detailed waveforms for comprehensive nonlinear model validation, including AM-AM, AM-PM, PAE, harmonic distortion in magnitude and phase, and much more, without the need for additional instruments, such as spectrum analyzers, that only give the magnitude of the generated spectra.

Examples of a prototype modeling system for parameter extraction and validation of empirical models from NVNA waveforms are shown in Figures 6.43 and 6.44 [87–90]. In general, the measured versus simulated waveforms provide great insight into where the models need enhancement or modification, and how tuning model parameter values can provide local models (parameter sets) optimized for distinct applications.

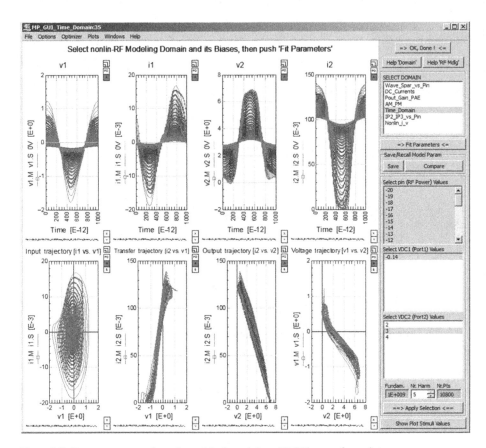

Figure 6.43 Parameter extraction of empirical models to NVNA waveform data.

6.7.2   Identification of Advanced FET Models from Large-Signal NVNA Data

The model identification problem means defining the detailed functional form of the terminal current and terminal charge functions at the gate and the drain, as functions of the many internal controlling state variables. The state variables are not directly controllable, and their values must be identified from the large-signal measurements.

Data is taken at several different ambient temperatures, power levels at each port, and relative phases between drive signals, all at several different quiescent bias points. Measurements at different fundamental frequencies may also be needed, although a method described in Section 6.7.3 can reduce the required NVNA data to a single fundamental frequency at the cost of neglecting the trap state dependence of the terminal charges. A great advantage of NVNA data is that the extreme regions of the device operation can be characterized with much less degradation of the transistor compared to static measurements. This is because the instantaneous voltages only enter the high-stress regions for sub-nanosecond periods as the device is stimulated with signals at one GHz or higher frequency. At these frequencies, the device dissipates much less energy at high instantaneous power regions than under DC conditions.

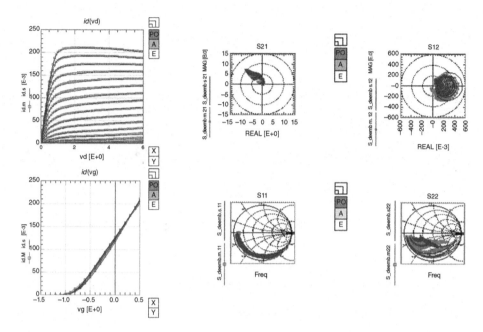

Figure 6.44 Parameter extraction of empirical models to NVNA waveform data enabling tradeoffs between small-signal and large-signal fits. The model used here is the Angelov Model.

The larger domain of device operation means that the need for the final model to extrapolate during large-signal simulation is dramatically reduced or even eliminated.

Just as for the flow described in the previous Section, 6.7.1, the NVNA data used for identifying this advanced time-domain model provides, as well, detailed waveforms, including two-tone intermodulation measurements with relative phase, for comprehensive model validation without the need for additional instrumentation.

The modeling identification process for the trap states is depicted in Figure 6.45. The identification from CW large-signal RF or microwave measurements is extremely simple, given the widely separated time-scales of the RF signal and the emission and capture rates of the traps, which is summarized by (6.75). Conditions (6.75) ensure that for the model identification process from CW large-signal data, the trap states $\phi_1(t)$ and $\phi_2(t)$, take *constant* values on any given trajectory. This can be seen from Figure 6.45, where the actual trap state values, ϕ_1 and ϕ_2, can be explicitly identified as the minimum $V_{GS}(t)$ and maximum $V_{DS}(t)$ values, respectively, over a given period of the waveforms for each experimental condition [36].

$$\frac{1}{\tau_{emit}} << f_{RF} << \frac{1}{\tau_{capt}} \tag{6.75}$$

We emphasize that it is only in the auxiliary dynamic state variable identification phase that we know the steady-state solution of the dynamical equations (6.72)–(6.74). It is only here that we appeal to the fact that the CW large-signal waveforms establish the steady-state behavior of the device so as to fix the dynamical variables under the

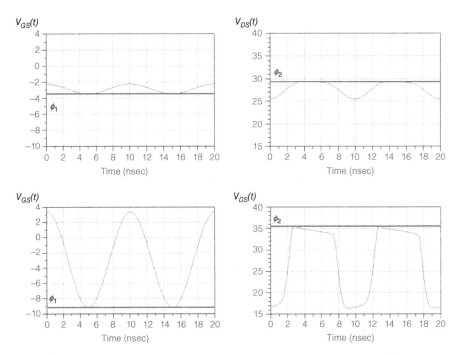

Figure 6.45 Identification of trap states from CW large-signal waveform data under the conditions of (6.75). Top row: small amplitude excitation. Bottom row: large amplitude excitation.

assumption (6.75). The full time-dependent equations (6.70)–(6.74) are evaluated based on the arbitrary time-dependent stimuli experienced by the device model during simulation. The model responds appropriately, in time, to any excitation and works like any other time-domain compact model in all simulation modes (e.g. transient analysis, harmonic balance, S-parameter analysis, etc.).

Similarly, the junction temperature, T_j, is essentially *constant* over the period of an RF or microwave waveform (as we demonstrated in simulation in Section 6.6) and can be calculated in terms of the average power dissipated along the load-line using the thermal resistance extracted as in [25]. Different load-lines will generally correspond to different temperatures.

The ensemble of waveforms over all measurement conditions therefore serves to "prepare" the device in all sets of possible states defined by the trap and thermal variables and the instantaneous terminal voltages. At each sample of the waveform, the currents and intrinsic terminal voltages are also measured. The result is a large collection of data for the terminal currents as functions of the trap states, junction temperature, and the instantaneous terminal voltages.

Prior to the availability of large-signal NVNA data, the behavior of the traps and self-heating effects were often separated by careful pulsed measurements [74–76]. The NVNA can probe the device at timescales several orders of magnitude faster than typical pulsed I–V systems. That is, NVNA data is often more indicative of the actual operating conditions of devices manufactured to operate at frequencies in the tens of GHz range.

$$I_{drain}(t) = \underbrace{I_D(V_{GS}(t), V_{DS}(t),}_{\text{Terminal voltages}} \underbrace{T_j(t), \phi_1(t), \phi_2(t))}_{\text{Auxiliary variables}} + \frac{d}{dt}\underbrace{Q_D(V_{GS}(t), V_{DS}(t),}_{\text{Terminal voltages}} \underbrace{T_j(t), \phi_1(t), \phi_2(t))}_{\text{Auxiliary variables}}$$

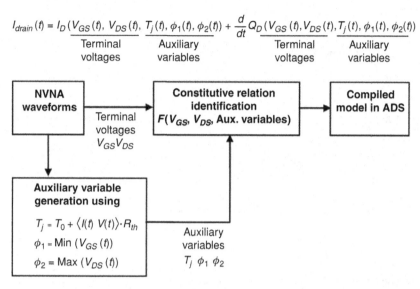

Figure 6.46 Model identification process for advanced FET model from waveform measurements.

The current and charge constitutive relations are identified according to the procedure outlined in Figure 6.46. The mathematical machinery used to represent the constitutive relations are the ANNs. It would be highly impractical to determine complicated nonlinear functional dependence on five independent variables without a powerful mathematical infrastructure for approximating the multivariate constitutive relations as functions of these variables. Moreover, the auxiliary (state) variables for each experiment are dependent variables (functionals of the waveforms) and assume values scattered in a multidimensional space. ANNs are easily trained on scattered data, and require no underlying structure for smooth, global approximation. The ANN representation couples the trap and thermal degrees of freedom to the currents and charges without the need to model any specifically proposed physical mechanism, such as self-backgating [74] or virtual gate resistance [73], which may vary from device to device. Whatever the data indicates about the coupling is modeled by the multivariate ANN constitutive relations. The final model can be compiled into a conventional nonlinear circuit simulator as with any compact model.

There is great insight that can be obtained by looking at the constructed constitutive relations based on large-signal steady-state waveforms. Two examples of generated intrinsic constitutive relations for different sets of trap states are shown in Figure 6.47. The model current constitutive relations corresponding to extreme trap states (Figure 6.47a) bears a striking resemblance to pulsed bias characterization from quiescent bias points associated with the trap state biases [26,74]. The advantage of the NVNA approach is that the model characteristics are inferred from DUT responses to signals typically three or more orders of magnitude faster than what can be measured with most pulsed systems that are typically limited to 0.1–1μs.

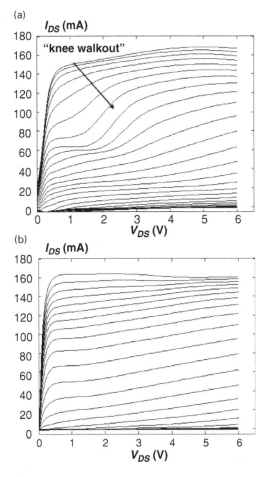

Figure 6.47 Advanced FET model intrinsic I–V constitutive relations at different trap state values: (a) fixed trap states corresponding to the extremes of large-signal trajectory, and (b) trap states following the DC bias conditions.

The complete model solves for the trap states, junction temperature, and currents self-consistently during simulation. When embedded back into the parasitic model, final comparison can be made to measured data. Figure 6.48 shows the validation with measured DC I–V curves. Note how different the static non-isothermal I–V curves of Figure 6.48 are from the intrinsic model constitutive relations of Figure 6.47. The dependence of the key constitutive relations on five state variables, some depending on slow dynamics like the junction temperature and trap states, provides sufficient degrees of freedom to fit the bias dependence of the small-signal model over the entire bias space at both DC and high frequencies. That is, frequency dispersion phenomena are predicted accurately under small-signal and large-signal conditions by having a model with both dynamic trapping and electro-thermal effects. Models with just electro-thermal effects are not capable of such good fits to both DC and high-frequency behavior at all biases.

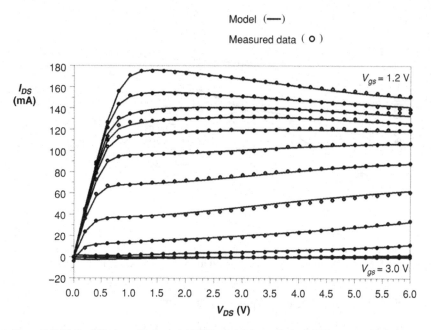

Figure 6.48 DC validation of advanced FET model for a GaAs pHEMT.

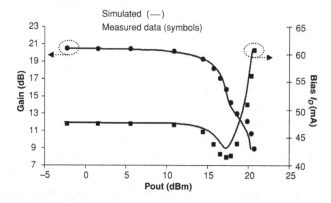

Figure 6.49 Large-signal validation of advanced FET model for a GaAs pHEMT. Gain and drain current versus output power.

Figure 6.49 shows the nonlinear validation results for the advanced FET model for power-dependent gain and bias current versus power. The distinctive car-shaped gain compression characteristic and significant non-monotonic dependence of the bias current with power is a result of the dynamics of drain-lag and the detailed constitutive relation obtained with the ANN training. Figure 6.50 shows the model validation of distortion versus power for this device, validating both the dynamical description and accuracy and robustness of the ANN approach to model the complicated constitutive relations.

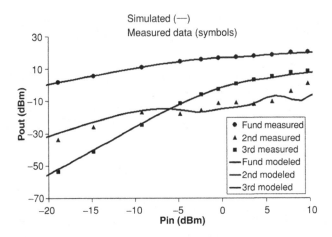

Figure 6.50 Large-signal validation of advanced FET model for a GaAs pHEMT. Output power at the fundamental frequency, second and third harmonics versus input power.

6.7.3   Simplifications: Lower Frequency Waveforms

The process above is all based on high-frequency NVNA-measured waveforms, typically at a fundamental frequency between 5 to 10 GHz. This is usually high enough in frequency to resolve the device capacitances under large-signal operation while being low enough to measure 13 or at least 6 harmonics, respectively, on a 67 GHz NVNA instrument. Data taken at GHz frequencies result in current waveforms having significant contributions from both the resistive and the reactive mechanisms of the device. This requires simultaneous self-consistent training of the ANNs for the voltage controlled current sources and the terminal charge sources, as in Figure 6.46, to properly separate and identify them as independent nonlinear real-valued functions from the data [36]. This can be a difficult and a time time-consuming computational task. Waveforms measured at more than one fundamental frequency are needed to help separate the controlled current source from that of the charge function, increasing the measurement time.

A practical simplification of the above method is to perform the NVNA measurements at only a single CW frequency that is sufficiently low that the contributions to the device response from the terminal charges can be neglected. This reduces the number of waveforms required and produces models with nearly the same accuracy as that described in Section 6.7.2. Low-frequency large-signal measurements have been used for a long time to characterize transistors at frequencies above the inverse thermal time constants and slow trapping constants for modeling purposes (see for example, [96–98].

The network analyzer is still used to take high-frequency linear S-parameters for characterizing the parasitic elements and for obtaining intrinsic capacitances for the charge model. Much of the following work has been done with dynamic load-lines measured at 100MHz. External high-power couplers were used to bypass the internal couplers of the PNA-X for high-power on-wafer measurements. The drain current ANN-based constitutive relation can therefore be directly trained from the measured current load-lines, from 100 MHz data, in much less time and with greater accuracy than the original approach of [36]. However, the terminal charge dependences on trap states are no longer available from such data, and in the following examples, [99–101], the terminal charges are constructed only from temperature-dependent high-frequency (e.g. 10 GHz) S-parameters. Specifically, we modify (6.70) and (6.71) to remove the trap state dependencies.

6.7.4   Results for GaN Transistors

6.7.4.1   GaN Device 1: 0.15 µm × 6 finger × 60 µm GaN HFET

This device is a 0.15 µm × 6 finger × 60 µm GaN HFET with individual source vias from Raytheon Integrated Defense Systems [99]. The device is optimized for 1–40 GHz operation for applications including power amplifiers, low noise amplifiers, and switch applications [99].

6.7.4.2   GaN Device 2: 0.5 um × 6 × 75 um GaN HFET

The modeled device was an on-wafer, 6 × 75 um (450um total) gate-width, 0.5um gate-length GaN HFET, from the RFMD (now Qorvo) GaN2C process. This technology is optimized for high linearity applications such as PMR and CATV in the 500 MHz to 4 GHz range. The transistor technology has a power density up to 4W/mm with a breakdown voltage over 300 V [100].

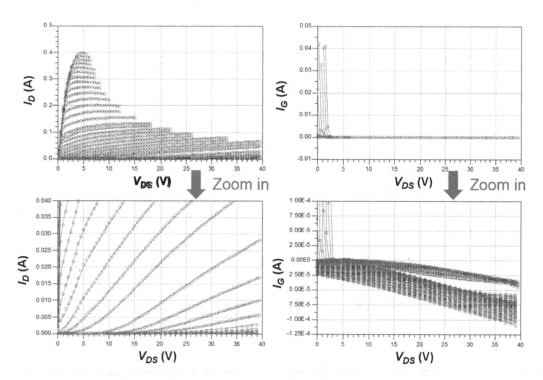

Figure 6.51 DC validation at 55°C. Measured (symbols), model (lines) for GaN HFET device 1.

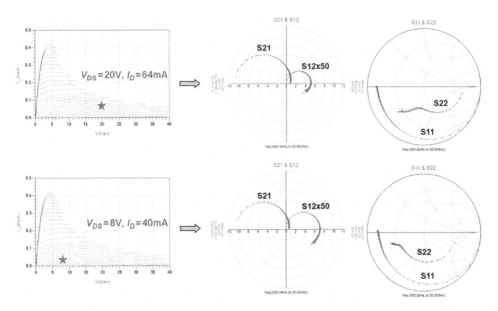

Figure 6.52 Broad-band S-parameter validation of GaN HFET device 1 at two different bias conditions. Measured (symbols), Model (lines). Frequency range 0.5 GHz to 50 GHz. The model was constructed from waveform data at the single fundamental frequency of 100 MHz.

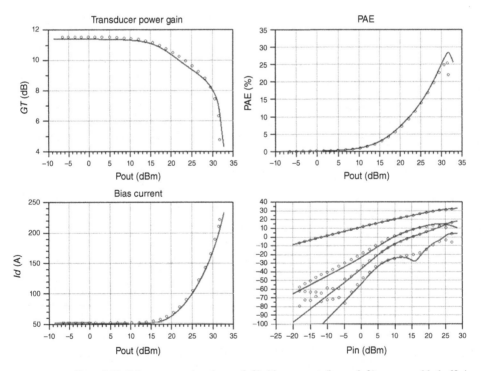

Figure 6.53 Gain compression (upper left), bias current (lower left), power added efficiency (upper right), and fundamental and first three harmonics versus input power (lower right), modeled (lines) and measured (symbols) for GaN device 1. Frequency is 10 GHz, $V_{DS} = 20$ V, $I_D = 54$ mA, Source & Load impedance = 50 Ω.

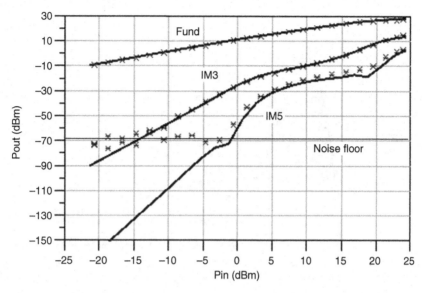

Figure 6.54 Two-tone intermodulation distortion (upper sidebands) versus input power per tone for $f_1 = 10.00$ GHz and $f_2 = 10.02$ GHz, for $V_{DS} = 20$ V, $I_D = 54$ mA. Simulation (lines) and measurements (symbols), for GaN HFET 1.

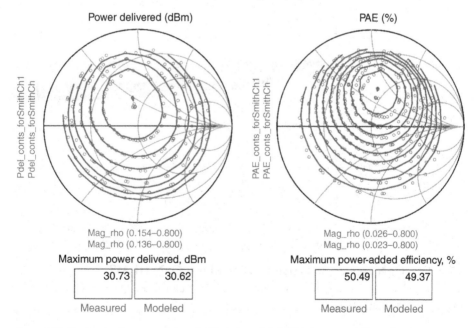

Figure 6.55 Contours of measured (symbols) and modeled (lines) power delivered and power added efficiency, at $V_{DS} = 12$ V, $I_D = 54$ mA, $Pin = 24$ dBm, at 10 GHz for GaN HFET device 1. Contours in 1 dB and 5% steps, respectively. The measured contours are plotted using load-dependent X-parameters to maintain matched harmonic load conditions at the second and third harmonics at all fundamental loads.

(a)

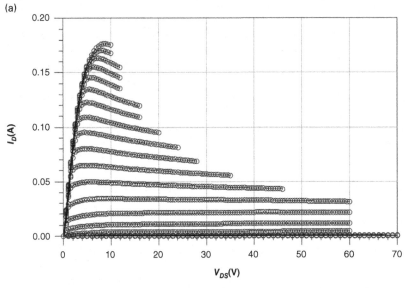

(b)

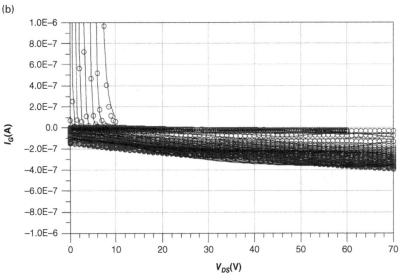

Figure 6.56 Drain current (top) and gate current (bottom) versus bias at an ambient temperature of 55°C. Modeled (solid lines) and measured (symbols) for GaN HFET device 2.

6.7.4.3   GaN Device 3: TriQuint 10 × 90 µm GaN HEMT

The device was from TriQuint Semiconductor (now Qorvo) with a total gate width of 900 um [101].

6.7.5   Discussion and Implications

The modeling process above has separated the basic physical mechanisms responsible for the device response. These are nonlinearities with respect to the intrinsic terminal voltages, dynamic self-heating and temperature dependence of the electrical character-istics, and the influence of two types of traps with asymmetric emission and capture times. The waveform excitations over bias, temperature, power, and load are rich enough to enable us to separately identify each of the above mechanisms, independently model them, and then recombine them (mutually couple them) in the model. Even the DC plots of Figures 6.48, 6.51, and 6.56 can be properly considered validation – not playback – because at each DC operating point the model is computing the junction temperature and the trap states (that move with bias) and computing the combined effects on the currents. The model does not blindly fit the DC measurements.

Unlike many models, there are enough dynamic degrees of freedom in this model, and the ANN-based constitutive relations are defined with such fidelity as function of so many variables, that it is possible to get an excellent broad-band fit at nearly any DC operating point. This benefit applies as well to the distortion results. The correct dynamical model and the detailed constitutive relations, with their infinite order differentiability from the ANNs, enable accurate harmonic and intermodulation simulation performance. The dependence of the model characteristics as a function of trap states does not need to be postulated to be caused by any particular physical mechanism, be it self-back gating or "virtual gate" formation [73], because the details are captured from the data and

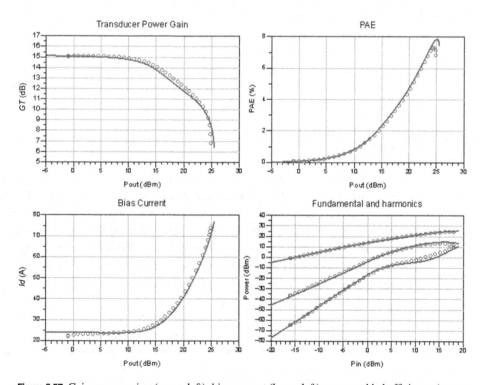

Figure 6.57 Gain compression (upper left), bias current (lower left), power added efficiency (upper right), and fundamental and first two harmonics versus input power (lower right), modeled (lines) and measured (symbols) for GaN HFET device 2. Frequency is 1 GHz, $V_{DS} = 48$ V, $I_D = 20$ mA, Source and Load impedance = 50 Ω.

Harmonic termination:

	Zs	Zl
H1	8.91+j34.73	44.17+j91.36
H2	432.64–j119.63	9.16–j26.86
H3	9.12–j25.2	25.31+j63.81

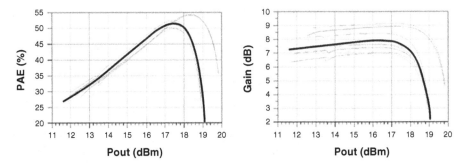

Figure 6.58 Nonlinear validation with specified source and load harmonic tuned impedances specified in the table for GaN device 3. The dark trace is the simulation with the model. The lighter traces are load-pull data from several devices from the same lot, but not the same device as used to construct the model. The bias condition is $V_{DS} = 15$ V, $I_D = 9$ mA, and the frequency was 10 GHz.[15]

represented in the model by the ANNs. The results shown in Figure 6.58 demonstrate that the model fits very well the device performance under tuned harmonic load conditions that were not controlled during the device characterization or the model identification.

6.8   Summary

Several theoretical foundations and practical applications of large-signal device modeling techniques for nonlinear circuit simulation have been presented. The nonlinear circuit theoretic foundation for device modeling was motivated by its historical origins from basic physical device operating principles. Large-signal models were classified as physically based, empirical, or measurement-based, depending upon the nature of their constitutive relations. The practical utility of artificial neural networks (ANNs) for smooth, multivariate approximation of nonlinear model constitutive relations was a recurrent theme throughout the chapter.

The correspondence between large-signal models and linear data, based on linearizing the intrinsic large-signal model equations, was described in detail. Nonlinear charge modeling was given much attention, and principles of terminal charge conservation were introduced formally and evaluated experimentally. An approach

[15] Acknowledgment to Dr. Charles Campbell and Maureen Kalinski (TriQuint), Prof. Zoya Popovic's group (Univ. of Colorado, Boulder), in particular David Sardin, UCB validation work funded under the DARPA MPC program (Dr. Dan Greene), through ONR (Dr. Paul Maki).

to separate depletion from drift charges in FETs for physical insight was reviewed. Practical trade-offs associated with constructing nonlinear charge models from bias-dependent linear data for table-based and ANN models were illustrated. Concepts of energy conservation and nonreciprocal capacitance matrices derived from terminal charge-based models were presented, even though all issues are not yet resolved. Quasi-static modeling assumptions and implications were elucidated and compared to experiment. Non-quasi-static mechanisms, necessary to explain observed device behavior, were introduced, specifically dynamic self-heating and charge capture and emission processes for III-V transistors, to explain and accurately model such phenomena.

Much attention was paid to symmetry principles, from a theoretical as well as a practical perspective, as a necessary tool for correct implementation of models of devices that possess such properties.

Finally, several applications of recently available large-signal microwave measurement instrumentation – specifically the NVNA – were presented, enabling three distinct advanced device modeling flows. These included X-parameter-based device models (mostly described in Chapter 4), empirical model parameter extraction and validation from NVNA waveform data, and the generation and extensive validation of an advanced ANN-based time-domain III-V FET nonlinear simulation model featuring dynamic self-heating and trapping mechanisms. These were cited as important future trends for successful and efficient state-of-the-art large-signal device modeling.

6.9   Exercises

Exercise 6.1   Derive (6.10) from (6.6). Relate the parameters V_T, V_{max}, I_{max} of (6.10) to the parameters $\{A_n\}$ in (6.6). Hint, expand (6.10) and collect powers of V_1.

Exercise 6.2   Derive (6.23) from (3.36).

Exercise 6.3   Consider the following proposed bias-dependent nonlinear capacitance equations, defined with respect to the conventional equivalent circuit diagram, defined for values of their arguments where the functions are nonsingular: $C_{GS}(V_{GS}, V_{DS}) = C_0\left(1 - \frac{V_{GS}}{V_{BI}}\right)^{-1/2} + \gamma\frac{V_{DS}}{V_{sat}}$ and $C_{GD}(V_{GS}, V_{DS}) = C_0\left(1 - \frac{(V_{GS}-V_{DS})}{V_{BI}}\right)^{-1/2} + \lambda\frac{V_{GS}}{V_{sat}}$

    a.   What are the conditions on the value of λ for these equations to satisfy terminal charge conservation at the gate?

    b.   Under what conditions are the above equations consistent with drain-source exchange symmetry?

    c.   Under the condition that terminal charge conservation is satisfied at the gate, compute the gate charge function up to a constant.

Exercise 6.4   Prove any large signal capacitance-based intrinsic three-node model based on two-terminal nonlinear capacitors with Q–V relations depending only on the voltage difference across the terminals, automatically satisfies the energy conservation principle as given by (6.33) and (6.34).

Exercise 6.5   Starting from energy-conservation equation (6.35), assuming port 1 is the gate and port 2 the drain of a FET, derive, in terms of the coefficients, A_{nm}, expressions for the resulting model capacitance functions. Hint: see (6.37) and Figure 6.22.

    a. C_{GD}

    b. C_{GS}

    c. C_{DS}

    d. C_m

Exercise 6.6   Express the capacitance and transcapacitance ECPs appearing in Figure 6.14 and defined in (6.26) in terms of the ECPs of Figure 6.13.

Exercise 6.7   The following is a proposed constitutive relation for the DC I–V curves of a drain-source symmetric FET: $I_D = \beta(V_{GS} - \lambda V_{DS} - V_T)^2 \tanh(\gamma V_{DS})$.

    a.   Assuming $I_G = 0$, is there a value of λ for which the above constitutive relation can be globally defined for a symmetric device?

    b.   Prove drain-source exchange symmetry transformations (6.61) and (6.62) are invertible and are their own inverses.

Exercise 6.8   Consider the following proposed formal expression for nonlinear terminal charge functions given by $Q_1(V_1, V_2) = \sum a_{n,m} V_1^n V_2^m$ and $Q_2(V_1, V_2) = \sum b_{n,m} V_1^n V_2^m$ where the sums are taken over all non-negative integer values of n and m.

    a.   Under what conditions are these functions consistent with energy conservation?

    b.   Under the conditions of (a), compute the stored energy.

Exercise 6.9   Is the Shockley model consistent with drain-source exchange symmetry?

Exercise 6.10   A diode is placed in the intrinsic FET equivalent circuit, parallel to C_{GS}, with anode at the intrinsic gate node and cathode at the intrinsic source node, to model positive current across the Schottky barrier. Another diode is placed in parallel to C_{GD} with anode at the drain node and cathode at the gate node, to model breakdown. Is this compatible with drain-source exchange symmetry?

References

[1] W. Shockley, "A unipolar 'field-effect' transistor," *Proceedings of the I.R. E.*, vol. 40, Nov. 1952, pp. 1365–1376.

[2] Y. P. Tsividis, *"Operation and Modeling the MOS Transistor,"* New York: McGraw-Hill, 1987, Appendix K.

[3] D. E. Ward, "Charge-based modeling of capacitance in MOS transistors," Technical Report G201–11, Integrated Circuits Laboratory, Stanford University, June 1981.

[4] J. Wood and D. E. Root, editors, *"Fundamentals of Nonlinear Behavioral Modeling for RF and Microwave Design,"* Boston: Artech House, 2005, chapter 7.

[5] Q. Xie, J. Xu, and Y. Taur, "Review and critique of analytical models of MOSFET short-channel effects in subthreshold," *IEEE Transactions on Electron Devices*, vol. 59, no. 6, June 2012, pp. 1569–1579

[6] G. Gildenblat, (Ed.) *"Compact Modeling, Principles, Techniques, and Applications,"* New York: Springer, 2010.

[7] G. Gildenblat, H. Wang, T. L. Chen, X. Gu, and X. Cai, "SP: An advanced surface-potential based compact MOSFET Model," *Custom Integrated Circuits Conference*, 2003.

[8] S. Khandelwal, Y. S. Chauhan, and T. A Fjeldly, "Analytical Modeling of Surface-Potential and Intrinsic Charges in AlGaN/GaN HEMT Devices," *IEEE Transactions on Electron Devices*, vol.59, no.10, pp. 2856, 2860, Oct. 2012.

[9] U. Radhakrishna, P. Choi, S. Goswami, L-S. Peh, T. Palacios, and D. Antoniadis, "MIT virtual source GaNFET – compact model for GaN HEMTs: from device physics to RF frontend circuit design and validation," *IEEE Electron Device Meeting (IEDM)*, Dec. 2014, pp. 11.6.1–11.6.4.

[10] W.R. Curtice, "A MESFET model for use in the design of GaAs integrated circuits, *IEEE Trans. Microw. Theory Techn.*, vol. 28, Issue 5, 1980, pp. 448–456.

[11] W.R. Curtice, "GaAs MESFET modeling and nonlinear CAD," *IEEE Trans. Microw. Theory Techn.*, *vol. 36*, issue 2, 1988, pp. 220–230.

[12] R. A. Pucel, H. A. Haus, and H. Statz, "Signal and noise properties of gallium arsenide microwave field-effect transistors," *Advances in Electronics and Electron Physics*, vol. 38, 1975, Elsevier, pp. 195–265.

[13] J. M. Golio, *Microwave MESFETs and HEMTs*, Norwood, MA: Artech House, Jan. 1991.

[14] EEHEMT Models: Keysight Advanced Design System (ADS 2016.01) Documentation, Nonlinear Devices, pp. 454–480.

[15] I. Angelov, H. Zirath, and N. Rosman, "A new empirical nonlinear model for HEMT and MESFET devices," *IEEE Trans. Microw. Theory Techn.*, vol. 40, no. 12, pp. 2258–2266, 1992.

[16] I. Angelov, L. Bengtsson, and M. Garcia, "Extensions of the Chalmers Nonlinear HEMT and MESFET model, *IEEE Trans. Microw. Theory Techn.*, vol. 46, no. 11, Oct. 1996, pp.1664–1674.

[17] W.R. Curtice, and M. Ettenberg, "A nonlinear GaAs FET model for use in the design of output circuits for power amplifiers," *IEEE Trans. Microw. Theory Techn.*, vol 33, Dec. 1985, pp.1383 – 1394.

[18] D. E. Root "Overview of microwave FET modeling for MMIC design, charge modeling and conservation laws, and advanced topics," 1999 Asia Pacific Microwave Conference Workshop Short Course on Modeling and Characterization of Microwave Devices and Packages, Singapore, Nov. 1999.

[19] D. E. Root, "Principles and procedures for successful large-signal measurement-based FET modeling for power amplifier design," Nov. 2000. Available: http://cp.literature.keysight .com/litweb/pdf/5989-9099EN.pdf

[20] S. Maas, "Fixing the Curtice FET Model" *Microwave Journal*, March 2001

[21] D. E. Root, J. Xu, J. Horn, and M. Iwamoto, "The large-signal model: theoretical foundations, practical considerations, and recent trends," chapter 5 in *Nonlinear Transistor Model Parameter Extraction Techniques*, Cambridge University Press, 2012.

[22] J. C. Pedro and N. B. Carvalho, *Intermodulation Distortion in Microwave and Wireless Circuits*, Norwood, MA: Artech House, 2003.

[23] S. Kirkpatrick, C. D. Gelett, and M. P. Vecchi, "Optimization by simulated annealing," *Science*, 220. May 1983, pp. 621–680.

[24] G. Antoun, M. El-Nozahi, and W. Fikry, "A hybrid genetic algorithm for MOSFET parameter extraction," *IEEE CCECE*, vol. 2, May 2003, pp. 1111–1114

[25] M. Iwamoto, J. Xu, and D. E. Root, "DC and thermal modeling for III-V FETs and HBTs," chapter 2 in *Nonlinear Transistor Parameter Extraction Techniques*, M. Rudolph, C. Fager, and D. E. Root, Eds., Cambridge Univ. Press.

[26] D. E. Root "Measurement-based mathematical active device modeling for high frequency circuit simulation," *IEICE Trans. on Electronics*, vol. E82-C June 1999, pp. 924–936.

[27] D. J. McGinty, D. E. Root, and J. Perdomo, "A Production FET modeling and library generation system," *IEEE GaAs MANTECH Conference Technical Digest*, San Francisco, CA, July 1997 pp. 145–148.

[28] S. Akhtar, P. Roblin, S. Lee, X. Ding, S. Yu, J. Kasick, and J. Strahler, "RF electro-thermal modeling of LDMOSFETs for power-amplifier design," *IEEE Trans. Microw. Theory Tech.*, vol. 50, Jun. 2002, pp. 1561–1570.

[29] W. M. Coughran, W. Fichtner, and E. Grosse, "Extracting transistor charges from device simulations by gradient fitting," *IEEE Trans. on Electron Devices*, vol. 8, pp. 380–394, 1989.

[30] V. Cuoco, M .P. van den Heijden, and L. C. N de Vreede, "The 'Smoothie' data base model for the correct modeling of non-linear distortion in FET devices," *IEEE Int. Microwave Symp. Dig.*, vol. 3, 2002, pp. 2149–2152.

[31] D. E. Root, S. Fan, and J. Meyer, "Technology independent non quasi-static FET models by direct construction from automatically characterized device data," 21st European Microwave Conf. Proc., Stuttgart, Germany, Sept. 1991, pp. 927–932.

[32] Keysight W8532EP IC-CAP manual.

[33] J. Xu; D. Gunyan, M. Iwamoto, A. Cognata, and D. E. Root, "Measurement-based non-quasi-static large-signal FET model using artificial neural networks," *IEEE Int. Microwave Symp. Dig.*, June 2006, pp. 469–472.

[34] J. Xu, D. Gunyan, M. Iwamoto, J. Horn, A. Cognata, D. E. Root; "Drain-source symmetric artificial neural network-based fet model with robust extrapolation beyond training data," *IEEE International Microwave Symposium Digest*, June 2007.

[35] J. Wood, P. H. Aaen, D. Bridges, D. Lamey, M. Guyonnet, D. S. Chan, and N. Monsauret; "A nonlinear electro-thermal scalable model for high-power RF LDMOS transistors," *IEEE Trans. Microw. Theory Techn.*, Feb. 2009, vol. 57, pp. 282–292.

[36] J. Xu, J. Horn, M. Iwamoto, and D. E. Root, "Large-signal FET model with multiple time scale dynamics from nonlinear vector network analyzer data," *IEEE Int. Microwave Symposium Digest*, May 2010.

[37] S. Haykin, *Neural Networks: A Comprehensive Foundation* (2nd ed.) Upper Saddle River, New Jersey: Prentice Hall; 1998.

[38] Q. J. Zhang and K. C. Gupta, *Neural Networks for RF and Microwave Design*, Boston: Artech House, 2000.

[39] Matlab Neural Network Toolbox™.

[40] J. Xu, M. C. E. Yagoub, D. Runtao, and Q. J. Zhang, "Exact adjoint sensitivity analysis for neural-based microwave modeling and design," *IEEE Trans. Microw. Theory Techn.*, vol. 51, Jan. 2003, pp. 226 – 237.

[41] D. E. Root, J. Xu, M. Iwamoto, "Nonlinear FET modeling fundamentals and neural network applications," International Microwave Symposium Workshop (WMA) Advances in Active Device Characterization & Modeling for RF & Microwave, Honolulu, Hawaii, June 2007.

[42] C. B. Barber, D. P. Dobkin, and H. T. Huhdanpaa, "The Quickhull Algorithm for Convex Hulls," *ACM Trans. on Mathematical Software*, vol 22, Dec. 1996, pp. 469–483.

[43] L. Zhang and Q. J. Zhang, "Simple and effective extrapolation technique for neural-based microwave modeling," *IEEE Microwave and Wireless Components Letters*, vol 20, June 2010, pp. 301–303.

[44] D. E. Root, "Measurement-based active device modeling for circuit simulation," European Microwave Conf. Advanced Microwave Devices, Characterization, and Modeling Workshop, Madrid, Sept. 1993 (available from author).

[45] J. Staudinger, M. C. De Baca, and R. Vaitkus, "An examination of several large signal capacitance models to predict GaAs HEMT linear power amplifier performance," *IEEE Radio and Wireless Conf.*, Aug. 1998, pp. 343–346.

[46] D. E. Root, "Nonlinear charge modeling for FET large-signal simulation and its importance for IP3 and ACPR in communication circuits," Proc. of the 44th IEEE Midwest Symp. on Circuits and Systems, Dayton, OH, August 2001, pp. 768–772 (corrected version available from author).

[47] M. Rudolph, *Introduction to Modeling HBTs*, Norwood, MA: Artech House, 2006.

[48] M. Rudolph, R. Doerner, K. Beilenhoff, P. Heymann, "Unified model for collector charge in heterojunction bipolar transistors," *IEEE Trans. Microw. Theory Tech.* vol. 50 Jul. 2002, pp. 1747–1751.

[49] M. Iwamoto, P. M. Asbeck, T. S. Low, C.P. Hutchinson, J. B. Scott, A. Cognata, X. Qin, L. H. Camnitz, and D. C. D'Avanzo, "Linearity characteristics of GaAs HBTs and the influence of collector design," *IEEE Trans. Microw. Theory Techn.* vol. 48, 2000, pp. 2377–2388.

[50] M. Iwamoto and D. E. Root, "Agilent HBT Model Overview," Compact Model Council Meeting, San Francisco, CA, Dec. 2006. Avaliable: https://community.keysight.com/docs/DOC-1201

[51] G. Gonzalez, *Microwave Transistor Amplifiers* (2nd ed.) Englewood Cliffs, NJ: Prentice Hall, 1984, pg. 61.

[52] A. Parker, "Getting to the heart of the matter," *IEEE Microw. Mag.*, April 2015, pp. 76–86.

[53] R. van der Toorn, J.C.J.. Paasschens, R.J. Havens, "A physically based analytical model of the collector charge of III-V heterojunction bipolar transistors," *IEEE Gallium Arsenide Integrated Circuit (GaAs IC) Symp.*, Nov. 2003, pp. 111–114.

[54] M. Iwamoto, J. Xu, J. Horn, and D. E. Root, "III-V FET High Frequency Model with Drift and Depletion Charges," *IEEE International Microwave Symposium*, June 2011.

[55] D. E. Root, J. Xu, D. Gunyan, J. Horn, and M. Iwamoto, "The large-signal model: theoretical and practical considerations, trade-offs, and trends," IEEE Int. Microwave Symp. Parameter extraction strategies for compact transistor models workshop (WMB), Boston, 2009.

[56] K. Kundert, "*The designer's guide to SPICE and Spectre*," Boston: Kluwer Academic Publishers, 1995.

[57] D. E. Root and S. Fan, "Experimental evaluation of large-signal modeling assumptions based on vector analysis of bias-dependent S-parameter data from MESFETs and HEMTs," *IEEE Int. Microwave Symp. Dig.*, 1992, pp. 255–259.

[58] M. Iwamoto, D. E. Root, J. B. Scott, A. Cognata, P. M. Asbeck, B. Hughes, and D. C. D'Avanzo, "Large-signal HBT model with improved collector transit time formulation for GaAs and InP technologies," *IEEE Int. Microwave Symp. Dig.*, Philadelphia, PA, June 2003, pp. 635–638.

[59] ADS FET, *Keysight Advanced Design System (ADS 2016.01) Manual, Nonlinear Devices*, pp. 379–383.

[60] D. E. Root "Elements of Measurement-Based Large-Signal Device Modeling," IEEE Radio and Wireless Conference (RAWCON) Workshop on Modeling and Simulation of Devices and Circuits for Wireless Communication Systems, Colorado Springs, August 1998.

[61] D. E. Root, "ISCAS tutorial/short course and special session on high-speed devices and modeling," Sydney, May 2001, pp. 2.71–2.78

[62] D. E. Root, M. Iwamoto, and J. Wood, "Device modeling for III-V semiconductors: an overview," IEEE Compound Semiconductor IC Symp., October 2004.

[63] A. C. T. Aarts, R. van der Hout; J. C. J. Paasschens, A. J. Scholten, M. Willemsen, and D. B. M. Klaassen; "Capacitance modeling of laterally non-uniform MOS devices," *IEEE IEDM Tech. Dig.*, Dec. 2004, pp. 751–754.

[64] A. E. Parker and S. J. Mahon, "Robust extraction of access elements for broadband small-signal FET models," *IEEE Int. Microwave Symp. Dig.*, 2007, pp.783–786.

[65] H. Statz, P. Newman, I.W. Smith, R. A. Pucel, and H. A. Haus, "GaAs FET device and circuit simulation in SPICE," *IEEE Trans. Electron Devices*, vol. 34, Feb. 1987, pp. 160–169.

[66] I. W. Smith, H. Statz, H. A. Haus, and R. A. Pucel, "On charge nonconservation in FETs," *IEEE Trans. Electron Devices*, vol. 34 Dec. 1987, pp. 2565 – 2568.

[67] C. A. Desoer and E. S. Kuh, *Basic Circuit Theory*, New York: McGraw-Hill, 1969, Table 19.1, p. 801.

[68] M. Iwamoto and D.E. Root, "Large-Signal III-V HBT Model with Improved Collector Transit Time Formulations, Dynamic Self-Heating, and Thermal Coupling," Int. Workshop on Nonlinear Microwave and Millimeter Wave Integrated Circuits (INMMIC), Rome, Nov. 2004.

[69] J. G. Leckey, "A new current dependent gate charge model for GaN HFET devices," 11th European Microwave Integrated Circuits Conference (EuMIC), London, UK, Oct. 2016, pp. 556–558.

[70] C. C. McAndrew, "Validation of MOSFET model source-drain symmetry," *IEEE Trans. Electron Devices*, vol. 53, no. 9, Sept. 2006, pp. 2202–2206.

[71] K. Yhland, N. Rorsman, M. Garcia, and H. F. Merkel, "A symmetrical nonlinear HFET/ MESFET model suitable for intermodulation analysis of amplifiers and resistive mixers," *IEEE Trans. Microw. Theory Techn.*, vol. 48, no. 1, January 2000, pp 15–22.

[72] K. Poulton et al, "Thermal design and simulation and design of bipolar integrated circuits," *IEEE Journal of Solid-State Circuits*, Vol. 27., No. 10, October 1992. pp. 1379 – 1387

[73] A. M. Conway, P. M. Asbeck, "Virtual gate large-signal model of GaN HFETs," *IEEE Int. Microwave Symp. Dig.*, June 2007, pp. 605–608.

[74] O. Jardel, F. DeGroote, T. Reveyrand, J. C. Jacquet, C. Charbonniaud, J. P. Teyssier, D. Floriot, and R. Quere "An electrothermal model for AlGaN/GaN power HEMTs including trapping effects to improve large-signal simulation results on high VSWR," *IEEE Trans. Microw. Theory Tech.*, vol. 55, Dec. 2007, pp. 2660–2669.

[75] O. Jardel, R. Sommet, J-P Teyssier, and R. Quere, "Nonlinear characterization and modeling of dispersive effects in high-frequency power transistors," chapter 7 in *Nonlinear Transistor Model Parameter Extraction Techniques*, Cambridge University Press, 2012.

[76] A. E. Parker, D. E. Root, "Pulse measurements quantify dispersion in pHEMTs," URSI Int. Symp. on Signals, Systems, and Electronics (ISSSE), Pisa, Sept. 1998, pp. 444–449.

[77] A. Santarelli *et al*, "A double-pulse technique for the dynamic characterization of GaN FETs," *IEEE Microwave Wireless Component Letters*, vol. 24, no. 2, pp. 132–134, February, 2014

[78] A. Santarelli *et al*, "GaN FET nonlinear modeling based on double pulse I/V characteristics," *IEEE Trans. Microw. Theory Techn.*, vol. 62, no. 12, December 2014, pp. 3262–3273.

[79] Keysight Technologies. Available: http://www.keysight.com/find/nvna

[80] P. Blockley D. Gunyan, and J. B. Scott, "Mixer-based, vector-corrected, vector signal/ network analyzer offering 300kHz – 20GHz bandwidth and traceable phase response," *IEEE Int. Microwave Symp. Dig. Long Beach*, Jun. 2005, pp. 1497–1500.

[81] J. Verspecht, "Calibration of a measurement system for high frequency nonlinear devices," Ph. D. Dissertation, Dept. ELEC, Vrije Universiteit Brussel, Nov. 1995.

[82] W. Van Moer and L. Gomme," NVNA versus LSNA: enemies or friends?" *IEEE Microw. Mag.*, volume: 11, issue 1, 2010, pp. 97–103.

[83] E. P. Vandamme, W. Grabinski, and D. Schreurs, "Large-signal network analyzer measurements and their use in device modeling," Proc. 9th Int. Conf. Mixed Design of Integrated Circuits and Systems (MIXDES) Wroclaw, 2002

[84] D. Schreurs, J. Verspecht, B. Nauwelaers, A. Van de Capelle, and M. Rossum, "Direct extraction of the non-linear model for two-port devices from vectorial nonlinear network analyzer measurements," 27th *European Microwave Conf. Proc.*, 1997, pp. 921–926.

[85] M.C. Curras-Francos, P. J. Tasker, M. Fernandez-Barciela, Y. Campos-Roca, and E. Sanchez, "Direct extraction of nonlinear FET Q-V functions from time domain large signal measurements," *IEEE Microwave and Guided Wave Letters*, vol. 10, 2000, pp. 531–533.

[86] P.J. Tasker, M. Demmler, M. Schlechtweg, M. Fernandez- Barciela, "Novel approach to the extraction of transistor parameter from large signal measurements," 24th European Microwave Conf. Sept. 1994, pp. 1301–1306.

[87] D. E. Root, J. Horn, J. Xu, M. Iwamoto, F. Sischka, and Y. Yanagimoto, "Time and frequency domain transistor modeling based on Nonlinear Vector Network Analyzer data," MOS-AK/GSA Workshop, San Francisco, CA, Dec. 2010.

[88] D. E. Root et al "NVNA Measurements for Behavioral and Compact Device Modeling," IEEE International Microwave Symposium Workshop (WMB) Device Model Extraction from Large-signal Measurements, Montreal, CA, June 2012.

[89] F. Sischka, "Nonlinear network analyzer measurements for better transistor modeling," 2011 IEEE Conference on Microelectronic Test Structures, Apr. 4–7, 2011 Amsterdam, The Netherlands.

[90] F. Sischka, "Improved compact models based on NVNA measurements," in *Proc. European Microwave Week 2010 Workshop*, Paris, France.

[91] D. E. Root, "Future Device Modeling Trends," *IEEE Microw. Mag.*, Nov./Dec. 2012, pp 45–59.

[92] W. Grabinski, E. P. Vandamme, D. Schreurs, H. Maeder, O. Pilloud, and C. C. McAndrew, "5.5 GHz LSNA MOSFET modeling for RF CMOS circuit design," in *60th IEEE ARFTG Conf. Dig.*, 2002, pp. 39–47.

[93] D. E. Root, "Compact and behavioral modeling techniques for GaN devices," IEEE CSICS Short Course on GaN Modeling, La Jolla, CA, 2014.

[94] I. Angelov, "Advanced GaN and GaAs transistor evaluation and transistor modeling using combined small – and large-signal VNA & NVNA microwave measurements," IEEE International Microwave Symposium Workshop: Direct Extraction of FET Circuit Models

from Microwave and Baseband Large-Signal Measurements for Model-Based Microwave Power Amplifier Design, 2015.

[95] I. Angelov, M. Thorsell, K. Andersson, N. Rorsman, and H. Zirath, "Recent results on using LSVNA for Compact modeling of GaN FET devices," *IEEE International Microwave Symposium Workshop (WMB)*: Device Model Extraction from Large-Signal Measurements, Montreal, June 2012.

[96] M. A. Smith, T. S. Howard, K. J. Anderson, and A. M. Pavio, "RF nonlinear device characterization yields improved modeling accuracy," *IEEE International Microwave Symposium Digest*, 1986. pp. 381–384.

[97] A. Raffo, S. D. Falco, V. Vadala, and G. Vannini, "Characterization of GaN HEMT low-frequency dispersion through a multiharmonic measurement system," *IEEE Trans. Microw. Theory Techn.*, vol. 58, no. 9, September 2010.

[98] A. Raffo, G. Bosi, V. Vadalà, and G. Vannini, "Behavioral modeling of GaN FETs: a load-line approach," *IEEE Trans. Microw. Theory Techn.*, vol. 62, no. 1, January 2014.

[99] J. Xu, R. Jones, S. A. Harris, T. Nielsen, and D. E. Root, "Dynamic FET model – DynaFET – for GaN transistors from NVNA active source injection measurements," *International Microwave Symposium Digest*, June 2014.

[100] J. Xu, S. Halder, F. Kharabi, J. McMacken, J. Gering, and D. E. Root, "Global dynamic FET model for GaN transistors: DynaFET Model validation and comparison to locally tuned models," 83rd IEEE ARFTG Measurement Conference, Tampa, FL., June 2014.

[101] J. Xu and D. E. Root, "NVNA Characterization enables DynaFET: an advanced compact time-domain FET model," NVNA Users Forum, Tampa, FL, June 2014.

7 Nonlinear Microwave CAD Tools in a Power Amplifier Design Example

Many important concepts regarding nonlinear microwave systems and an extensive discussion of nonlinear device modeling techniques were introduced in previous chapters. Now, to summarize what we have learned and to facilitate the comprehension of the presented concepts, this chapter illustrates how accurate nonlinear device models and circuit simulators can be used in the design of microwave circuits. For these purposes, the design of a typical medium-power amplifier (PA) circuit will be used as an example.

We show that the effort of considering nonlinear circuit simulation and active device models is justified as these tools reduce the complexity, and thus the time needed for practical circuit designs and their characterization. Moreover, these device and circuit representations enable the analysis of the circuit's operation, not only via the input-output observation, as is done in the laboratory, but also via looking into every internal equivalent circuit node of the active device.

The chapter is organized in two major thematic parts.

The first part, consisting of Section 7.1, recalls some results of active device modeling from previous chapters, explaining the importance and impact they have on the predicted power amplifier behavior. Special attention will be paid to the model features that determine the most important PA characteristics: output power, drain efficiency, and AM/AM and AM/PM distortion, i.e., gain amplitude and phase profiles.

The second part, Section 7.2, details the PA computer-aided design and verification procedures. It first illustrates the roles that the nonlinear device model and a simulator can play in the various circuit design steps, constituting an aid, or even a substitute, to the laborious and expensive measurement-intensive conventional PA design procedure. Then, we will also show how simulations can be used to estimate and interpret the performance characteristics of the implemented circuit.

7.1 Nonlinear Device Modeling in RF/Microwave Circuit Design

Contrary to a low-noise amplifier, or a low-power PA design, which, nowadays, often use MMIC implementation, medium to high-power PAs rely on packaged devices mounted in a PCB. Therefore, not only do the equivalent circuit models of Chapters 5 and 6 have to be embedded in a more or less complex package model, but the designer may also face the situation of having to design a PA with a device for which there is still no available model provided by the device vendor. This is quite common in mobile communications base station design, and is aggravated in high-power (hundreds of watts peak-power) circuits whose active devices are often internally prematched. This is one field in which the frontier between active device modeling and circuit design is blurred, and so where the microwave engineer has to make a prudent choice of the best compromise between a model's complexity (and, thus, difficulty of extraction) and accuracy, i.e., mostly the model's capability to predict the actual device's output power and efficiency load-pull contours and AM/AM and AM/PM profiles. As we will see, this goes all the way from the judicious choice of the equivalent circuit topology to the formulations of the $i_{DS}(v_{GS},v_{DS})$, $Q_{gs}(v_{GS},v_{DS})$, $Q_{gd}(v_{GS},v_{DS})$ and $Q_{ds}(v_{GS},v_{DS})$ nonlinear constitutive relations.

To obtain a representation as close as possible to the real device, the model topology is frequently established according to our knowledge of the device's structure and physical operation. Actually, Figure. 7.1(a) is an illustration of the geometry of an RF MOSFET, which is then translated into the equivalent circuit model of Figure 7.1(b). As discussed in Chapters 5 and 6, this model can be divided into intrinsic and extrinsic subcircuits. The former ones are associated with the channel, while the latter are the representation of the parasitic effects associated to the metallization of the die and the package.

7.1.1   The Extrinsic Subcircuit Model

Since the extrinsic elements are bias-independent, or linear, we could think that they may not be as important as the intrinsic elements for the design of a nonlinear circuit such as a power amplifier. However, these extrinsic elements play a fundamental role in the accurate prediction of output power and efficiency load-pull contours. To illustrate this, Figure 7.2 shows three different sets of load-pull contours for a commercial 240 W packaged Si LDMOS device. Figure 7.2(a) depicts the obtained load-pull predictions in which the used model is only composed of the adopted FET $i_{DS}(v_{GS},v_{DS})$ nonlinear model; (b) shows the updated load-pull predictions – reported at the intrinsic reference plane – when the intrinsic $Q_{gs}(v_{GS},v_{DS})$, $Q_{ds}(v_{GS},v_{DS})$ and $Q_{gd}(v_{GS},v_{DS})$ nonlinear voltage-dependent charge models or their equivalent capacitances, $C_{gs}(v_{GS},v_{DS})$, $C_{ds}(v_{GS},v_{DS})$, and $C_{gd}(v_{GS},v_{DS})$, were also included; and, (c) illustrates the final load-pull predictions of the entire model of Figure 7.1, i.e., when also the extrinsic sub-circuit is considered.

Comparing the contours of Figure 7.2(b) and (c) with those determined by the nonlinear FET channel current of (a), first we should note a non-uniform stretching of the contours, in particular in high impedance values, where the output voltage excursion is higher, producing a large variation of the nonlinear output capacitance. Thus, after de-embedding the small-signal capacitance, the optimum impedances become more reactive. Finally, with the inclusion of the extrinsic elements in (c), the load-pull contours moved to a different Smith chart region.

Measuring the load-pull contours at the extrinsic reference plane, so that they can be used as a practical power amplifier design tool, usually requires a very complicated and

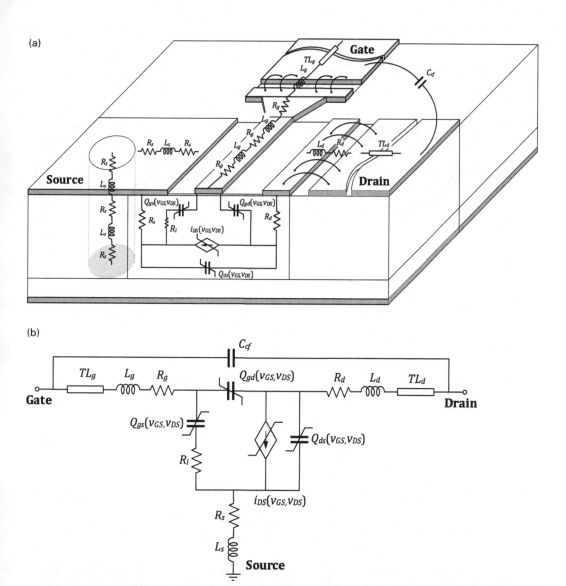

Figure 7.1 (a) Illustration of the physical structure of a MOSFET and (b) its corresponding equivalent circuit model.

time-consuming task of laboratory load-pull characterization. However, if we had access to a complete equivalent circuit model, we could obtain these load-pull contours through simulation, or simply determine their estimates at the intrinsic level [1,2] and then calculate their extrinsic counterparts, embedding the impedance transformation effects of the extrinsic subcircuit elements. Alternatively, the power amplifier design could be done at the intrinsic reference plane incorporating all extrinsic components into the matching networks.

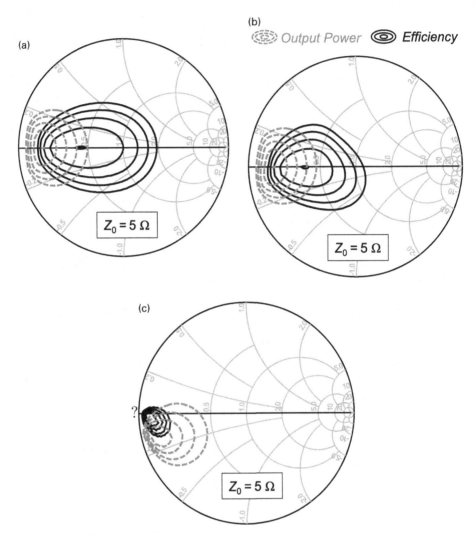

Figure 7.2 Output power and efficiency load-pull contours of a 240 W packaged Si LDMOS device simulated at 1800 MHz in three different situations: (a) considering only the adopted FET $i_{DS}(v_{GS}, v_{DS})$ nonlinear model, (b) considering the FET $i_{DS}(v_{GS}, v_{DS})$ and $C_{gs}(v_{GS}, v_{DS})$, $C_{ds}(v_{GS}, v_{DS})$ and $C_{gd}(v_{GS}, v_{DS})$ intrinsic nonlinear capacitances at the intrinsic reference plane, and (c) considering the entire model, i.e., also including the extrinsic linear elements.

    The extraction of the package parasitic network requires S-parameter characterization of a dummy device, i.e., an empty package, and/or measurements of a complete device biased at extreme quiescent states, such as cut-off, or gate-channel junction forward bias at the FET's triode region.

    In high-power devices, as represented by the illustration presented in Figure 7.3(a), the intrinsic optimum impedances are so low that in-package prematching networks are often needed to transform them into manageable impedance values

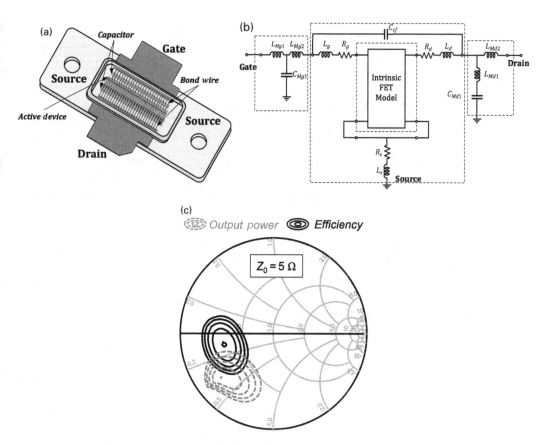

Figure 7.3 (a) Illustration of a high-power device with in-package prematching networks, (b) its equivalent circuit model representation, and (c) the respective output power and efficiency load-pull contours at the package reference plane.

at the package reference plane. Unfortunately, the physical dimensions of these multifinger devices and of their connection pads, bond-wires, leads and prematching structures are often an appreciable fraction of the wavelength. Therefore, their accurate modeling cannot be accomplished with lumped elements, requiring, instead, complex electromagnetic simulations [3]. The obtained S-matrices – or their time-domain equivalent circuit representation [4] – are then included into the entire high-power device model, as illustrated in Figure 7.3(b). With these intricate prematching structures, the external reference plane moves further away from the intrinsic one, which produces a larger transformation of the load-pull contours, as shown in Figure 7.3(c). Moreover, the narrowband nature of these prematching networks strongly influences the harmonic impedances at the intrinsic reference plane, regardless of the harmonic terminations provided at the package terminals; and this significantly limits the use of harmonic manipulation for optimized PA performance.

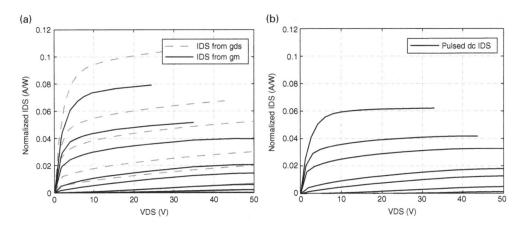

Figure 7.4 I/V curves of a GaN HEMT device (a) obtained by $G_m(v_{GS},v_{DS})$ and $G_{ds}(v_{GS},v_{DS})$ integration – referred to as "IDS from gm" and "IDS from gds," respectively – and (b) pulsed dc I/V curves, when the measurements are not isodynamic.

7.1.2 Nonlinear $i_{DS}(v_{GS},v_{DS})$ Current Model

After prescribing the equivalent circuit topology, we now need to discuss the intrinsic elements. Of these, a realistic voltage-dependent $i_{DS}(v_{GS},v_{DS})$ drain current model is key for an accurate prediction of the AM/AM profile and the output power and drain efficiency load-pull contours. As explained in Chapter 6, various model formulations can be adopted, from more or less compact formulae, to complex, but systematic, neural networks. However, more important than the actual $i_{DS}(v_{GS},v_{DS})$ formulation is the measurement dataset from which it is extracted.

7.1.2.1 $i_{DS}(v_{GS},v_{DS})$ Current Model Extraction

In devices that suffer from obvious thermal, or even trapping, effects, such as power FETs, dc and RF data are not consistent [5]. Hence, either we have an NVNA to measure the device in a wide set of operating conditions [6] and then rely on the simultaneous nonlinear optimization of dc and drive-level and load dependent RF data described in Chapter 6, or we try more conventional data collection procedures such as bias-dependent S-parameter data. In that case, i_{DS} current is obtained by integration of the transconductance $G_m(v_{GS},v_{DS})$ and output admittance $G_{ds}(v_{GS},v_{DS})$ profiles measured under pulsed bias conditions.

As an example, in Figure 7.4 we can observe pulsed dc I/V curves, from a GaN HEMT device, and the ones obtained from the G_m and G_{ds} integration over v_{GS} and v_{DS}, when the applied pulses were incapable of producing isodynamic measurements. As expected, if the thermal and trapping states are not the same for all tested bias points, the obtained I/V curves will be different, being impossible to fit these measurements to any single $i_{DS}(v_{GS},v_{DS})$ model expression.

On the contrary, if a prepulse is used to set the traps' state, guaranteeing that the thermal and trapping states do not change for all measured bias points [7] – obtaining,

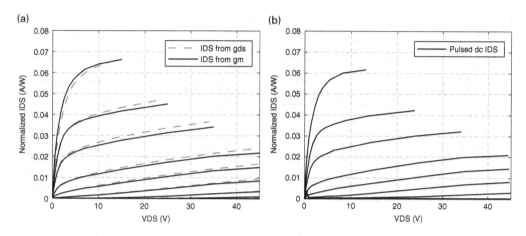

Figure 7.5 I/V curves of a GaN HEMT device (a) obtained by $G_m(v_{GS},v_{DS})$ and $G_{ds}(v_{GS},v_{DS})$ integration – referred to as "IDS from gm" and "IDS from gds", respectively – and (b) with the direct pulsed I/V method, when the measurements are guaranteed to be isodynamic.

this way, isodynamic measurements – the I/V curves will be the same, either directly measured from the i_{DS} pulses, or obtained from the G_m and G_{ds} integration, as shown in Figure 7.5.

7.1.2.2 Impact of the $i_{DS}(v_{GS},v_{DS})$ Current on the PA Output Power and Efficiency

The knowledge obtained from modeling these thermal and trapping effects also helps us to understand how they affect the I/V curves, which may guide us later during our PA design.

For instance, it is known that, when a high v_{DS} peak voltage is applied to the device, the maximum i_{DS} current, I_{MAX}, decreases and a threshold voltage, V_T, displacement is observed due to the trapping effects [5–8]. Consequently, the $i_{DS}(v_{GS},v_{DS})$ curves will be different if we consider the I/V curves pulsed from 0 V or from a high peak quiescent voltage, which, in turn, will modify the optimum power and efficiency load terminations, as show in Figure 7.6(a).

Thermal effects will also produce an I_{MAX} drop when the device operates in large-signal mode, because of the increased device's dissipation and temperature, and thus decreased channel conductivity [9]. Therefore, if we only consider the I/V curves measured at room temperature, the estimation of the optimum power and efficiency impedances could also be different from the real ones obtained when the high-power signal is applied to the device, and thus its temperature is increased, as illustrated in Figure 7.6(b).

7.1.2.3 Impact of the $i_{DS}(v_{GS},v_{DS})$ Current on the PA AM/AM Distortion

Having discussed the implications of the $i_{DS}(v_{GS},v_{DS})$ drain current model on the maximum output power and efficiency, we will now take a look at the distortion arising from this nonlinear voltage-controlled current source. For that, we will remove the

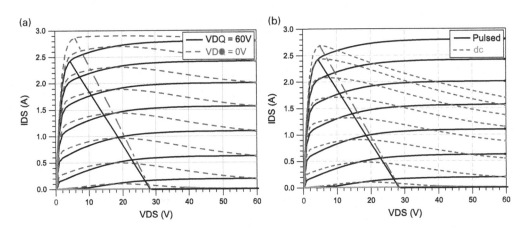

Figure 7.6 Illustration of the (a) trapping and (b) thermal effects' impact on the optimum power load estimation.

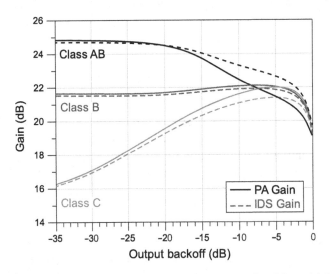

Figure 7.7 Comparison between the simulated complete PA model (PA Gain) and the predictions obtained when the nonlinear device model is only composed by the $i_{DS}(v_{GS},v_{DS})$ nonlinearity (IDS Gain).

nonlinear intrinsic capacitances – i.e., replace them by their linear versions – and compare the simulation results with the ones obtained when the full nonlinear model is considered.

Figure 7.7 shows this comparison for three different PA operation classes corresponding to shallow class AB, class B, and shallow class C. Looking into these results, we can conclude that the nonlinear $i_{DS}(v_{GS},v_{DS})$ current is, indeed, the main contributor to the AM/AM characteristic of a PA, because only slight differences are observed due to the input and output mismatches produced by the nonlinear capacitances. Actually, in [10] it is explained how each of these curve shapes can be related to particular $i_{DS}(v_{GS},v_{DS})$ model features such as the FET's soft turn-on and its saturation to triode region transition.

7.1.3 Nonlinear Intrinsic Capacitance Models

7.1.3.1 Nonlinear Capacitance Models Extraction

The problems experienced with the $i_{DS}(v_{GS},v_{DS})$ model extraction, which were related to the thermal and trapping effects, are also present in the identification of the intrinsic nonlinear voltage-dependent charge sources. Therefore, to guarantee an isodynamic extraction, the double-pulse bias-dependent S-parameter measurements we adopted for the i_{DS} model extraction should still be used for the capacitances.

Figure 7.8 shows typical $C_{gs}(v_{GS},v_{DS})$, $C_{ds}(v_{GS},v_{DS})$, and $C_{gd}(v_{GS},v_{DS})$ profiles. A high variation with v_{GS} is observed in the C_{gs} profile around the threshold voltage, whereas C_{gd} and C_{ds} only present a significant variation – with v_{GD} and v_{DS} voltages, respectively – when the device becomes operated in the triode region.

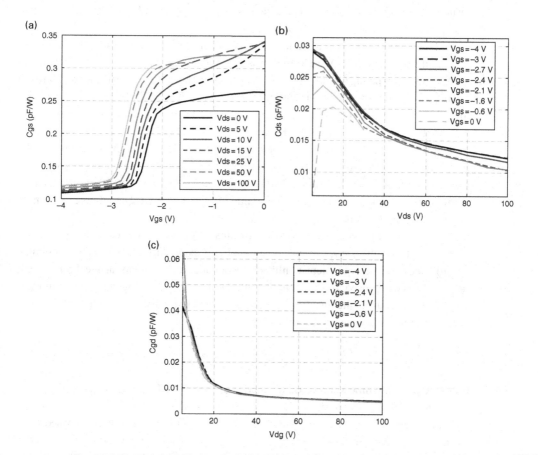

Figure 7.8 Extracted (a) $C_{gs}(v_{GS},v_{DS})$, (b) $C_{ds}(v_{GS},v_{DS})$, and (c) $C_{gd}(v_{GS},v_{DS})$ normalized profiles obtained from pulsed bias S-parameters of a GaN HEMT device.

These variations will detune the output efficiency load-pull contours predicted when only the I/V curves are considered, as was seen in Figure 7.2. In addition, for advanced PA architectures, such as the Doherty PA, where it is necessary to compensate the delay between the main and auxiliary amplifiers [11], the variation of these capacitances will make it more difficult to determine the correct delay that should be used for large-signal operation. Moreover, these variations will have a severe impact on the AM/PM PA characteristic, as we will explain next. Consequently, the nonlinearity of the intrinsic capacitances and their accurate modeling should be also taken into account for a successful PA design.

7.1.3.2   Impact of the Intrinsic Capacitances on the PA AM/AM and AM/PM Distortion

With respect to the nonlinear distortion induced by the intrinsic capacitances, we have to distinguish the PAs based on the Si LDMOS and GaN HEMT, the two main RF power transistor technologies used for mobile communication base stations, since they present significant differences in their normalized capacitances. Si LDMOS devices usually have higher C_{gs} and C_{ds} capacitances and lower C_{gd} feedback capacitance, when compared with those of GaN HEMTs [12].

Another aspect that we should take into consideration when assessing the impact of C_{gd} nonlinearity in the PA performance is that this can be viewed as two independent effects. On one hand, C_{gd} is a nonlinearity in itself because, as seen Figure 7.8(c), it manifests a noticeable variation with the v_{GD} voltage. But, on the other hand, it should be noted that, even if C_{gd} were linear (i.e., constant or bias independent), it would still constitute a source of nonlinearity. Because C_{gd} is reflected, through the Miller effect, to the input as $C_{gd}(1 - A_v)$, and to the output as $C_{gd}(A_v - 1)/A_v$, and the voltage gain, A_v, will have to vary due to the inevitable PA gain compression, the effect of C_{gd} will also be nonlinear. Measurements and simulations of microwave FET amplifier circuits have shown that, perhaps unexpectedly, the input Miller reflected C_{gd} nonlinearity is more significant than the direct $C_{gd}(v_{GD})$ nonlinearity [10].

To understand the underlying reasons for this surprising result, equation (7.1) gives the value of the intrinsic v_{gs} voltage when the PA is excited by a sine wave of amplitude $V_s(\omega)$. Because the equivalent nonlinear capacitance that appears at the FET's input terminals is composed by this input Miller reflected C_{gd} along with the nonlinear $C_{gs}(v_{GS})$ capacitance, the intrinsic V_{gs} phasor voltage will vary, in amplitude and phase, with the amplitude drive, A, according to

$$V_{gs1}(A) \approx \frac{V_s(\omega)}{1 + j\omega Z_s(\omega)\left[C_{gs0}(A) + C_{gd}(1 - A_v(A))\right]} \tag{7.1}$$

where C_{gs0} is the time-varying $C_{gs}(t)$ averaged over one RF period [10].

Figures 7.9 and 7.10 show the normalized (in a per watt basis) C_{gs0} and the $C_{gd,Miller}$ dependence with the amplitude for Si LDMOS and GaN HEMT based PAs, respectively. The selected V_{GS} bias points were the ones corresponding to the same three operation classes used for the $i_{DS}(v_{GS}, v_{DS})$ distortion analysis. Although both PAs present similar capacitance profile shapes, the input Miller reflected C_{gd} value for Si

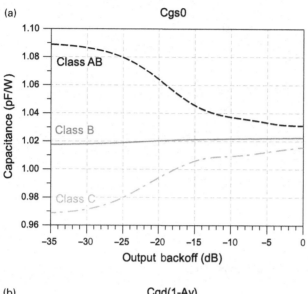

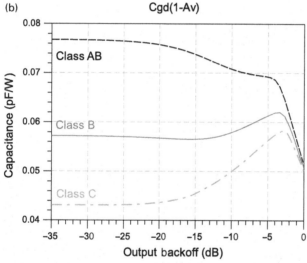

Figure 7.9 Normalized (per Watt) profiles of (a) the C_{gs0} component and (b) the input Miller reflected C_{gd} for a Si LDMOS based PA.

LDMOS PAs is almost insignificant when compared with the C_{gs0}, whereas in the GaN HEMT PAs it is the major contributor to the input capacitance variation. This means that in GaN HEMT PAs, the input capacitance decreases whenever the gain compresses or it increases when a gain expansion is observed.

As far as the PA output node is concerned, since the voltage gain is normally very high, the output Miller reflected C_{gd} variation $[C_{gd}(A_v - 1)/A_v]$ is almost negligible. Thus, the C_{ds} variation will be the main contributor to the phase shift on the fundamental v_{DS} voltage, according to equation (7.2).

(a)

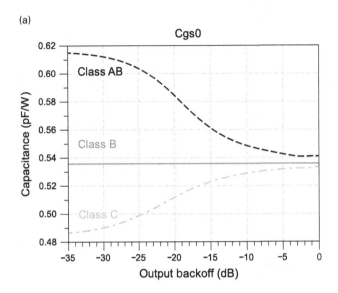

(b)

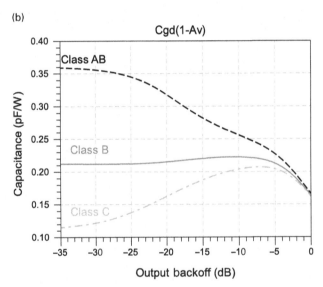

Figure 7.10 Normalized (per watt) profiles of (a) the C_{gs0} component and (b) the input Miller reflected C_{gd} for a GaN HEMT based PA.

$$V_{ds1}(A) \approx \frac{-Z_L(\omega)I_{ds1}(A)}{1 + j\omega Z_L(\omega)C_{ds0}(A)} \qquad (7.2)$$

The C_{ds0} capacitance variation for both Si LDMOS and GaN HEMT based PAs is depicted in Figure 7.11. Again, although the capacitance profile shapes of both devices are very similar, their values are completely different, with C_{ds0} being much higher for Si LDMOS based PAs. Actually, for GaN HEMT based PAs, the C_{ds0} variation is so

(a)

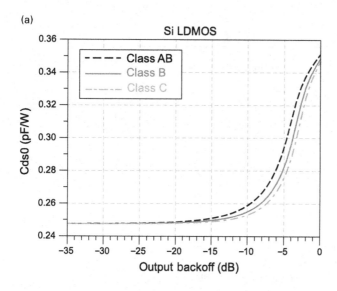

(b)

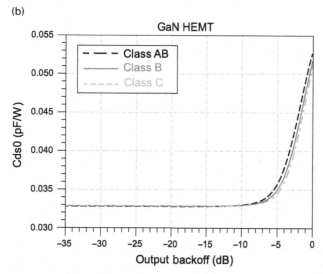

Figure 7.11 Normalized (per Watt) profiles of the C_{ds0} component of (a) Si LDMOS and (b) GaN HEMT based PAs.

small when compared with the input capacitance variation (see Figures 7.10 and 7.11) that it will be insignificant for the overall amplitude nonlinearity (AM/AM) or phase shift (AM/PM) of the PA.

Because the $i_{DS}(v_{GS},v_{DS})$ nonlinearity is the dominant contributor to the PA amplitude, AM/AM, distortion, the main effect of capacitance nonlinearities is on the amplitude dependent phase shift, AM/PM. But, contrary to what happens with the $i_{DS}(v_{GS},v_{DS})$ nonlinearity, and thus with the amplitude distortion, these two technologies present quite different intrinsic capacitance variations. Hence, Si LDMOS and GaN

(a)

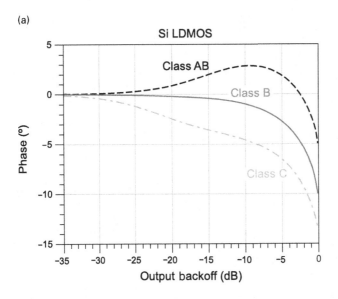

(b)

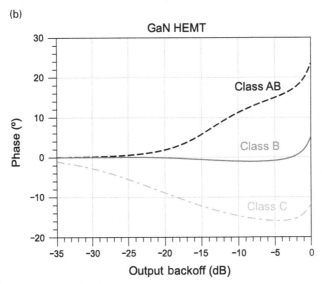

Figure 7.12 AM/PM characteristics for (a) Si LDMOS and (b) GaN HEMT based PAs obtained from harmonic-balance simulations.

HEMT based PAs evidence significantly distinct AM/PM characteristics, as shown in Figure 7.12. For Si LDMOS, the C_{ds0} variation is the major contributor to the AM/PM distortion, producing an almost phase-lagging behavior for all operation classes; only a slight phase-leading behavior is presented at the mid-power region for class AB due to the initial C_{gs0} reduction. Conversely, the AM/PM distortion of GaN HEMT PAs is mostly determined by the phase shift imposed on the V_{gs} phasor voltage, due to C_{gs0} and input Miller reflected C_{gd} variation. Therefore, since this capacitance variation is tied to the voltage gain, the AM/PM characteristic presents an opposite behavior to the AM/AM characteristic, as can be seen when we compare the AM/AM behavior shown in Figure 7.7 and the AM/PM of Figure 7.12(b).

7.1.4 Device Model Implementation in Commercial Microwave Circuit Simulation Platforms

To finalize this section devoted to illustrate the use of nonlinear models and computer-aided design tools in predicting the major RF characteristics of microwave devices, we now show how the above nonlinear models can be implemented in the available commercial simulators, such as the Keysight Technologies Advanced Design System, ADS [13], and the Applied Wave Research Microwave Office, MWO [14]. For that, we will use simple model formulations of the bidimensional nonlinear $i_{DS}(v_{GS}, v_{DS})$ current and the assumed one-dimensional gate-source charge, $Q_{gs}(v_{GS})$, or capacitance, $C_{gs}(v_{GS})$.

For the $i_{DS}(v_{GS}, v_{DS})$ formulation we will use a model (7.3a) consisting in the product of (7.3b) and (7.3c), which describe the i_{DS} dependence on the v_{GS} and v_{DS} voltages, respectively. Figure 7.13 shows the I/V curves of this simple model and the respective derivatives, G_m and G_{ds}.

$$i_{DS}(v_{GS}, v_{DS}) = f_g(v_{GS})f_d(v_{DS}) \tag{7.3a}$$

in which

$$f_g(v_{GS}) = \frac{1}{2K_g}\beta \cdot [K_g(v_{GS} - V_T) + \ln(2\cosh(K_g(v_{GS} - V_T)))] \tag{7.3b}$$

and

$$f_d(v_{DS}) = \tanh(\alpha \cdot v_{DS}) \tag{7.3c}$$

As far as the $Q_{gs}(v_{GS})$ formulation is concerned, we will adopt the model (7.4) – that we have used in the core of $i_{DS}(v_{GS})$ model (7.3b) – leading to the $C_{gs}(v_{GS})$ hyperbolic tangent expression of (7.5). The profiles for these $Q_{gs}(v_{GS})$ and $C_{gs}(v_{GS})$ models can be shown in Figure 7.14(a) and (b), respectively.

$$Q_{gs}(v_{GS}) = C_{gs0}v_{GS} + 0.5A_{Cgs}\left(v_{GS} - V_{Cgs} + \frac{\ln(2\cosh(K_{Cgs}(v_{GS} - V_{Cgs})))}{K_{Cgs}}\right) \tag{7.4}$$

$$C_{gs}(v_{GS}) = C_{gs0} + 0.5A_{Cgs}(1 + \tanh[K_{Cgs}(v_{GS} - V_{Cgs})]) \tag{7.5}$$

The model implementation itself in ADS can be readily done with *symbolically defined device*, SDD, components. These components allow us to define the desired equations as a function of the voltages, currents, or their derivatives. Figure 7.15

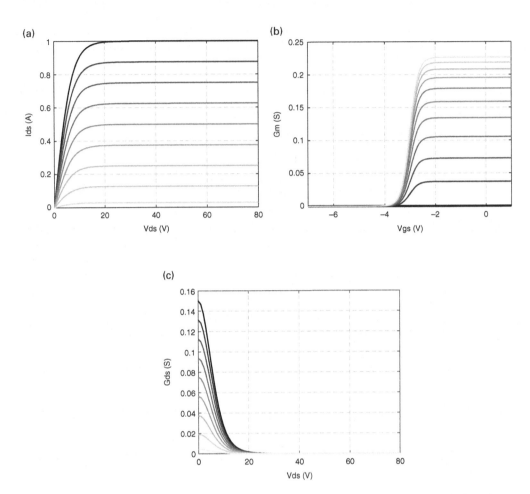

Figure 7.13 (a) $i_{DS}(v_{GS},v_{DS})$, (b) $G_m(v_{GS})$ and (c) $G_{ds}(v_{DS})$ profiles of the example formulae used to illustrate how to implement nonlinear static models in commercial nonlinear circuit simulators.

illustrates the implementation of our $Q_{gs}(v_{GS})$ charge and $i_{DS}(v_{GS},v_{DS})$ current models.

To implement a model in MWO, we can use APLAC netlists. There, we define not only the charge and current equations, but also their derivatives to improve the simulator convergence. Tables 7.1 and 7.2 show the APLAC-MWO netlist implementation of our $Q_{gs}(v_{GS})$ and $i_{DS}(v_{GS},v_{DS})$ models, respectively. After implementation of these netlists, we need to compile and import them to a schematic, as is shown in Figure 7.16.

7.2   Computer-Aided Power Amplifier Design Example

After having concluded our explanation of the impact of some of the active device model parameters in the performance of the circuits in which these devices operate, we

(a)

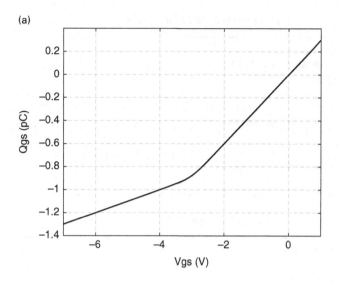

(b)

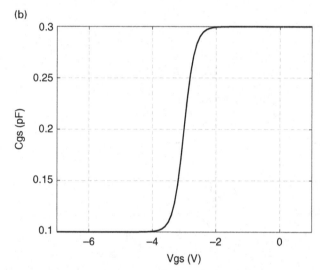

Figure 7.14 (a) $Q_{gs}(v_{GS})$ and (b) $C_{gs}(v_{GS})$ profiles of the model used to illustrate the implementation of nonlinear capacitance/charge models in commercial nonlinear circuit simulators.

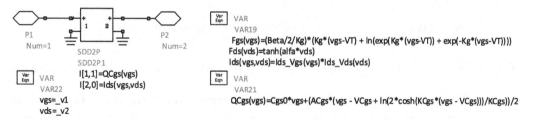

Figure 7.15 $Q_{gs}(v_{GS})$ and $i_{DS}(v_{GS},v_{DS})$ model implementation in the ADS simulator with a symbolically defined device component.

Table 7.1 $Q_{gs}(v_{GS})$ model implementation in the MWO with an APLAC netlist.

```
DefModel Cgs 2 n1 n2
  + PARAM 4 Cgs0 Acgs Vcgs Kcgs
  Function vgs = CV(0)
  Function Qgs_vgs {qv cv} [
      +qv = (Cgs0*vgs + (Acgs*(vgs + Vcgs + ln(2*cosh(Kcgs*(vgs − Vcgs)))/Kcgs))/2);
      +cv = (Cgs0 + 0.5*Acgs*(tanh(Kcgs*(vgs − Vcgs)) + 1));
      +qv,
      +cv]
  VCCS CgsVccs n1 n2 1 n1 n2 Qgs_vgs C
EndModel
```

Table 7.2 $i_{DS}(v_{GS}, v_{DS})$ model implementation in the MWO with an APLAC netlist.

```
DefModel Ids 4 n1 n2 n3 n4
+PARAM 13 Beta
Function vgs = CV(0)
Function vds = CV(1)
Function ids_vgs_vds {Fgs, Fds, dFgs_dvgs, dFds_dvds} [
      +Fgs = (0.5*Beta/Kg)*(Kg*(vgs − VT) + ln(exp(Kg*(vgs − VT)) + exp(−Kg*(vgs − VT))));
      +Fds = tanh(alfa*vds);
      +dFgs_dvgs = 0.5*Beta*(1 + tanh(Kg*(vgs − VT)));
      +dFds_dvds = alfa*(1 − tanh(alfa*vds)^2);
      +Fgs*Fds,dFgs_dvgs*Fds,Fgs*dFds_dvds]
      VCCS Idsvccs n3 n4 2 n1 n2 n3 n4 ids_vgs_vds
EndModel
```

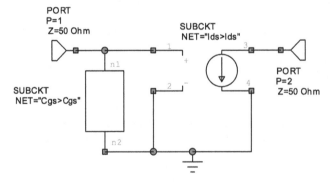

Figure 7.16 $Q_{gs}(v_{GS})$ and $i_{DS}(v_{GS}, v_{DS})$ schematic of the imported APLAC netlists in the MWO simulator.

will now move on to discuss a particular application example: a GaN HEMT based medium-power amplifier for mobile communications. For that, we will show how computer-aided design tools can help with the bias point and input and output termination selection and how they can help in the performance evaluation.

7.2.1 Bias Point Selection for High Efficiency

Starting the PA design flow with the bias point selection, we will use harmonic-balance (HB) simulation and our previously discussed GaN HEMT model to help us search for the quiescent point that presents the best compromise between efficiency and linearity.

For maximized output voltage excursion, and so linear output power capability, we consider a typical current mode design where $v_{ds}(t)$ is assumed nearly sinusoidal, and the V_{DS} quiescent point is selected as the middle point between the device's knee and breakdown voltages, V_K and V_{BR}, or $V_{DS} = 28\,\text{V}$, as is recommended by the device's manufacturer.

On the other hand, maximized efficiency requires a gate voltage V_{GS} close to the threshold voltage, V_T, which, in an idealized piecewise linear device, would correspond to class B operation with a flat gain. In a real FET, the V_{GS} quiescent point that we should use to obtain a maximally flat gain is not well-defined, since the turn-on is not abrupt but smooth. However, we can still define it as the V_{GS} quiescent point where small-signal input 3rd-order nonlinearity vanishes [15]. This could be thought of as the point of vanishing $G_{m3} \equiv \partial^3 i_{DS}/\partial v_{GS}^3$ obtained from a small-signal two-tone test, or, alternatively, from small-signal harmonic distortion (see [15–17]) or of vanishing $\partial^2 G_m/\partial v_{GS}^2$, i.e., the first inflection point of the transconductance, $G_m(V_{GS})$, characteristic, derived from small-signal linear S-parameter data.

Unfortunately, this traditional methodology is not appropriate for our GaN HEMT as its small-signal V_T is different from the actual large-signal V_T, due to the different thermal and trapping states determined by the dc quiescent, and the large-signal operating points. To illustrate this, Figure 7.17(a) and (b) present two gain versus input power drive level characteristics respectively measured with a CW and with a two-tone signal whose period of the envelope is much smaller than the device's time constants associated with its thermal or trapping phenomena. Although Figure 7.17(a) seems to correspond to a class B biased PA while Figure 7.17(b) to a PA biased in class C, they were measured on the same device with exactly the same dc quiescent point.

These two curves clearly reveal the threshold voltage shift caused by the trapping effects of GaN HEMTs [18]. At low drive levels, the v_{DS} peak voltage is not much different from the V_{DD} bias and the associated drain-lag traps are charged to their reference level. The threshold voltage is also at its reference level, V_{T0}, and $V_{GS} \approx V_{T0}$ is such that the device shows a typical flat small-signal gain class B behavior. However, when the drive level increases, the new v_{DS} peak voltage sets these traps to a different state, increasing V_T to V_{T1}, so that, now, the device is no longer biased at $V_{GS} \approx V_{T1}$ but at $V_{GS} < V_{T1}$, i.e., in class C. If now the drive level is decreased so slowly that the trap state dynamics can follow the decrease of v_{DS} peak voltage, the device will show again its flat small-signal gain, as seen in Figure 7.17(a). If, on the other hand, this decrease is much faster than the traps' discharging time, the device will keep its apparent class C bias, exhibiting an increasing small-signal gain as shown in Figure 7.17(b).

Therefore, a different strategy has to be adopted. Instead of the usual sinusoidal HB simulations, which mimic the CW lab tests, two-tone harmonic-balance (or envelope transient harmonic-balance, ETHB) simulations were used to determine the V_{GS}

(a)

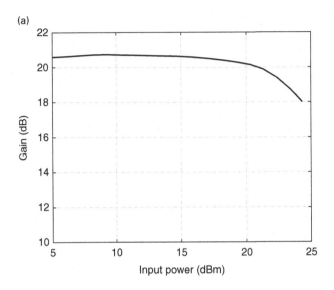

(b)

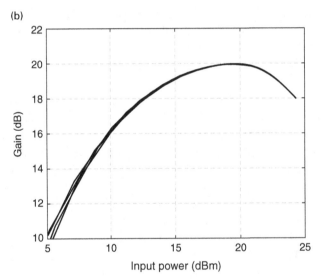

Figure 7.17 Gain versus input drive level of the same device and quiescent point, but driven with (a) a CW signal of increasing amplitude and (b) a two-tone signal of 1 kHz frequency separation. Note how the curve of (a) could misleadingly indicate a class B operation, while the one of (b) reveals a class C gain characteristic imposed by the trapping induced V_T shift.

quiescent point of best efficiency-linearity trade-off. This is what is shown in Figure 7.18(a) in terms of power added efficiency, PAE, Figure 7.18(b) gain, $Gain$, and 7.18(c) carrier-to-intermodulation-ratio, IMR, for three different V_{GS} biases corresponding to shallow class AB, class B and shallow class C operation.

Since the simulated results indicate similar values of PAE but considerably different ones for IMR and gain profile flatness, the class B bias point of $V_{GS} = -3.58\,\text{V}$ was selected.

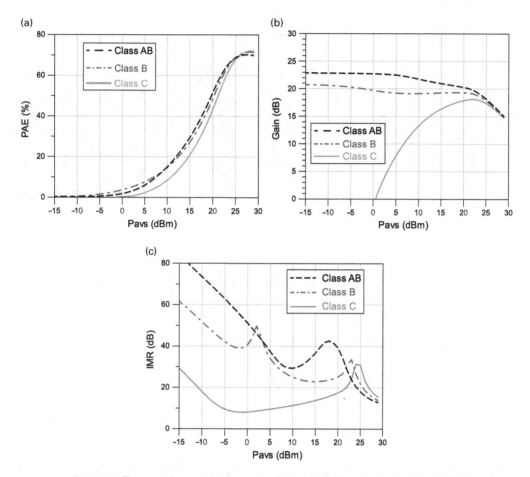

Figure 7.18 Two-tone harmonic-balance simulations of (a) power-added efficiency, PAE, (b) gain, Gain, and (c) carrier-to-intermodulation-ratio, IMR, versus input drive level for three different V_{GS} quiescent points ($V_{GS} = -3.3$ V, $V_{GS} = -3.58$ V and $V_{GS} = -3.9$ V) corresponding to shallow class AB, class B and shallow class C operation, respectively.

7.2.2   Source- and Load-Pull Simulations

After having selected a convenient quiescent point for the best trade-off between efficiency and linearity, we will now select proper source and load terminations.

Again, this will be done using the circuit simulator, avoiding the need for tedious source/load-pull measurements and expensive equipment. Actually, if good models are available (which is not necessarily the case when dealing with very high power devices), we have an advantage in the simulator over what we can do in the lab. Indeed, having access to the intrinsic device's voltages and currents (something particularly useful in prematched FETs), we can simulate source/load-pull with ideal lossless components directly at very low impedances, i.e., without the need to use impedance

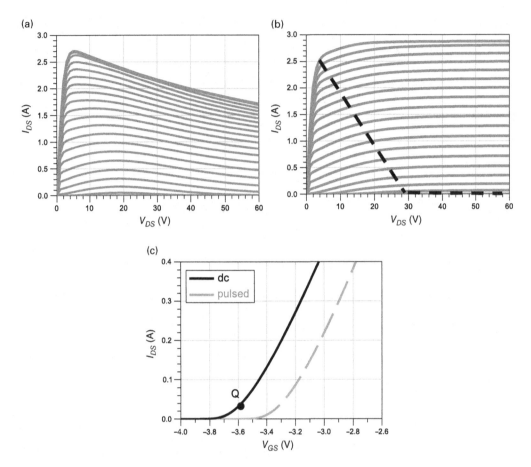

Figure 7.19 (a) Conventional dc I/V curves obtained from a dc simulation and (b) pulsed I/V curves simulated via a transient analysis in which the quiescent point was set at $V_{DS} = v_{DS\_Max}$ and $V_{GS} < V_T$, whose pulse duration was significantly smaller than the device's thermal and de-trapping time-constants and the duty cycle was 0.1%. (c) Comparison between the dc and pulsed I_{DS} versus V_{GS} for a $V_{DS} = 28$ V to show the shift in the threshold voltage due to trapping and thermal effects.

transformers to convert them to the normalized 50 Ω found in the real lab. Furthermore, with a good model, we may even start by calculating a first estimate of the required load.

7.2.2.1   Estimation of the PA Required Load

In order to obtain a first estimate of the PA required load, we will use the Cripps load-line method [1] for the intrinsic load resistance that maximizes linear output power capability (from now on designated as the power load, R_{L_pwr} or Z_{L_pwr}) and its extension to optimum efficiency provided by [2] (which we will designate the efficiency load, R_{L_eff} or Z_{L_eff}). We need to start by drawing the intrinsic $i_{DS}(v_{GS}, v_{DS})$ output curves. However, what we would obtain from a typical dc simulation would be the curves of Figure 7.19(a), which suffer from thermal and trapping phenomena. To overcome this, we need to first set the temperature and the trap states at values close to the ones we expect the amplifier to operate in steady-state, and then dynamically test the device from that quiescent state.

From the device's datasheet, such a device, driven by a WiMAX signal, is supposed to deliver about 2 W of average power with 26% of drain efficiency. So, since its thermal resistance is on the order of 8 K/W [19], we obtain a rough estimate of 46°C of temperature increase over room temperature. Hence, we set the device model temperature control node to 71°C. As far as the trap states are concerned, we will assume that they will be imposed by the maximum v_{DS} voltage, and that this value is given by: $v_{DS\_Max} = V_{DS} + (V_{DS} - V_K) \approx 51$ V. Consequently, the I/V curves we are interested in are the ones obtained from a transient pulsed I/V simulation in which the quiescent point is set at $V_{DS} = v_{DS\_Max}$ and $V_{GS} < V_T$, whose pulse duration is significantly smaller than the device's thermal and detrapping time-constants and the duty cycle is some 0.1% or lower. Figure 7.19(b) depicts the output I/V curves obtained from such a transient pulsed I/V simulation.

Superimposed to these pulsed I/V curves is the ideal class B dynamic load line, the selected quiescent point, Q, and its correspondent position after the discussed shift determined by the thermal and trap steady-states. According to the Cripps load-line method, such a dynamic trajectory requires an intrinsic termination of $R_{L\_pwr} = 19\,\Omega$, which, transformed by the device's output parasitic network, leads to a fundamental termination of $Z_{L\_pwr} = 14.4 - j7.2\,\Omega$ at the device's drain terminal.

As far as the optimum intrinsic termination for maximized drain efficiency is concerned, the approximate estimate of [2] led to an $R_{L\_eff} = 79\,\Omega$ and a corresponding extrinsic termination of $Z_{L\_eff} = 72.9 + j15\,\Omega$.

7.2.2.2 Simulated Load-Pull Contours

Although these two Z_L values could already be used to get an estimate of a fundamental load impedance capable of providing a good compromise between output power capability and efficiency, a set of CW HB simulations produced the load-pull contours shown in Figure 7.20. Commercial nonlinear microwave circuit simulators already include preprogrammed aids to perform automatic load-pull simulations as shown in Figure 7.20(b) given a set of predefined test loads distributed as in Figure 7.20(a). As a matter of fact, such a simulation requires also the definition of a particular source impedance (we selected one that could produce a reasonable input match for the maximum output power load) and load and source impedances at all harmonics. In the present case, we selected a short circuit at all even and odd harmonics, as is the condition required for current-mode class B operation.

These conditions led to a maximum output power of 43 dBm (about 20 W), a maximum drain efficiency of 76% and to an optimum load of $Z_{L\_opt} = 29.3 - j3.3\,\Omega$, which represents a good compromise between these maximum output power and efficiency figures. Such a load provides an output power of 41 dBm (12.6 W) and a drain efficiency of 74%.

An alternative to the current-mode class B harmonic termination set previously used could also be chosen. Specifically, we now use the set determining class F operation,

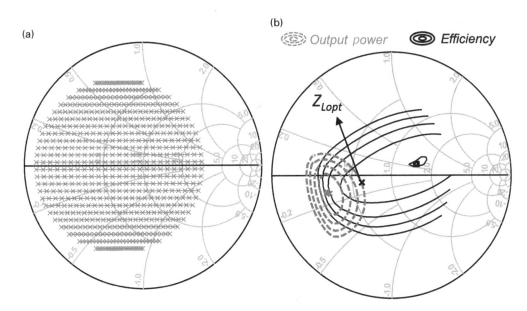

Figure 7.20 (a) Set of predefined load impedances used to simulate the output power (dashed lines) – $P_{Max}=43.1$ dBm and $P_{Step}=0.5$ dB – and drain efficiency (solid lines) – $\eta_{Max}=76.6\%$ and $\eta_{Step}=5\%$ – load-pull contours shown in (b).

i.e., a short circuit to all intrinsic even order harmonics and an open circuit to all odd order ones. In this case, the obtained load-pull contours are the ones shown in Figure 7.21, from which a maximum power of nearly 43.5 dBm (22.4 W) and maximum efficiency of 93% are predicted. As shown in Figure 7.21, the optimum compromise between output power and efficiency now corresponds to an improved output power of 42.5 dBm (22.6 W) and efficiency of 89%, reason why this was the adopted configuration.

The dynamic load-line and the intrinsic voltage and current waveforms shown in Figure 7.22 (a) and (b), respectively, confirm the expected class F operation.

7.2.2.3   Small- and Large-Signal Stability Check

Before selecting the source impedance, a stability check (both at the dc operating point and the large-signal nominal excitation amplitude) was conducted and, as shown in Figure 7.23(a), it was verified that the selected load impedance was dangerously close to the region that produces a negative input resistance ($|\Gamma_{in}| > 1$, indicating a potentially unstable design). Therefore, following the advice of the GaN HEMT manufacturer, a stabilizing resistor of 5 Ω was introduced in series with the gate terminal, resulting in the new stability region shown in Figure 7.23(b).

To be complete, such a stability check would have to be done at all possible excitation amplitudes, with the aid of a small-signal perturbation of the LSOPs, and extended to all frequencies at which the device is active, something that is theoretically complicated, and quite cumbersome in practice. Alternatively, a single stability check

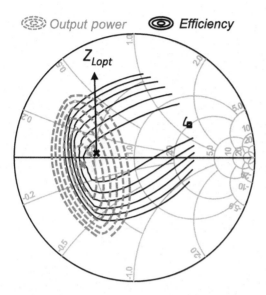

Figure 7.21 Output power (dashed lines) – $P_{Max} = 43.5$ dBm and $P_{Step} = 0.5$ dB – and drain efficiency (solid lines) – $\eta_{Max} = 93\%$ and $\eta_{Step} = 5\%$ – load-pull contours obtained when the harmonic terminations were selected to determine a class F operation and the set of tested loads were again the ones shown in Figure 7.20(a).

may be performed with the aid of a transient simulator. Nevertheless, one would still need to repeat this transient simulation for, at least, the cases of infinitesimal and nominal excitation levels.

7.2.2.4 Simulated Source-Pull Contours

With the selected impedance for the fundamental, and the other ideal source and load harmonic terminations for class F operation, we then conducted a source-pull for the fundamental resulting in the contours of Figure 7.24. From these, a maximum output power of 42.5 dBm (17.8 W) and a maximum efficiency of 85.6%, were obtained for the source impedance of $Z_S = 7.6 + j11.9\ \Omega$.

After several iterations of source and load tuning, the found source and load terminations were the ones listed in Table 7.3.

Note that, because of the transistor package parasitics and the design of the matching networks, only the first three harmonics could be considered, which led to a predicted design performance of 42 dBm (15.8 W) of output power and 81.8% of efficiency.

7.2.3 Input and Output Bias and Matching Network Design

Having selected the input and output termination impedances, initial matching and biasing networks were designed. This process started by using ideal transmission line elements, which were then converted into microstrip elements. This led to a first draft of the layout, which was then retuned, representing these planar structures by S-parameter matrices obtained from an electromagnetic simulator. Such a tuning

(a)

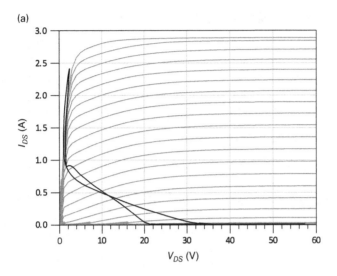

(b)

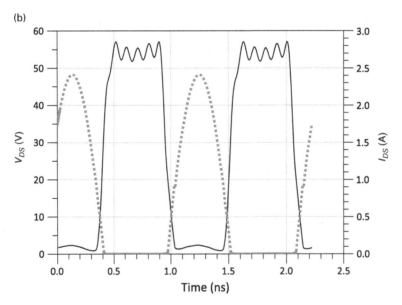

Figure 7.22 (a) Dynamic load-line and (b) intrinsic $v_{DS}(t)$ (——) and $i_{DS}(t)$ ($\cdots$) waveforms of the device terminated for class F operation.

process should be made for both the desired frequency response around the carrier and harmonics as well as around dc, since this determines the so-called video-band-width, the main characteristic responsible for the amplifier's envelope memory effects. This electromagnetic simulation of the microstrip layout is particularly important for high-power devices that usually require transmission lines of very low characteristic impedance. In this scenario, the obtained layout tends to have microstrip lines or discontinuities (such as microstrip steps, T and cross junctions,

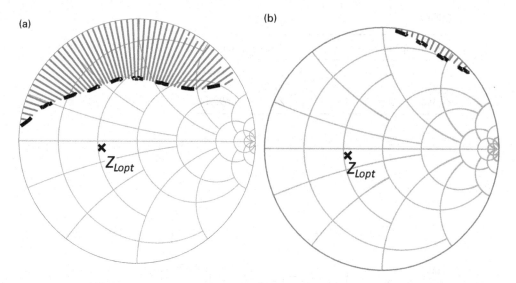

Figure 7.23 (a) Load terminations of negative input resistance ($|\Gamma_{IN}| > 1$) simulated at the large-signal excitation amplitude, and the selected load termination, which shows that such a design is too close to instability. (b) Same simulation but now when the device includes a 5 Ω stabilizing resistor at the gate terminal.

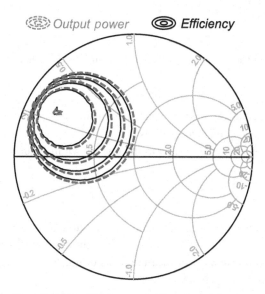

Figure 7.24 Simulated source-pull contours of output power (dashed lines) – P_{Max} = 42.5 dBm and P_{Step} = 0.5 dB – and efficiency (solid lines) – η_{Max} = 85.6% and η_{Step} = 5% –, for constant source available power equal to 24 dBm, when the device is terminated at the output to guarantee optimum class F operation.

Table 7.3

Source impedances	Load impedances
$Z_S(f_0) = 7.6 + j11.9\,\Omega$	$Z_L(f_0) = 27 + j3.4\,\Omega$
$Z_S(2f_0) = -j8.3\,\Omega$	$Z_L(2f_0) = -j14.5\,\Omega$
$Z_S(3f_0) = -j12.6\,\Omega$	$Z_L(3f_0) = +j69.6\,\Omega$

(a)

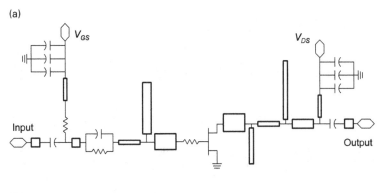

(b)

Figure 7.25 (a) Final circuit schematic and (b) photograph of the implemented medium-power amplifier example.

bends, etc.) whose geometrical dimensions may fall outside of the validity of the simulator's embedded microstrip element models.

After this layout design process, the result was the one shown in Figure 7.25.

7.2.4   Performance Evaluation

After the use of the described nonlinear simulating techniques for a typical circuit design, we will now illustrate the utilization of these tools in the context of performance evaluation. For that, we will make a set of measurements of our PA, to compare with results obtained from the nonlinear circuit simulator.

Because of the specific responses a nonlinear circuit offers to distinct excitations, the illustrated measurements and simulations will be grouped according to the tested class of stimuli. Therefore, as is quite usual in the test of PAs, these classes of excitations will be a CW signal and a modulated carrier. However, contrary to the CW simulations, which are quite straightforward, that is not the case of the simulations conducted with the modulated stimuli because the time-domain description of the aperiodic modulation requires that every component (even the EM simulated blocks of the matching networks) is represented by an appropriate lumped equivalent circuit, e.g., a broadband spice model, BBSM [4]. So, before we proceed to simulate the circuit driven by a modulated signal with, e.g., envelope transient harmonic-balance, ETHB, it is always advisable to test the generated equivalent circuit models comparing HB responses of the circuit with similar results obtained via ETHB. This requires a test whose results can be obtained from both HB and ETHB, like periodic multitone excitations.

7.2.4.1   Continuous-Wave Testing

Continuous-wave measurements and simulations are the simplest tests we can make in our circuit because most RF instrumentation and simulators were especially conceived for that purpose. These tests have been grouped into linear and nonliner measurements, according to the considered excitation amplitude and required instrumentation or simulation tools. However, since the circuit's small-signal response should naturally arise in the limit of the large-signal tests when the excitation amplitude is sufficiently small (so as not to generate harmonics or change the dc quiescent point), we will treat them in a unified manner. Hence, we will place the PA in a matched source and load environment and then sweep the excitation amplitude, from small signal up to a few dB of gain compression, and measure the input–output transfer characteristic in amplitude and phase. These amplitude and phase profiles are known as the quasi-static AM/AM and the AM/PM and are shown in Figure 7.26(a) and (b).

If these tests were accompanied by measuring the input, output, and dc supplied powers, an efficiency versus drive level plot like the one shown in Figure 7.27 could also be reproduced.

7.2.4.2   Multitone Testing

As said in the introduction to this section, multitone testing constitutes a performance evaluation procedure, but it also serves as a means to test the accuracy of the equivalent circuit models used in the envelope following simulators that operate in the time domain. This is the reason why two-tone and four-tone tests will be treated in this section, and their measurement results compared with multitone harmonic-balance and envelope transient harmonic-balance simulations.

Starting with two-tone tests, Figure 7.28 is a repetition of the AM/AM plots discussed above for the CW excitation but now measured capturing the envelope time evolution (with a vector signal generator and a vector signal analyzer instead of a network analyzer) for a two-tone stimulus of constant frequency separation ($\Delta f = 50\,\text{kHz}$) and increasing peak power level.

Note the progressive reduction of small-signal gain verified for the increasing $v_{DS}(t)$ voltage peak, an indication of the GaN HEMT long-term memory effects [18].

Figure 7.29 describes another set of two-tone tests, but now with a constant input peak envelope power ($PEP = 24.7\,\text{dBm}$) and increasing separation frequency to evidence the PA's long-term dynamic behavior.

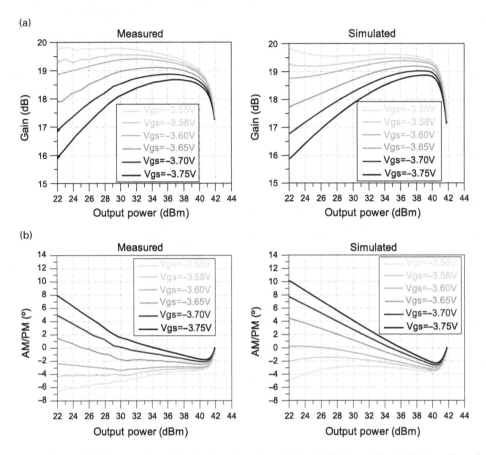

Figure 7.26 Measured and HB simulated quasi-static (a) AM/AM and (b) AM/PM profiles of the PA prototype, for various V_{GS} bias points corresponding to class AB, class B and class C.

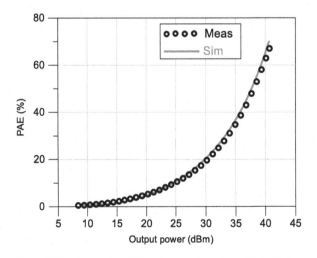

Figure 7.27 Measured and HB simulated power-added-efficiency of the tested PA circuit for the selected bias ($V_{GS} = -3.58$ V).

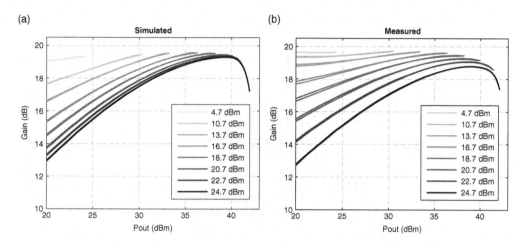

Figure 7.28 Dynamic gain (or dynamic AM/AM) profiles of our PA example obtained with two-tone tests of constant frequency separation ($\Delta f = 50$ kHz) and increasing peak envelope power level: (a) measurements and (b) two-tone HB simulated results.

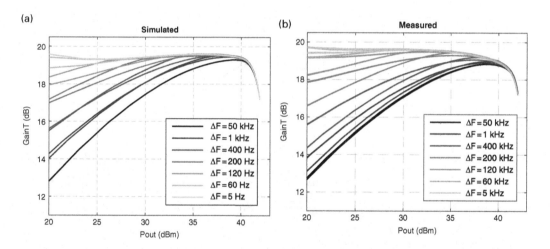

Figure 7.29 Dynamic gain (or dynamic AM/AM) profiles of our PA example obtained with two-tone tests of constant input peak envelope power level ($PEP = 24.7$ dBm) and increasing frequency separation (a) measurements and (b) two-tone HB simulated results.

To close this discussion regarding two-tone tests, Figure 7.30 is an illustrative spectrum of the measured and simulated results obtained for the input PEP of 24.7 dBm and a separation frequency of 0.8 MHz.

Finally, Figures 7.31 and 7.32 summarize the measurement and simulation tests conducted with a 4-tone excitation in order to check the accuracy of the time-domain envelope simulation models. These figures compare the dynamic AM/AM and AM/PM profiles (Figure 7.31) and spectra (Figure 7.32) obtained with measurements, multitone HB simulation and ETHB envelope following simulation at the same conditions.

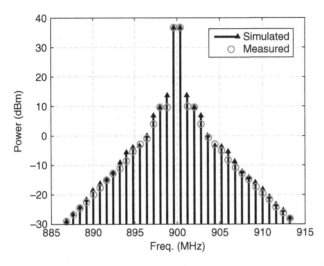

Figure 7.30 Illustrative measurements and HB simulated results of a two-tone test performed with an input *PEP* of 24.7 dBm and a separation frequency of 0.8 MHz.

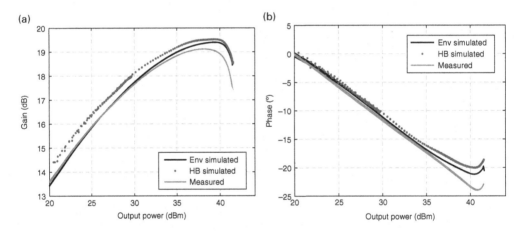

Figure 7.31 (a) Dynamic AM/AM and (b) AM/PM profiles directly obtained from laboratory measurements and multitone HB and ETHB simulations of our PA circuit example excited with a 4-tone stimulus.

7.2.4.3   Modulated Signal Testing

As a final example of the nonlinear measurement and simulation tests on our PA circuit example, we excited it with a 64-QAM signal of 24.1 dBm of available input peak envelope power and 10 MHz of bandwidth. Figure 7.33–7.35 describe the obtained results in several different forms.

Similarly to what was done before, Figures 7.33 and 7.34 report the measured and ETHB simulated time-domain AM/AM and AM/PM and corresponding frequency-domain spectra, respectively.

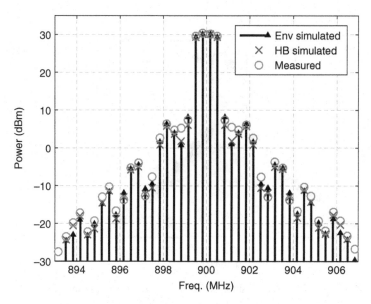

Figure 7.32 Measured and simulated (with both HB and ETHB) spectra corresponding to the 4-tone tests reported in figure Figure 7.31.

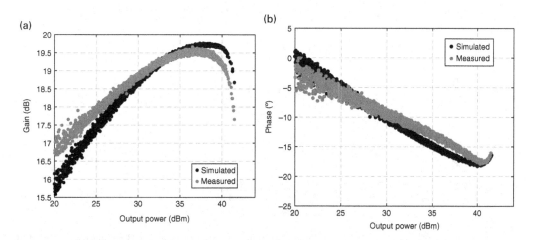

Figure 7.33 (a) Dynamic AM/AM and (b) AM/PM profiles directly obtained from laboratory measurements and ETHB simulations of our PA circuit example excited with a 64-QAM signal of 10 MHz bandwidth.

As a final illustration of presently available nonlinear simulation capabilities, Figure 7.35 depicts a system level constellation diagram of the PA response to the 64-QAM signal, from which error vector magnitude, EVM, performance figures could be drawn.

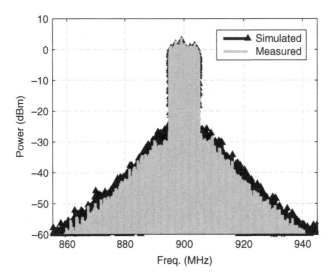

Figure 7.34 Measured and ETHB simulated spectra corresponding to the time-domain dynamic AM/AM and AM/PM plots shown in figure Figure 7.33.

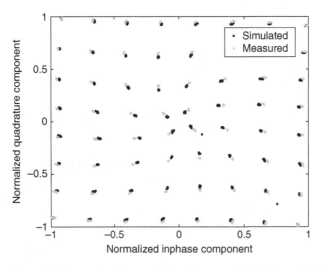

Figure 7.35 System-level constellation diagram obtained from the ETHB simulations reported in Figures 7.33 and 7.34.

7.3   Summary

The authors hope that this closing chapter provided an illustration of some of the concepts addressed in the book's preceding parts, putting them in an application context, so that their significance in nonlinear microwave circuit design can be better understood.

We started by revisiting some features of the active device modeling presented in the previous chapters discussing their impact on the predicted circuit behavior. Then, a step-by-step design and verification procedure for a wireless medium-power amplifier based on a GaN HEMT device was presented. This exemplified how a nonlinear device model, connected to some commercial nonlinear simulation tools, can be used to aid, first, the design and, then the test of the selected circuit.

This way, we showed how these modeling and simulation techniques can lead to a more profound understanding of the circuit operation, and thus to an optimized circuit design. The consequent reduction in the number of iterations necessary for the circuit design/implementation justifies the effort spent on device characterization and modeling.

Actually, the modeling and simulation techniques addressed in this book, and experimented in this last chapter, have become indispensable in modern nonlinear microwave circuit designs, enabling monolithic implementations, higher operating frequencies, wider bandwidths, or architectures of increased levels of complexity.

References

[1] S. C. Cripps, "A theory for the prediction of GaAs FET load-pull power contours," in *IEEE MTT-S Int. Microw. Symp. Dig.*, Boston, MA, 1983, pp. 221–223.

[2] J. C. Pedro, L. C. Nunes, and P. M. Cabral, "A simple method to estimate the output power and efficiency load–pull contours of class-B power amplifiers," *IEEE Trans. on Microw. Theory Techn.*, vol. 63, no. 4, April 2015, pp. 1239–1249.

[3] J. Wood, P. H. Aaen, D. Bridges, D. Lamey, M. Guyonnet, D. S. Chan, and N. Monsauret, "A Nonlinear electro-thermal scalable model for high-power RF LDMOS transistors," *IEEE Trans. on Microw. Theory Techn.*, vol. 57, no. 2, Feb. 2009, pp. 282–292.

[4] B. Gustavsen and A. Semlyen, "Rational approximation of frequency domain responses by vector fitting," *IEEE Trans. on Power Delivery*, vol. 14, no. 3, July 1999, pp. 1052–1061.

[5] O. Jardel, F. De Groote, T. Reveyrand, J. C. Jacquet, C. Charbonniaud, J. P. Teyssier, D. Floriot, and R. Quere, "An electrothermal model for AlGaN/GaN power HEMTs including trapping effects to improve large-signal simulation results on high VSWR," *IEEE Trans. Microw. Theory Techn.*, vol. 55, no. 12, Dec. 2007, pp. 2660–2669.

[6] P. Roblin, D. E. Root, J. Verspecht, Y. Ko, and J. P. Teyssier, "New trends for the nonlinear measurement and modeling of high-power RF transistors and amplifiers with memory effects," *IEEE Trans. Microw. Theory Techn.*, vol. 60, no. 6, June 2012, pp. 1964–1978.

[7] A. Santarelli, R. Cignani, G. P. Gibiino, D. Niessen, P. A. Traverso, C. Florian, D. M. M. P. Schreurs, and F. Filicori, "A double-pulse technique for the dynamic I/V characterization of GaN FETs," *IEEE Microwave Wireless Comp. Lett.*, vol. 24, no. 2, Feb. 2014, pp. 132–134.

[8] L. C. Nunes, J. M. Gomes, P. M. Cabral, and J. C. Pedro, "A new nonlinear model extraction methodology for GaN HEMTs subject to trapping effects," in *IEEE MTT-S Int. Microwave Symp. Dig.*, Phoenix, AZ, 2015, pp. 1–4.

[9] P. Aaen, J. A. Plá and J. Wood, *Modeling and Characterization of RF and Microwave Power FETs*, New York: Cambridge University Press, 2011.

[10] L. C. Nunes, P. M. Cabral and J. C. Pedro, "AM/AM and AM/PM distortion generation mechanisms in Si LDMOS and GaN HEMT based RF power amplifiers," *IEEE Trans. Microw. Theory Techn.*, vol. 62, no. 4, April 2014, pp. 799–809.

[11] B. Kim, I. Kim and J. Moon, "Advanced Doherty Architecture," *IEEE Microw. Mag.*, vol. 11, no. 5, Aug. 2010, pp. 72–86.

[12] S. J C H Theeuwen and J. H. Qureshi, "LDMOS technology for RF power amplifiers," *IEEE Trans. on Microw. Theory Techn.*, vol. 60, no. 6, June 2012, pp.1755–1763.

[13] Keysight Technologies Advanced Design System (ADS) Software, 2016, url: http://www .keysight.com/en/pc-1297113/advanced-design-system-ads.

[14] Applied Wave Research Microwave Office (AWR - MWO), 2016, url: http://www.awrcorp .com/products/microwave-office.

[15] J. Pedro and N. Carvalho, *Intermodulation Distortion in Microwave and Wireless Circuits*, Norwood, MA: Artech House, 2003.

[16] S. Maas and A. Crosmun, "Modeling the gate I/V characteristic of a GaAs MESFET for Volterra-series analysis," *IEEE Trans. Microw. Theory Techn.*, vol. 37, no. 7, July 1989, pp. 1134–1136.

[17] J. Pedro and J. Perez, "Accurate simulation of GaAs MESFET's intermodulation distortion using a new drain-source current model," *IEEE Trans. Microw. Theory Techn.*, vol. 42, no. 1, Jan. 1994, pp. 25–33.

[18] J. Pedro L. Nunes and P. Cabral, "Soft compression and the origins of nonlinear behavior of GaN HEMTs," in *44th Europ. Microwave Conf. Proc., 29th European Microw. Conf. Proc.*, 2014, pp. 1297–1300.

[19] CGH35015 Datasheet, Wolfspeed, 2015, url: http://www.wolfspeed.com/media/downloads/ 231/CGH35015.pdf

Appendix

Introduction

This appendix shows how to convert among several different equivalent linear two-port descriptions of a three-terminal device. The method can be directly applied to more general networks with an arbitrary number of terminals.

The Indefinite Admittance Matrix

For a three-terminal device, we start with the "indefinite admittance matrix," a 3×3 matrix, shown in (1), with each row and column associated with one of the three terminals in the same but arbitrary order. The matrix is defined as the coefficients of the linear relationships between the terminal currents and terminal voltages (rather than port relationships) given by (2). In (2), e_j is the voltage at the jth node (terminal) of the three-terminal device.

$$Y^{indef} = \begin{bmatrix} Y_{11} & Y_{12} & Y_{13} \\ Y_{21} & Y_{22} & Y_{23} \\ Y_{31} & Y_{32} & Y_{33} \end{bmatrix} \tag{1}$$

$$I_i = \sum_{j=1}^{3} Y_{ij}^{indef} e_j \tag{2}$$

The circuit law KCL can be applied by summing (2) over the index i to obtain zero, which must hold for any possible sets of e_j. Therefore, we obtain the result (3) that the sum of the elements in each column of (1) must be zero:

$$\sum_{i=1}^{3} I_i = 0 = \sum_{i=1}^{3} \sum_{j=1}^{3} Y_{ij}^{indef} e_j \Rightarrow \sum_{i=1}^{3} Y_{ij}^{indef} = 0 \tag{3}$$

The circuit law KVL has as a consequence that only voltage differences are physically significant, so we can subtract a fixed but arbitrary voltage from each of the node voltages in (2) and must still get the same output currents. This means we have the result that the sum of the elements of each row of (1) equals zero.

$$I_i = \sum_{j=1}^{3} Y_{ij}^{indef} (e_j - e_k) = \sum_{j=1}^{3} Y_{ij}^{indef} e_j - \left(\sum_{j=1}^{3} Y_{ij}^{indef} \right) \cdot e_k \Rightarrow \sum_{j=1}^{3} Y_{ij}^{indef} = 0 \tag{4}$$

Equations (3) and (4) reduce the number of independent elements of the 3×3 matrix (1) from 9 to 4, the latter is perfect for a 2×2 port matrix!

We now choose an ordering for the rows and columns of (1) to correspond with the terminals of the device. Without loss of generality, we choose the order gate = 1, drain = 2, and source = 3.

We now suppose we know (e.g., from measurements) the device 2×2 common source admittance matrix with index 1 for the gate and 2 for the drain, written according to (5).

$$Y^{CS} = \begin{bmatrix} Y_{11}^{CS} & Y_{12}^{CS} \\ Y_{21}^{CS} & Y_{22}^{CS} \end{bmatrix} \tag{5}$$

The task is to compute any one of the other possible 2×2 admittance matrices (e.g. common gate or common drain) of which each has two distinct possible port orderings. We illustrate the general procedure by calculating the common gate admittance matrix with the drain port 1 and the source port 2.

We first place, as a submatrix, the known common source two-port matrix (5) within the 3×3 indefinite matrix (1) and apply the conditions (3) and (4) to obtain (6).

$$Y^{indef} = \begin{bmatrix} Y_{11}^{CS} & Y_{12}^{CS} & -\left(Y_{11}^{CS} + Y_{12}^{CS}\right) \\ Y_{21}^{CS} & Y_{22}^{CS} & -\left(Y_{21}^{CS} + Y_{22}^{CS}\right) \\ -\left(Y_{11}^{CS} + Y_{21}^{CS}\right) & -\left(Y_{12}^{CS} + Y_{22}^{CS}\right) & Y_{11}^{CS} + Y_{12}^{CS} + Y_{21}^{CS} + Y_{22}^{CS} \end{bmatrix} \tag{6}$$

The common gate 2×2 admittance matrix corresponding to the same device can now be directly obtained from (6) simply by "crossing out" the first row and column (top row and left-most column) – corresponding to the gate – and reading out the 2×2 submatrix that remains. The result is (7). This gives the explicit relationships of each element of the common gate admittance to simple combinations of the known common source admittance matrix elements.

$$Y^{CG} \equiv \begin{bmatrix} Y_{11}^{CG} & Y_{12}^{CG} \\ Y_{21}^{CG} & Y_{22}^{CG} \end{bmatrix} = \begin{bmatrix} Y_{22}^{CS} & -\left(Y_{21}^{CS} + Y_{22}^{CS}\right) \\ -\left(Y_{12}^{CS} + Y_{22}^{CS}\right) & Y_{11}^{CS} + Y_{12}^{CS} + Y_{21}^{CS} + Y_{22}^{CS} \end{bmatrix} \tag{7}$$

Example

As an example, we start with the common source admittance matrix of the intrinsic linear equivalent circuit model shown in Chapter 5, Figure 5.20, and derive the corresponding common gate matrix. For convenience, the common source admittance matrix, equation 5.11, is repeated in (8). We note $G_{DS} \equiv \frac{1}{R_{DS}}$.

$$Y^{CS} = \begin{bmatrix} 0 & 0 \\ G_m & G_{DS} \end{bmatrix} + j\omega \begin{bmatrix} C_{GS} + C_{GD} & -C_{GD} \\ -C_{GD} & C_{GD} + C_{DS} \end{bmatrix} \tag{8}$$

Plugging the values of (8) into the right-hand side of (7) results in (9).

$$jw \begin{bmatrix} C_{GS}+C_{GD} & -C_{GD} \\ -C_{GD} & C_{GD}+C_{DS} \end{bmatrix} \quad \begin{array}{l} \text{Start from CS matrix} \\ \text{Port 1 Gate} \quad \text{Port 2 Drain} \end{array}$$

$$\begin{array}{ccc} & G & D & S \end{array}$$

$$\begin{array}{l} G \\ \Rightarrow D \quad jw \\ S \end{array} \begin{bmatrix} C_{GS}+C_{GD} & -C_{GD} & -C_{GS} \\ -C_{GD} & C_{GD}+C_{DS} & -C_{DS} \\ -C_{GS} & -C_{DS} & C_{GS}+C_{DS} \end{bmatrix} \quad \begin{array}{l} \text{Augment to 3×3 "indefinite} \\ \text{admittance matrix" by} \\ \text{adding terms such that the} \\ \text{sum of each row \& col. =0} \end{array}$$

$$\begin{array}{ccc} & G & D & S \end{array}$$

$$\begin{array}{l} G \\ \Rightarrow D \quad jw \\ S \end{array} \begin{bmatrix} C_{GS}+C_{GD} & -C_{GD} & -C_{GS} \\ -C_{GD} & \boxed{C_{GD}+C_{DS}} & -C_{DS} \\ -C_{GS} & -C_{DS} & C_{GS}+C_{DS} \end{bmatrix} \Rightarrow \begin{array}{l} D \\ S \end{array} jw \begin{bmatrix} C_{GD}+C_{DS} & -C_{DS} \\ -C_{DS} & C_{GS}+C_{DS} \end{bmatrix}$$

$$C_{GD}(V_{GS},V_{DS}) = \frac{\mathrm{Im}Y_{11}^{(CG)}(V_{GS},V_{DS},\omega) + \mathrm{Im}Y_{12}^{(CG)}(V_{GS},V_{DS},\omega)}{\omega}$$

Figure A.1 Common source to common gate admittance transformation process.

$$Y^{CG} \equiv G^{CG} + jwC^{CG}$$

$$= \begin{bmatrix} G_{DS} & -G_m - G_{DS} \\ -G_{DS} & G_m + G_{DS} \end{bmatrix} + jw \begin{bmatrix} C_{GD}+C_{DS} & -C_{DS} \\ -C_{DS} & C_{GS}+C_{DS} \end{bmatrix} \qquad (9)$$

As a check, we can identify the ECP C_{GD} from the following entries of the common gate 2×2 admittance representation. That is, as claimed in Equation 5.2, we have the explicit formula for C_{GD} in terms of the common gate matrix elements given in (10).

$$C_{GD} = C_{11}^{CG} + C_{12}^{CG} \qquad (10)$$

The process is summarized graphically for the imaginary part of the admittances in Figure A.1.

Had we wanted the imaginary part of the common drain admittance (with port 1 = gate and port 2 = source), we would have crossed out (or removed) the second row and second column of the 3×3 matrix in Figure A.1. The resulting 2×2 matrix would be the common drain admittance in terms of the common source elements. Rows and columns are simply exchanged to obtain results with different port orderings.

Exercises

Exercise A.1   a.   Derive the common gate representation of the augmented linear model corresponding to Figure 5.27 that was expressed in common source admittance representation by 5.18.

b.   Derive an explicit expression for the ECP C_{GD} of Figure 5.27 in terms of the intrinsic common gate admittance parameters derived in part (a).

Exercise A.2   Use (1), (3), (4), and relation 5.3, to define the following properties of the 3×3 indefinite scattering matrix: $\sum_{i=1}^{3} S_{ij} = \sum_{j=1}^{3} S_{ij} = 1$.

Index

Printed in the United States
by Baker & Taylor Publisher Services